GEOLOGICAL MODELS OF PETROLEUM ENTRAPMENT

GEOLOGICAL MODELS OF PETROLEUM ENTRAPMENT

KINJI MAGARA

Professor of Petroleum Geology, Faculty of Earth Sciences, King Abdulaziz University, Jeddah, Saudi Arabia

ELSEVIER APPLIED SCIENCE PUBLISHERS
LONDON and NEW YORK

ELSEVIER APPLIED SCIENCE PUBLISHERS LTD
Crown House, Linton Road, Barking, Essex IG11 8JU, England

Sole Distributor in the USA and Canada
ELSEVIER SCIENCE PUBLISHING CO., INC.
52 Vanderbilt Avenue, New York, NY 10017, USA

WITH 15 TABLES AND 246 ILLUSTRATIONS

British Library Cataloguing in Publication Data

Magara, Kinji
Geological models of petroleum entrapment.
1. Oil reservoir engineering —
Mathematical models
I. Title
553.2′82′0724 TN871

Library of Congress Cataloging-in-Publication Data

Magara, Kinji.
Geological models of petroleum entrapment.

Bibliography: p.
Includes index.
1. Petroleum—Geology. I. Title.
TN870.5.M33 1986 553.2′82 86-4520

ISBN 0-85334-439-6

Photoset in Malta by Interprint Ltd.

Printed in Great Britain by Page Bros. (Norwich) Limited

Foreword

The principal controls of oil and gas accumulation have long been known: source, generation and maturation, migration, reservoir, trap, and seal. Certain of these controls are parts of essentially static geologic systems; with varying amounts of data, these elements may be reconstructed and interpreted. However, other critical controls—most notably sourcing, generation, maturation, and migration—are elements of dynamic geologic systems; direct evidence of these systems over time may be elusive to non-existent.

Even with static systems for which numerous data are available, a considerable part of the geologic reconstruction and interpretation may be on inference exclusively. The value of our interpretations, especially of dynamic systems, should not be assessed in absolute terms of right or wrong, but rather as logical or illogical.

In exploration, our efforts to understand geologic systems are chiefly for one paramount goal—the ability to predict. To the extent that predictive capability can be enlarged, inherent risks in exploration are reduced.

In this treatise, Dr Magara distinguishes static and dynamic elements and attempts to integrate evidence and inference fundamental to the practical exploration for oil and gas. His essential predication is that even with limited data, or where, by definition, the record no longer exists, marginal predictive capability is superior to no prediction at all.

Our accumulated data base in exploration permits such synthesis; the geologic reality of future exploration in many parts of the world demands it.

WILLIAM L. FISHER

Morgan J. Davis Professor of Petroleum Geology and Chairman, Department of Geological Sciences Director, Bureau of Economic Geology The University of Texas at Austin Texas, USA

Preface

For the assessment of petroleum potential of a sedimentary basin, sophisticated computer models have been developed and used by many organizations in the world. Such a computer model would provide us with answers on timing, efficiency, and volume of petroleum migrating and accumulating into traps in the basin. Modelling can be made either one-, two- or three-dimensionally.

Sophisticated computer modelling became popular in many organizations not only because of the recent development of advanced computer technology but also owing to improvements in our understanding of such basic problems as hydrocarbon generation and maturation, primary and secondary migration, sealing and hydrocarbon entrapment, reservoir properties, etc. For an effective application of a computer model, of course, several important data such as those from seismic, wire-line logs, drill stem tests, production tests, core analysis, both organic and inorganic geochemical analyses, paleontological study, etc., must be used.

The only drawback in using such a computer model is that, unless a geologist can himself carry out the programming, he would need help by a computer programmer to write a program. This can be done easily if he works for a large organization, but is not so easy in a relatively small organization.

A simple (one-dimensional) geological model can, however, be made by a petroleum geologist using the methods described in this book. These methods basically use graphs and charts with some calculations on a calculator. Chapters 3, 5 and 6 discuss applications of such simple geological models.

An effective model demands a large amount of reliable data which may become available only at later stages of exploration. However, such a model would be more useful at earlier stages of exploration than at the later stages. The chance of discovering new and significant fields in a basin would decrease gradually with time (or with more drilling), because the remaining unexplored area gets smaller with time.

In some cases, the most critical data may never become available even at the latest stages of exploration. For example, if hydrocarbons were generated mostly in synclinal areas where sedimentary rocks are usually more mature than those of the anticlinal areas, and if they have migrated laterally into traps, the true source rock evaluation must be made on the basis of data from the syncline. However, such synclinal areas may never be drilled, and thus there would be almost no chance for us to obtain reliable source rock data.

To partially solve this problem, I propose to use the methods described in Chapters 2 and 5, which estimate both the concentration of organic matter and its maturity from burial history plots. A burial history plot (Chapter 6) can be made using seismic data only, so that it has a very wide application at almost any exploration stage after the seismic survey.

To assist in understanding the primary hydrocarbon migration and in evaluating volumes of hydrocarbons expelled from source rocks, the subsurface data shown in Chapter 4 may be of value. In improving the maturity estimate of a source rock, both the thickness of sedimentary rocks removed by erosion (Chapter 7) and heat flow and geothermal gradient (Chapter 8) can be the critical factors.

If hydrocarbons migrate vertically upward, instead of laterally (stratigraphy-controlled), vertical charging of fluids, which may take place along the fracture planes, must be understood. This problem is discussed in Chapter 9.

An example of the practical application of some of the above-mentioned methods for the Arabian Gulf region is described in Chapter 10.

Chapters 11 to 14, on the other hand, explain the results of my own and my coworkers' research works on reservoir properties which could be critical in improving the geological modelling.

Finally, in Chapter 15, I intend to describe the size distribution and other important attributes of the world's giant oil and gas fields.

In summary, this book is made of three main parts as follows:

(1) simple geological models (Chapters 3, 5 and 6);
(2) improvement of the basic input data for modelling (Chapters 2, 4, 7, 8, 9, 11, 12, 13 and 14); and

(3) others—introduction, summary, and regional application (Chapters 1, 10 and 15).

Following the significant developments of the concepts and technology of petroleum geology in the last several decades, many ideas are still quite controversial. We would need more time to sort them out based on the increased amount of new data and experience in the future. In this book I chose a set of concepts which I favour from my own scientific standpoint and experience, and which are not seriously contradictory to each other. In other words, this book presents my çurrent thoughts on the present state of petroleum geology.

KINJI MAGARA
Jeddah, Saudi Arabia

Acknowledgements

Most of the research works described in this book were completed at both the Faculty of Earth Sciences, King Abdulaziz University, Jeddah, Saudi Arabia and the Bureau of Economic Geology, the University of Texas at Austin, Texas, USA. I wish to express my deepest appreciation to these organizations for providing the opportunities for me to conduct such research works and to complete this book. I have been particularly happy with the drafting, typing and photo reproduction staff of these two universities.

I wish to thank Professor W. L. Fisher, Director of the Bureau of Economic Geology, for writing the foreword of this book and my former colleagues, Professor W. E. Galloway of the University of Texas and Dr D. K. Hobday of Bridge Oil Ltd., Australia, for co-writing Chapter 12 of this book.

Even though most of the works included in this book were completed at these two universities, many of the original and fundamental concepts had been developed while I was working for Imperial Oil Ltd. (Exxon affiliate in Canada). My utmost gratitude goes out to all the pioneering scientists who worked in the late 1960s and early 1970s in Imperial's Geological Research Departments headed by Drs C. R. Evans and F. J. Moretti.

With respect to the reproduction of figures and data published in various journals and books, the following authors, academic societies and publishers kindly granted me their special permission, which is deeply appreciated; Dr A. Chiarelli; Dr W. G. Dow; Professor J. D. Haun; Dr. J. A. Momper; American Association of Petroleum Geologists; American Petroleum Institute; Bureau of Economic Geology, the University of Texas at Austin; Canadian Society of Petroleum Geologists; Elsevier Science Publishers, Amsterdam; Geological Society of America; Geological Survey of Canada; Gulf Publishing Company; Japanese Association of Petroleum Technologists (presently Japanese Association for Petroleum Technology); *Oil and Gas Journal*; Petroconsultants Ltd; Plenum Publishing Corp.; Prentice-Hall, Inc.; Scientific Press Ltd; Society of Petroleum Engineers; Springer-Verlag, Berlin; United States Geological Survey; and Wyoming Geological Association.

Contents

Chapter 1

Introduction

Hydrocarbons generated in fine-grained clastic rocks were probably disseminated at first, but later they moved from their host rocks to more porous and permeable beds to form a petroleum accumulation. Levorsen (1967) defined the movement of hydrocarbons from nonreservoir rocks to reservoir rocks as *primary migration*, thereby distinguishing it from their concentration and accumulation within the reservoir rocks, which he termed *secondary migration*.

PRIMARY AND SECONDARY HYDROCARBON MIGRATION

Primary migration of hydrocarbons is one of the most controversial matters in petroleum geology. It involves at least four major problems:

1. The time when hydrocarbons migrate.
2. The form in which they migrate, such as molecular solution, micellar solution or separate hydrocarbon phase.
3. The cause of migration.
4. The source of the motive water and the cause of its movement.

In discussing primary migration, it is important to make a clear distinction between liquid and gaseous hydrocarbons, because their respective physical and chemical properties, which are important in primary migration, are significantly different.

The time of primary oil migration has also been a controversial matter. Wilson (1975) stated that, 'Many geochemists advocate that oil is expelled from source rocks through the heat and pressure imposed by deep burial. Conversely, many experienced petroleum geologists point out that empirical data on the habitat of oil accumulations call for early, short-lived expulsion of hydrocarbons shortly after the burial of source and reservoir sediments'. In other words, many geochemists believe in relatively late migration while many petroleum geologists still consider migration to have taken place relatively early.

Whether hydrocarbons migrate from source to reservoir rocks dissolved in water or as separate entities has been another controversial matter. Even if most hydrocarbons migrate in solution in water, whether they are borne along with the moving water or move primarily by diffusion (the water can stay static while the hydrocarbons themselves move from a more to a less concentrated position by molecular solution) is a different kind of problem.

If most hydrocarbons migrate in their own phase separate from water, then we wish to know whether they move as tiny droplets or as films. These forms may influence whether the primary direction of movement is vertical or horizontal with reference to the bedding planes of the fine-grained source rocks. As with hydrocarbons in solution, we do not yet understand whether hydrocarbons in their separate phase move by themselves while most of the formation water remains static, or move along with the formation water.

As possible causes of primary migration, forces from sediment compaction, aquathermal effect, osmosis, clay-mineral conversion, capillary pressure, buoyancy, etc., are considered. The significance of each cause may be influenced by other factors such as the time and the form of hydrocarbon migration.

Even after significant generation of hydrocarbons, the concentration in source rocks is only in the order of a few hundred ppm or less. On the other hand, the source rocks contain a significant amount of water, which can move by the hydraulic gradient even though the rate of movement would have been quite slow. Excellent evidence of such water movement in shales are porosity–depth curves from different parts of the world, such as those shown in Fig. 1.1.

All the curves in Fig. 1.1 indicate a tendency for shale porosity to decrease with increased burial depth, which suggests that water was expelled during burial by one means or another.

Because most source rocks contain such a large volume of water, and because that water moves, we sometimes assume that the movement of such compaction water is important in moving the generated hydrocarbons. If this assumption is valid, then the source of the water and the cause of its movement must be examined carefully for a better understanding of primary migration. Regarding the movement of water, one

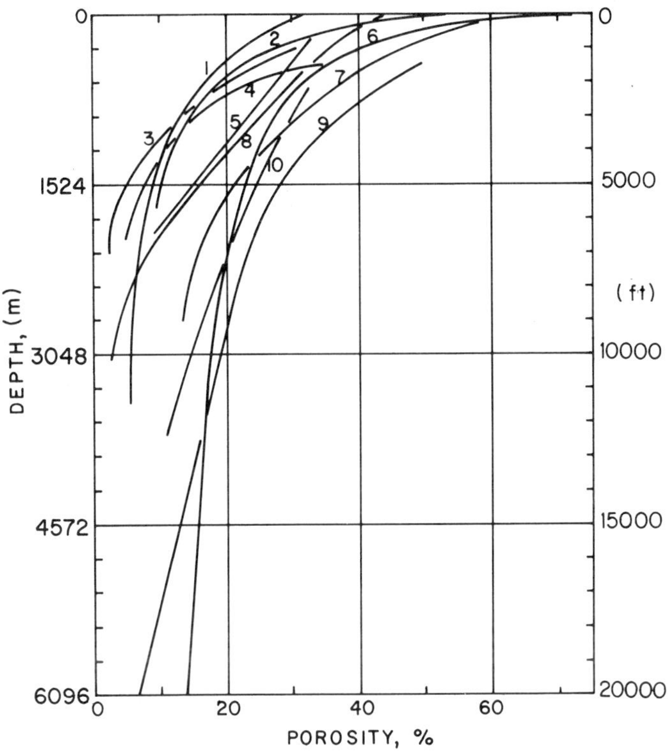

FIG. 1.1. Relationship between porosity and depth of burial for shales and argillaceous sediments. (From Rieke and Chilingarian, 1974, Fig. 17) 1, Proshlyakov (1960); 2, Meade (1966); 3, Athy (1930); 4, Hosoi (1963); 5, Hedberg (1936); 6, Dickinson (1953); 7, Magara (1968); 8, Weller (1959); 9, Ham (1966); 10, Foster and Whalen (1966). (Reproduced by permission of Elsevier Science Publishers, Amsterdam.)

may wish to consider the volume, the direction (vertical or horizontal) and the timing of the water movement.

There are two different kinds of water moving in a sedimentary basin. Each has specific characteristics as follows:

(a) Meteoric Water

1. Movement of meteoric water is important in the relatively shallow intervals of a sedimentary basin.
2. The direction of small-scale movement can be either from sand-

stone to shale or from shale to sandstone. But most movement of this type of water may occur in sandstones only.

3. The direction of large-scale water movement is from the basin's edges to its centre, or from shallow to deep. However, the water that entered the basin centre must eventually move out of the basin to maintain a material (water) balance.
4. The volume of water moving is unlimited, because the amount of rain is unlimited.
5. Movement of this type of water is probably unimportant in primary migration, but it may affect the trapping condition of hydrocarbons in a pool. Meteoric water could alter the quality of hydrocarbons by biodegradation and water washing.
6. Movement of this type of water is a present event and may or may not have developed in the geological past.

(b) Sediment-source Water

1. The movement of this type of water takes place in any part of a sedimentary basin (deep or shallow).
2. The principal direction of small-scale movement is from a shale or a clay to a sandstone or other permeable bed.
3. The direction of large-scale movement is from the basin's centre to its edges, or from the deeper parts to the shallower.
4. The volume of water is limited because the amount of sediment in a basin is usually limited.
5. Movement of this type of water is probably important in the primary migration of hydrocarbons.
6. Most such movement occurred in the geological past.

Item 6 in the above lists suggests important differences in the timing of movements of these different types of water, which may cause a significant difference in petroleum assessment.

Figure 1.2, by Hacquebard (1977), may be used to demonstrate this difference. It shows the temperature and coalification history of the Mannville Formation in Alberta, Canada. This formation, which is of Cretaceous age, experienced a relatively uniform and continuous burial and thermal history for the first 72 million years after deposition, as the downward trend shows.

In the latter half of this 72-million-year period, the organic matter within the formation reached significant stages of thermal maturation, as shown by the changing levels of vitrinite-reflectance values (R_0).

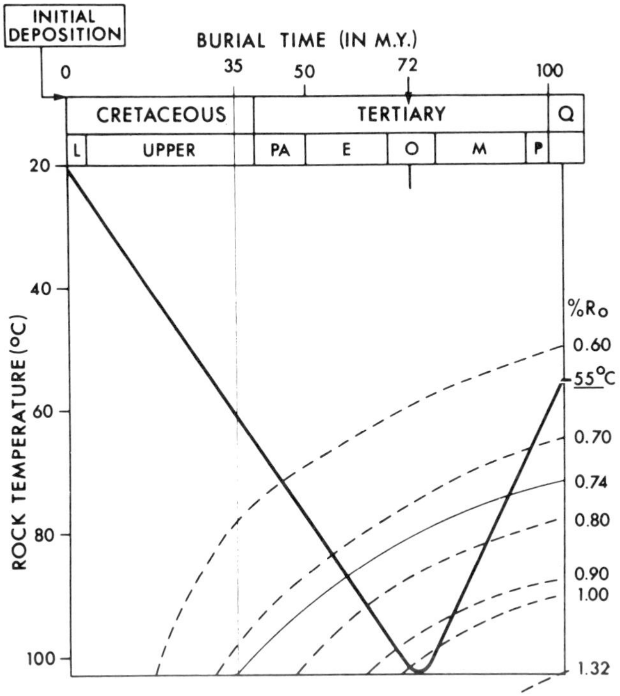

FIG. 1.2. Coalification history of Mannville coal with a rank of 0·74% R_0 (vitrinite reflectance) and present subsurface temperature of 55°C. Contour lines (dashed and solid) show R_0 values with uniform burial rates. (From Hacquebard, 1977.) (Reproduced by permission of the Geological Survey of Canada.)

Sediment-source water, or compaction water, was probably expelled primarily from shales. The principal direction of movement of such water is from a point of greater loading to one of less loading, or from the centre of a sedimentary basin to its edges. The water would also have moved from a hotter place to a cooler by aquathermal effect. This type of water movement would have caused significant movement of the generated hydrocarbons from source rocks to reservoirs (primary migration).

During this period of continuous deposition and burial, the area was largely covered by water. There may have been some scattered topographical highs from which some meteoric water penetrated into deeper sections. But the influence of such penetration would have been relatively insignificant, because at this stage most areas were under water with relatively little topographical relief above the surface.

An important point to make is that the present hydrodynamic condition of the meteoric water, which results primarily from the present topography, has nothing to do with the paleohydrodynamic conditions discussed above.

The final 30 million years (from 72 to 102 million years after deposition, see Fig. 1.2) is the period of significant uplift and erosion, as shown by an upward trend. During this later period, the movement of sediment-source water was probably insignificant, because there was no further compaction of the sedimentary rocks. However, there may have been some late-stage fluid movement caused by the difference in potential drops within the area, due to varying amounts of erosion and the associated temperature drops.

During and after the erosional period, fluid potential would have dropped more significantly in areas of greater erosion, because of greater temperature decline and more contraction of the pore-fluid. The reduced overburden stress after erosion may also have caused a slight expansion of most sedimentary rocks, thus creating more pore spaces. This also may have led to further reduction of pore-fluid potential.

Such unbalanced fluid-potential distribution due to the unequal erosion and fluid contraction must have caused late-stage fluid movement in the lateral direction or some secondary migration.

During this final period of uplift and erosion, hydrocarbons probably continued to be generated, but at a much slower rate because of the reduced formation temperature.

As uplift progressed, significant topographical highs were developed, resulting in some hydrodynamic conditions of meteoric origin. From the study of subsurface fluid potentials, one would be able to construct a present potentiometric map and to indicate possible fluid-flow directions at present. Figure 1.3 is an example of such a potentiometric map of the Viking sandstone in the Western Canada Basin, which merely indicates the present fluid environment (Hill *et al.*, 1961).

Migration and accumulation of petroleum, on the other hand, took place in the geological past. These paleo-events were controlled largely by the paleofluid environments, so that detailed study of the present fluid environment may help only a little in understanding the migration and accumulation of petroleum in this area.

Movement of hydrocarbons in the reservoir (secondary migration) could also have been influenced by most of the agents considered to have been active at the primary-migration stages. Certain other factors, however, are also involved. Once hydrocarbons from source rocks have

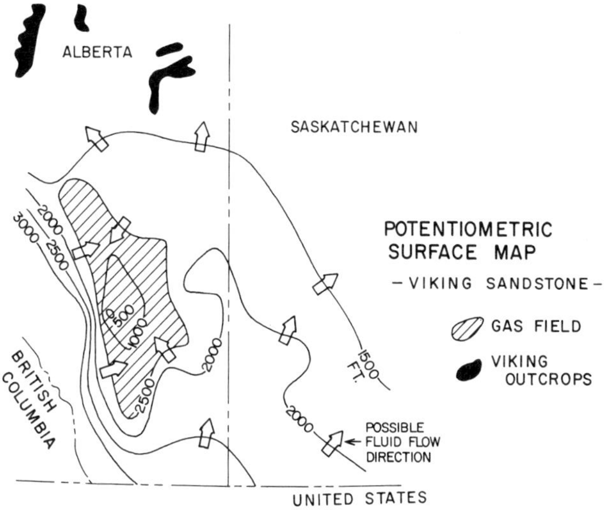

FIG. 1.3. Potentiometric map of Viking sandstone in Western Canada Basin. (From Hill *et al.*, 1961.) (Reproduced by permission of Prentice–Hall, Inc.)

entered reservoir rocks, they encounter several different physical conditions: (1) larger pore spaces, (2) fewer capillary restrictions, (3) less fluid pressure, (4) less structured water. (Water attached to clay surfaces is usually charged electrically and is believed to have a higher viscosity than normal water. The amount of such structured water in reservoirs is less than in shales.)

These new conditions would help to enlarge and connect the hydrocarbon globules. The vertical connection of globules would produce a significant buoyant force. The reduced capillary restrictions and reduced amount of structured or semisolid water would allow this buoyancy to move the interconnected globules to a higher structural position in the reservoir.

SEVEN FACTORS CONTROLLING OIL ACCUMULATION

Although primary and secondary migration are important in controlling petroleum accumulation, they are not the only agencies involved. Figure 1.4 shows seven important factors we must consider in petroleum exploration:

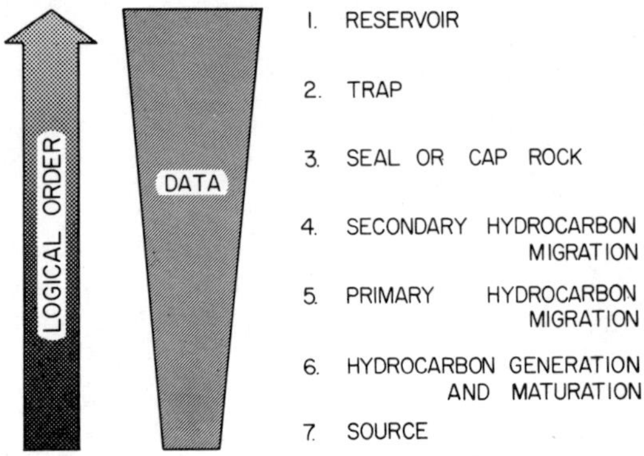

FIG. 1.4. Seven important factors in forming a petroleum accumulation (Magara, 1981). (Reproduced by permission of the American Association of Petroleum Geologists.)

1. Reservoir and effective pore spaces.
2. Trap.
3. Seal or cap rock.
4. Secondary hydrocarbon migration.
5. Primary hydrocarbon migration.
6. Hydrocarbon generation and maturation.
7. Source.

Of this list, items 4–6 are physical and chemical processes related to a petroleum accumulation, while items 1, 2, 3 and 7 are rock (or material) itself or certain forms of sedimentary rocks.

The logical order of occurrence of these seven factors or events is from item 7 to item 1. In other words, hydrocarbons were generated from source rocks, moved from source to reservoir by the primary migration agent, and migrated within the reservoir by the secondary migration process. Before and after such hydrocarbons reach a trap, they must be sealed by cap rocks. At the final trapping position, the reservoir must have sufficient porosity and permeability for petroleum production.

Although from 7 to 1 is the order in terms of timing, the abundances of available subsurface data do not usually follow it. We usually have good control of data on reservoir (1), trap (2) and seal (3) once a first well has been drilled in a structure. There are many unsolved problems regarding primary (5) and secondary (4) migration, and we usually have very little information on these factors. Source (7) and hydrocarbons (6) generated from it can be studied by geochemical methods if the source rock exists at the location where the well was drilled. However, if hydrocarbons were generated in the synclinal areas and migrated into the anticline where the well was drilled, the data from the anticline have little value in evaluating source potential and level of hydrocarbon generation.

In summary, although it would be ideal to have sufficient information on all the factors listed in Fig. 1.4, the amount of available data usually decrease from item 1 to item 7, especially when only a few wells have been drilled in an area.

Most petroleum accumulations are considered to have been accompanied by all the factors listed in Fig. 1.4. However, the presence of all of these factors in a certain area today does not necessarily indicate a petroleum accumulation. It could be that the timing of their occurrence in this area was not appropriate, so that petroleum either could not accumulate properly or was lost later.

Most existing textbooks on petroleum geology have discussed all the factors listed in Fig. 1.4 in terms of their current presence or absence, their extent and their effectiveness. However, none of these books has stressed the importance of the geological timing of their development or destruction. I believe that a proper interpretation of the relative timing of these geological factors will help us to discover more petroleum accumulations.

Geology, or petroleum geology, is a historical science. Two basic methods express the historical nature of a rock. The first method analyses the geologic age of the rock at the present time. For example, a rock may be estimated to be 100 million years old—Cretaceous age. Sedimentologists may study the environment in which this rock was deposited 100 million years ago, and if there is differential deposition of various rock types, geologists may be able to analyse the change in depositional environments through geologic time. However, for a given rock formation, only one geologic age and depositional environment can be inferred. Such a method of historical analysis presents a static view of geology since the only variables considered are those of geologic age and depositional environment.

The second type of historical analysis is based on the concept of successive historical changes with burial of a given rock of a certain geologic age. The study of sediment diagenesis may be a typical example of this second type of geological analysis. For example, what a Devonian reservoir rock, with its present geologic age dated to 380 million years, looked like during the Triassic time (180–225 million years ago) is a different kind of question. And it is one that can be quite important in petroleum geology because most petroleum migration occurred neither at the stage of deposition of the rock nor at the present stage. The migration probably took place between these stages.

This second type of historical analysis may be called a dynamic analysis of the rock, since for a given sample of rock, a series of historical interpretations with regard to changing rock properties may be made. The dynamic analysis is particularly important in petroleum exploration. Our judgement on whether a Devonian reservoir retained sufficient porosity and permeability during the Triassic time, when significant generation and migration of petroleum may have taken place, could have a significant effect on our assessment.

In natural science, it is always desirable or necessary to provide direct evidence to support theories, hypotheses, and interpretations. In the static analysis of geology, it is usually possible to provide direct or indirect evidence for the interpretations given by a geologist. However, if the dynamic analysis or interpretation is made, it is more difficult to provide solid evidence to support the analysis, because the events interpreted by a geologist occurred in the geologic past and no longer exist at the present time. In some cases, only inferred evidence may be provided to support the interpretation, however.

An excellent example of the difference between past and present events or situations may be observed in the hydrostatically pressured and the geopressured intervals of a relatively young sedimentary basin. Based on the concepts and observations of sediment compaction, there would be almost no doubt among geologists that hydrostatically pressured and normally compacted intervals were made by the proper balance of sediment loading and compaction, and the resultant fluid expulsion. In other words, these intervals experienced significant internal fluid movement in the geologic past.

The geopressured and undercompacted intervals are, on the contrary, believed to have been caused by the imbalance of sediment loading and fluid expulsion. In other words, the fluids did not move in these intervals as much as they did in the hydrostatically pressured intervals in the geologic past.

At the present time, if the excess fluid pressures (above hydrostatic pressure) or the fluid potentials are measured and if the values are plotted three-dimensionally, the geopressured sections would show more variations of the potentials, horizontally and vertically, suggesting a greater tendency of fluid motion at present, than the hydrostatic pressure sections. This means that in the hydrostatic pressure sections the most active fluid motion due to sediment compaction ceased before present, whereas in the geopressured sections some fluid motion still does exist. Note that even if the sediment permeability is relatively large, the fluid would not move significantly, if there is no or only a very little variation of fluid potential in the former sections.

The above example suggests that the situations observed presently are not necessarily identical to those in the geologic past. In other words, there is no effective way, using the data available at present, to document what happened in the geologic past.

Many of the dynamic interpretations presented in this book may not be checked by the presently available data, due to the reasons mentioned above. However, such interpretations are desirable or even necessary in some cases, since the processes of petroleum generation, migration, and accumulation were dynamic. The conceptual difference between the static and dynamic analysis of geology can be understood relatively easily, but the practical and dynamic analysis of a rock cannot easily be made in most cases.

DYNAMIC INTERPRETATION

If a geologist studies subsurface rocks for oil exploration, the data available are those describing present conditions. For example, it is not difficult to obtain porosity or permeability values for a reservoir that has been buried to a certain depth and is of a certain geologic age. By plotting the porosity or permeability values of reservoirs at different depth points, it may be possible to see a trend of changing porosity or permeability with increase of present burial depth. Such a plot is, therefore, a profile of porosity or permeability on the basis of present burial conditions. This type of information may be called 'static', because it indicates only the present condition.

To make a realistic historical analysis of the porosity or permeability of each reservoir, the geologist must extract 'dynamic' information from the static information. The difference between information that is present or static and that which is historical or dynamic may be explained by following simple examples.

Let us assume that a porosity–depth profile such as the one shown in Fig. 1.5 is established in an area, on the basis of the porosity measurements of reservoirs at different present burial depths. The porosity of reservoir A at 20 000 ft, which is 50 million years old, is 10%. The

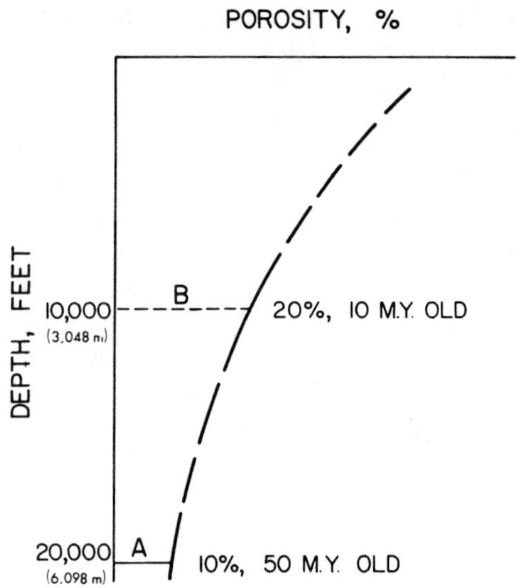

FIG. 1.5. Schematic diagram showing sandstone porosity–depth plot of a well (Magara, 1981). (Reproduced by permission of the American Association of Petroleum Geologists.)

porosity of another reservoir B at 10 000 ft, which is 10 million years old, is 20%.

Let us also assume that the geothermal gradient during active sedimentation and burial in the geological past was almost identical to the present geothermal gradient. Another assumption to be made is that the reservoir porosity is controlled primarily by three factors—pressure (or depth of burial), geologic age and temperature.

On the basis of the above information and assumptions, let us discuss the porosity of reservoir A when it was buried to 10 000 ft. An easy answer would be 20%, based on the present porosity of reservoir B whose burial depth is 10 000 ft. This answer is, however, not necessarily the true solution, because the geologic age of reservoir A by the time it was buried to 10 000 ft was probably older than 10 million years, on the basis of the age difference between these two reservoirs 10 000 ft apart (50 – 10 = 40 million years). Therefore, the porosity of reservoir A at that time would have been less than 20%. A realistic estimate of the porosity of reservoir A in the geological past can be made only by a sophisticated mathematical model that takes into account many geological, physical and chemical factors. Stephanson (1977) recently discussed the dynamic relationship among geologic age, pressure, and temperature on sandstone porosity.

Figure 1.6 shows a schematic example of shale porosity versus depth relationship for a well. Shales above about 12 000 ft (3658 m) are normally compacted, while those below 12 000 ft (3658 m) are undercompacted. The level of compaction of these undercompacted shales when they reached a depth of 11 000 ft (3353 m) is in question.

Assuming these shales experienced a normal compaction history (as shown by the solid line up to 12 000 ft (3658 m) or at the top of undercompacted zone), and were rebounded to the present porosity level, the compaction level at that time may be interpreted to be at A (Fig. 1.6). The rebound of shales by smectite (or montmorillonite) dehydration was suggested by Powers (1967), but was more recently rejected as the principal cause of undercompaction (Magara, 1975).

Therefore, the most likely level of compaction of these undercompacted shales, when they reached 11 000 ft (3353 m), may be shown by B (Fig. 1.6) on the dashed, curved line. Porosity at B is significantly higher than that at A, suggesting that B shales probably never reached the level

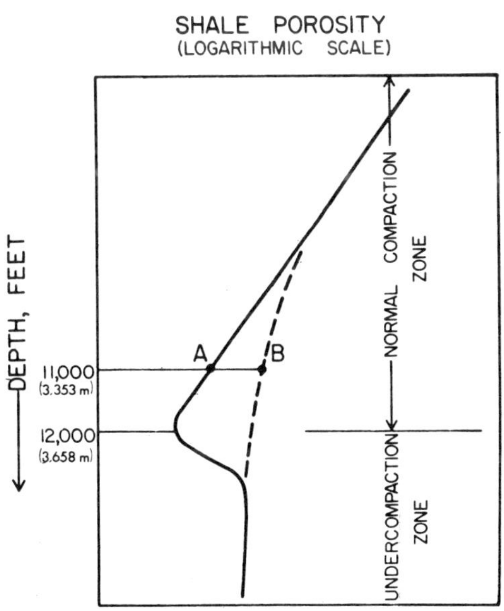

FIG. 1.6. Schematic diagram showing shale porosity–depth plot of a well. (Reproduced by permission of the American Association of Petroleum Geologists.)

of compaction illustrated at A. A more realistic historical change of porosity of these undercompacted shales may be shown by the dashed, curved line in Fig. 1.6.

The purpose of the above-mentioned porosity examples is to demonstrate the difference between the present profile which is static information, and the history of a changing geologic factor, which is dynamic information.

To make a realistic historical interpretation of any of the geologic factors listed in Fig. 1.4, such dynamic information is necessary. However, in most subsurface conditions, the only information available for geologists is static, and it is therefore necessary to deduce the dynamic information from it. A computer can be used quite effectively in this regard, but certain interpretation can be made manually as well.

THEME OF THIS BOOK

The theme of this book is, therefore, analysis of the timing of the development or destruction of the factors listed in Fig. 1.4. As pointed out previously, many problems related to petroleum generation, migration and accumulation are still controversial. A reference book would describe all of these conflicting ideas and hypotheses, so that the reader of such a book, though he would become knowledgeable in the various problems and arguments, might also find himself a bit confused.

For the practical application of geological methods to oil exploration, a geologist, at least at a given time, must have a set of basic geologic ideas that are in harmony if he is to have a chance to reach a conclusion from his interpretation. Otherwise he may arrive at a number of conflicting conclusions. If he has no idea which conclusion is most realistic, then his results are useless for practical purposes.

Because the principal objective of this book is practical application, I have been at pains to avoid describing too many conflicting opinions. Rather I have stressed my own ideas and the ideas of others that do not conflict seriously. I have presented many subsurface data that support these ideas. In short, the ideas and methods in this book have a common basis or consistency, so that none should be in serious conflict with another.

This book, then, is essentially a manual or guidebook based on the ideas I favour for practical application, rather than a reference book. I sincerely hope that other books based on other ideas and methods will be written by other petroleum geologists with different backgrounds, so that an operations geologist can choose the book or ideas that he prefers for practical application.

PRESENCE VS NONPRESENCE IN SOURCE ROCKS

In most prolific sedimentary basins, modern geochemical techniques have been used to indicate the level of petroleum maturation and the potential of source rocks. Most conclusions have been made on the basis of the concentration of hydrocarbons remaining in the source rocks at present. If the amount of hydrocarbons expelled from the source rock is always a certain proportion or percentage of the total hydrocarbons generated, as some geochemists assume, a study of the concentration of hydrocarbons now remaining in the source rocks would provide a meaningful estimate of source potential. However, if the expulsion of hydrocarbons is unrelated to the total hydrocarbons generated, then a higher hydrocarbon concentration in a rock does not necessarily mean a good source potential. There is a possibility that the rock generated some amount of hydrocarbons but that most of them may have stayed within the original rock. In other words, there was no effective expulsion of the hydrocarbons generated. Such a rock cannot be considered a good source rock, even though it may now have a relatively high concentration of hydrocarbons.

Suppose, on the other hand, that the same amount of hydrocarbons was generated in another rock but a larger proportion has been expelled. In this case, geochemical analysis of this rock would indicate a lower concentration of hydrocarbons at present. A geochemist might conclude that this is a poor source rock because of the current relatively low concentration of hydrocarbons. His conclusion, however, would not be warranted and such a rock, in my opinion, could be an excellent source rock because it flushed the generated hydrocarbons more effectively.

Tissot and Pelet (1971) analysed the amounts of hydrocarbons, resins and asphaltenes in shales adjacent to a reservoir in Algeria. Figure 1.7 shows the results of their analyses in mg/g organic carbon. While the amounts of resins and asphaltenes in the shale remain relatively constant, the amount of hydrocarbons decreases toward the reservoir. This suggests the presence of preferential expulsion of hydrocarbons over resins and asphaltenes in the shales closer to the reservoir.

The difference in hydrocarbon contents at the 14-m point and at the

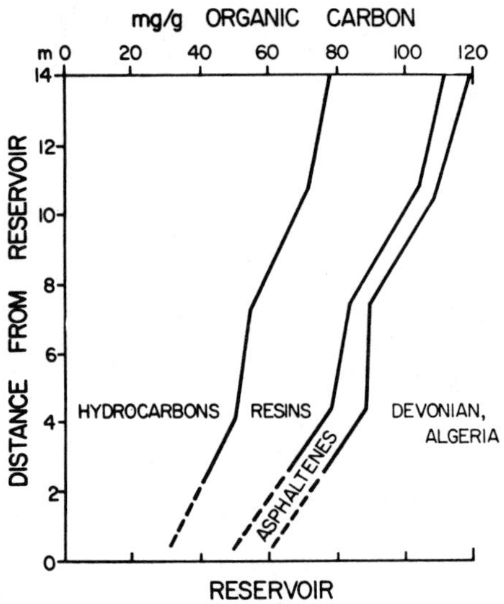

FIG. 1.7. Diagram showing amounts of hydrocarbons, resins, and asphaltenes of Devonian shales close to a reservoir in Algeria (Magara, 1981). (Data from Tissot and Pelet, 1971.) (Reproduced by permission of the American Association of Petroleum Geologists.)

near-reservoir point is about 40 mg/g organic carbon. Since the level of thermal maturation of organic matter within this short interval of 14 m is considered to be almost constant, this 40 mg represents the lowest possible amount of hydrocarbons expelled per gram of organic carbon from the shale closest to the reservoir.

In any situation, if geochemical source-rock analyses were made for the samples at the 14-m point and at the near-reservoir point, the former may be classified as a better source rock because of its higher hydrocarbon concentration.

I judge, however, that the latter shale has undergone more hydrocarbon expulsion than the former; that, in other words, the latter has been a better source rock although it now has less hydrocarbon concentration.

Figure 1.8 shows plots of shale transit time (A), remaining hydrocarbons in shales (C15+, B), estimated average rate of hydrocarbon expulsion from shales (C), and estimated rate of hydrocarbon generation (D) versus depth of a well in northern Japan. The type of organic matter in the shales is predominantly woody with some herbaceous organic

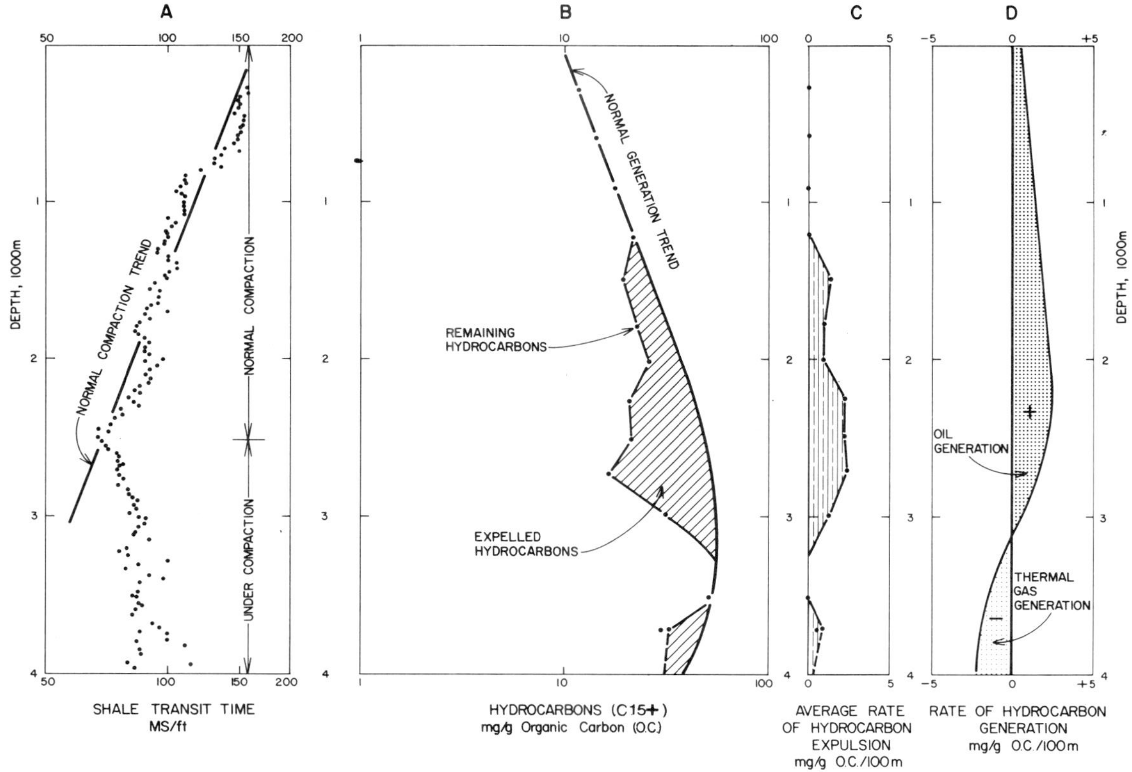

FIG. 1.8. Diagram showing plots of shale transit time (A), C15+ hydrocarbons (B), average rate of hydrocarbon expulsion (C), and rate of hydrocarbon generation (D) against depth of a well in northern Japan. (From Magara, 1981.) (Reproduced by permission of the American Association of Petroleum Geologists.)

matter. Relative concentration of these organic matters stays relatively constant throughout the interval shown in Fig. 1.8.

Petroleum generation from a given organic matter may be expressed by the Arrhenius equation (an exponential function) relating geologic time and temperature (Connan, 1974). Since geologic time and temperature normally increase with depth, it is expected that the petroleum generation trend can be expressed by a near straight line ('Normal Generation Trend') until a peak is reached on a semi-log plot, such as shown in Fig. 1.8B. The difference between this estimated generation trend and the actually measured value of C15+ may provide an estimate of amount of hydrocarbons expelled from shales (shaded areas in Fig. 1.8B).

The estimated amount of the expelled hydrocarbons is relatively large in the interval between 1250 and 3250 m (Fig. 1.8B), which covers most of the normal compaction zone (Fig. 1.8A). This suggests that the expulsion of the generated hydrocarbons was relatively effective in the normally compacted interval where the expulsion of pore-water was also effective.

Although a minor quantity of hydrocarbons was generated above 1250 m, most of the generated hydrocarbons seem to have stayed in the source rocks. The effective expulsion seems to have started below about 1250 m. As discussed by Magara (1979), the effective hydrocarbon expulsion in their own phase may start at a point in time when the viscosity of the water remaining in shales exceeds that of hydrocarbons. The viscosity of water in shales would increase continuously with compaction, as the relative proportion of structured or semi-solid water in shales increases with the expulsion of relatively mobile water (Magara, 1979). On the other hand, the viscosity of hydrocarbons normally decreases with burial and increasing temperature (Holmquist, 1965; Levorsen, 1967). Therefore, there must be a point in time or in depth beyond which less viscous hydrocarbons were preferentially expelled over the remaining structured water. This critical level of shale compaction seems to have been reached at about 1250 m in this well.

In the deepest interval below 3250 m, which is undercompacted (Fig. 1.8A), the expulsion efficiency of the generated hydrocarbons is believed to have been relatively low.

Figure 1.8C shows the estimated average rate of hydrocarbon expulsion since burial to 1250 m (mg/g organic carbon/100 m of burial). The rate of hydrocarbon expulsion peaked in the interval between 2200 and 2700 m.

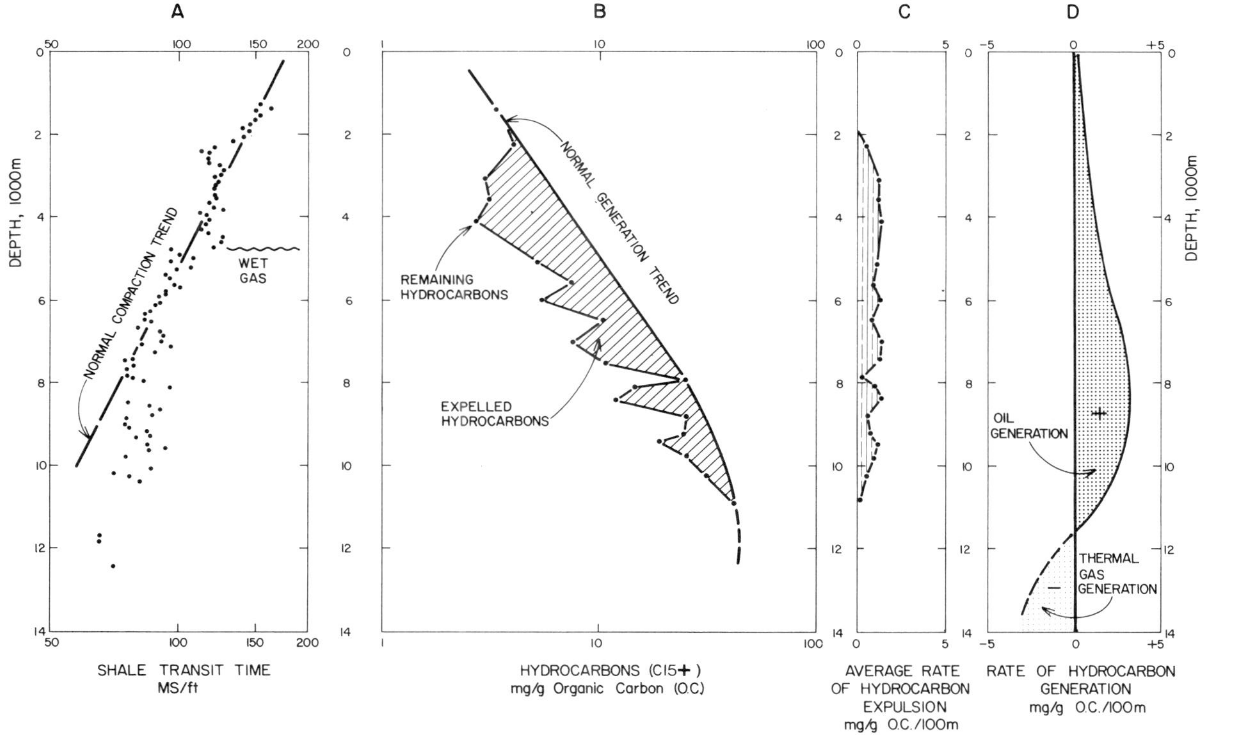

FIG. 1.9. Diagram showing plots of shale transit time (A), C15+ hydrocarbons (B), average rate of hydrocarbon expulsion (C), and rate of hydrocarbon generation (D) against depth of a well in Canadian Arctic region (from Magara, 1981). (Geochemical data from Powell, 1978.) (Reproduced by permission of the American Association of Petroleum Geologists.)

Using Magara's (1978) concept of rate of hydrocarbon generation, Fig. 1.8D was constructed. The rate of hydrocarbon generation increases from the surface to about 2300 m and then decreases rapidly. Below 3100 m, the rate has negative values, suggesting the destruction of the generated hydrocarbons (thermal gas generation, Fig. 1.8D). By comparing the plots in Figs. 1.8C and 1.8D, we may conclude that the timings of both generation and expulsion of hydrocarbons seem to have been concurrent in this area.

Figure 1.9 is a similar example from Canadian Arctic region. Geochemical data are derived from Powell (1978). There is an excellent match between the expulsion (Fig. 1.9C) and generation (Fig. 1.9D) of hydrocarbons. Organic matter of this well is predominantly herbaceous.

The original data for constructing Figs. 1.7, 1.8, and 1.9 are static. Using these static data, one is able to interpret the rate of changes with burial, which is dynamic information. Even without a computer, such a dynamic interpretation is possible.

In summary, then, to understand the amount and the mechanisms of petroleum migration, we must obtain data on the hydrocarbons that

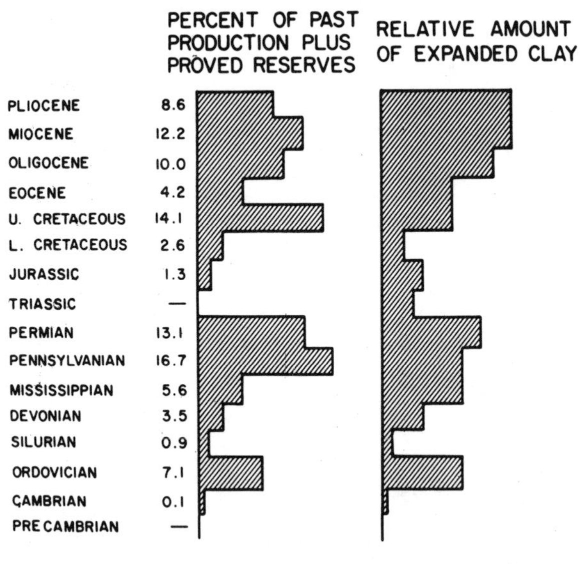

FIG. 1.10. Relationship of oil production to age and relative expanded clay content for United States. (From Weaver, 1960, as adapted by Price, 1976.) (Reproduced by permission of the American Association of Petroleum Geologists.)

have migrated from the source rocks. These hydrocarbons are no longer in the source rocks. However, most geochemical methods based on source rocks provide information on the hydrocarbons that have not moved out and hence are still in the source rocks. Thus, the objective of a geological interpretation for petroleum migration is to deduce the amount of migrated hydrocarbons on the basis of those that remain.

Figure 1.10 presents another example of possible confusion between presence and nonpresence of a geologic factor. Constructed by Weaver (1960), it shows the close relationship between the abundance of expandable clay (or smectite) and the occurrence of accumulated oil in the United States. Figure 1.10 suggests that more oil can be accumulated in formations containing higher concentrations of expandable clay.

The same diagram has been used to demonstrate the significance of

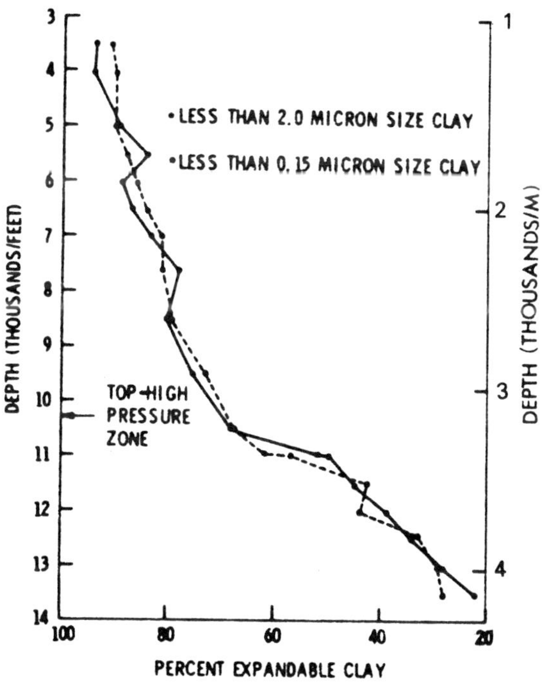

FIG. 1.11. Plot of % expandable clay versus depth showing accelerated increase in diagenesis of smectite to mixed-layer clay and illite. (From Schmidt, 1973.) (Reproduced by permission of the American Association of Petroleum Geologists.)

smectite–illite conversion (Price, 1976), in that the conversion of the clay minerals by the elevated temperature is considered to be the principal mechanism for moving hydrocarbons from source rocks to reservoir rocks (Burst, 1969). The conversion of the clay minerals seems to be quite significant in the Gulf Coast, as shown in Fig. 1.11 (Schmidt, 1973); the percentage of expandable clay seems to have changed from about 90 at 4000 ft to about 20 at 14 000 ft.

A question may arise: should the oil occurrence be associated with the low quantity of expandable clay rather than the high quantity, if clay–mineral conversion is the principal migration mechanism for the generated hydrocarbons—since most of the expandable clay would have become non-expandable after the conversion.

The association of the accumulated oils and the expandable clay may simply mean that the presence of the expandable clay, not its conversion, is a favourable factor in the migration and accumulation of petroleum. Expandable clays commonly contain a large quantity of water which is electrically charged by the clay surfaces and has a significantly higher viscosity than normal water (Low, 1976). Such water may be called 'structured' or 'semi-solid' water (Dickey, 1975). In summary, then, Weaver's plot (Fig. 1.10) suggests that the presence of such semi-solid water, rather than clay–mineral conversion, may have some bearing on the migration of petroleum.

In conclusion, what is needed for a historical analysis of each of the geological factors listed in Fig. 1.4 is dynamic information, but what is available in most cases is merely the static information that applies to the present. It is, therefore, recommended that petroleum geologists use these static data to reconstruct the dynamic data that will give them a clearer picture of the geologic past.

REFERENCES

Athy, L. F. (1930) Density, porosity and compaction of sedimentary rocks *A.A.P.G.*, **14**, 1–24.

Burst, J. R. (1969) Diagenesis of Gulf Coast clayey sediments and its possible relation to petroleum migration. *A.A.P.G.*, **53**, 73–93.

Connan, J. (1974) Time–temperature relations in oil genesis. *A.A.P.G.*, **58**, 2516–21.

Dickey, P. A. (1975) Possible primary migration of oil from source rock in oil phase. *A.A.P.G.*, **59**, 337–45.

Dickinson, G. (1953) Geological aspects of abnormal reservoir pressures in Gulf Coast Louisiana. *A.A.P.G.*, **37**, 410–32.

Foster, J. B. and Whalen, H. E. (1966) Estimation of formation pressure from electrical surveys—offshore, Louisiana. *Petrol. Technol. J.*, **18**, 165–71.

Hacquebard, P. A. (1977) Rank of coal as an index of organic metamorphism for oil and gas in Alberta. In *The Origin and Migration of Petroleum in the Western Canadian Sedimentary Basin*, Ch. 3, Alberta, Bulletin 262, Geol. Survey of Canada, p. 136.

Ham, H. H. (1966) New charts help estimate formation pressures. *Oil Gas J.*, **64**, 58–63.

Hedberg, H. D. (1963) Gravitational compaction of clays and shales. *Am. J. Sci.*, **31**, 241–87.

Hill, G. A., Colburn, W. A. and Knight, J. W. (1961) Reducing oil finding costs by use of hydrodynamic evaluations. In *Petroleum Exploration, Gambling Game or Business Venture*, Institute of Economic Petroleum Exploration, Development and Property Evaluation, Prentice-Hall, Englewood, N. J., pp. 38–69.

Holmquist, H. J. (1965) Deep pays in Delaware and Val Verde Basins. In Young, A. and Galley, J. E. (eds), *Fluids in Subsurface Environments*, A.A.P.G. Memoir 4, pp. 257–79.

Hosoi, H. (1963) First migration of petroleum in Akita and Yamagata prefectures. *Japan. Assoc. Mineral. Petrol. Econ. Geology J.*, **49**, 43–55, 101–14.

Levorsen, A. I. (1967) *Geology of Petroleum*, 2nd edn, Freeman, San Francisco, Calif., p. 724.

Low, P. F. (1976) Viscosity of interlayer water in montmorillonite. *Soil Sci. Soc. Am. Proc.*, **40**, 500–505.

Magara, K. (1968) Compaction and migration of fluids in Miocene mudstone, Nagaoka Plain, Japan. *A.A.P.G.*, **52**, 2466–2501.

Magara, K. (1975) Reevaluation of montmorillonite dehydration as cause of abnormal pressure and hydrocarbon migration. *A.A.P.G.*, **59**, 293–302.

Magara, K. (1978) Time of oil generation and migration and oil window. *Canadian Petrol. Geology Bull.*, **26**, 152–5.

Magara, K. (1979) Structured water and its significance in primary oil migration. *C.P.G. Bull.*, **27**, 87–93.

Magara, K. (1981) Static versus dynamic interpretation in petroleum geology. *A.A.P.G.*, **65**, 531–9.

Meade, R. H. (1966) Factors influencing the early stages of compaction of clays and sands—review. *Sediment. Geology J.*, **36**, 1085–1101.

Powell, T. G. (1978) An assessment of the hydrocarbon source rock potential of the Canadian Arctic Islands, Geol. Survey of Canada, Paper 78-12, 82p.

Powers, M. C. (1967) Fluid-release mechanisms in compacting marine mud rocks and their importance in oil exploration. *A.A.P.G.*, **51**, 1240–54.

Price, L. C. (1976) Aqueous solubility of petroleum as applied to its origin and primary migration. *A.A.P.G.*, **60**, 213–44.

Proshlyakov, B. K. (1960) Reservoir properties of rocks as a function of their depth and lithology. *Geol. Neft. Gaza.*, **12**, 24–9.

Rieke III, H. H. and Chilingarian, G. V. (1974) *Compaction of Argillaceous Sediments*, Elsevier, Amsterdam, 424p.

Schmidt, G. W. (1973) Interstitial water composition and geochemistry of deep Gulf Coast shales and sandstones. *A.A.P.G.*, **57**, 321–37.

Stephanson, L. P. (1977) Porosity dependence on temperature: limits on maximum possible effect. *A.A.P.G.*, **61**, 407–15.

Tissot, B. and Pelet, R. (1971) Nouvelles données sur les mécamosmes de genèse et de migration du petrole: simulation mathématique et application á la prospection. *Proc. 8th World Petrol. Congress*, 35–46.

Weaver, C. E. (1960) Possible uses of clay minerals in search for oil. *A.A.P.G.*, **44**, 1505–18.

Weller, J. M. (1959) Compaction of sediments. *A.A.P.G.*, **43**, 273–310.

Wilson, H. H. (1975) Time of hydrocarbon expulsion, paradox for geologists and geochemists. *A.A.P.G.*, **61**, 407–15.

Chapter 2

Organic Carbon Concentration in Source Rocks

INTRODUCTION

Among organic geochemists working in the oil industry and academia, it is widely accepted that the concentration of organic matter (or organic carbon) is one of the most important factors controlling the quality of source rocks. There would be a certain lower limit of the concentration in source rocks to form significant petroleum accumulations. Another factor influencing the source rock quality is the type of organic matter, which is more or less a geochemical problem. Refer to the books written by Tissot and Welte (1978) and by Hunt (1979) for most geochemical problems including the type of organic matter.

Among seven controlling factors for petroleum accumulations, such as source, maturation, primary and secondary migration, trap, seal, and reservoir (Fig. 1.4), the source rock may effectively rank first, because nothing else matters if there is no source (Momper, 1978).

On the basis of 25 742 samples from the Russian Platform, Ronov (1958), (see Fig. 2.1) indicated a significant difference between the petroliferous and the nonpetroliferous regions in concentration of organic carbon which consists of the main proportion of organic matter. Sediments (clays, sands, and carbonate) of the petroliferous region contain 0·5–1·4 average weight % organic carbon, whereas those of the nonpetroliferous region contain only 0·1–0·4 average weight %. Based on Ronov's (1958) data and on other known evidence, an arbitrary empirical lower limit of 0·4 weight % organic carbon (about 0·5 weight % organic matter) is generally accepted as the minimum kerogen concentration required for significant hydrocarbon generation (Dow, 1978). Figure 2.2 is a similar plot of organic carbon percentage vs grain size of sediments in the Bering Sea (from Bordovskiy, 1965). The highest

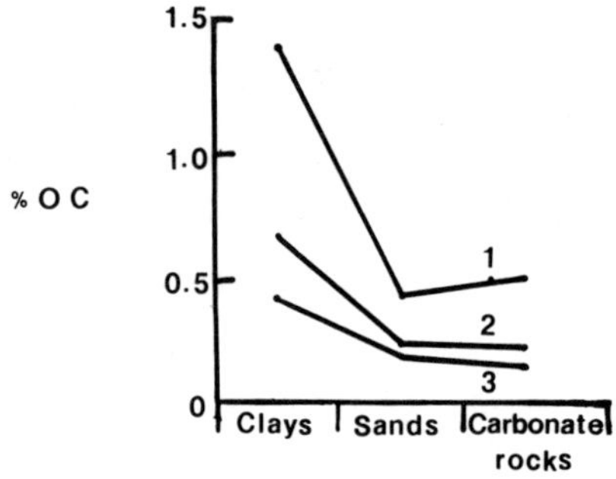

FIG. 2.1. Average weight % of organic carbon (OC) in Russian Platform; (1) petroliferous areas, (2) whole platform, and (3) non-petroliferous areas. (Based on Ronov's 25 742 samples, 1958.)

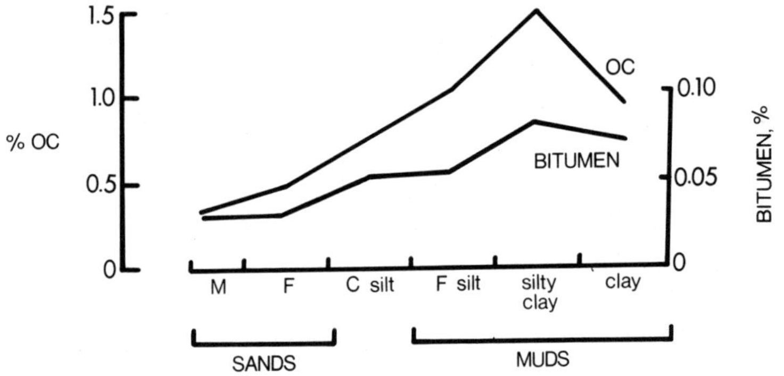

FIG. 2.2. Average weight % of organic carbon (OC) and bitumen in sediments of different grain sizes of Bering Sea. M, medium, F, fine, C, coarse. (From Bordovskiy, 1965.) (Reproduced by permission of Elsevier Science Publishers, Amsterdam.)

concentration of organic carbon and bitumen are observed in the silt-clay mud.

A relationship may exist between depositional environment and concentration of organic carbon or organic matter because production, transportation, and destruction or preservation of organic matter can be influenced by or may be controlled by the environmental factors (Dow, 1978). Assuming that such an empirical relationship is established in a region, it is possible to predict the organic carbon concentration for undrilled areas within the region if the depositional environment is known. However, prediction of the depositional environment particularly before drilling is not an easy task, even if some seismic records were available. I believe that predicting the organic matter concentration from depositional environment is a logical approach but is generally a difficult and unreliable method if there is no drilled data.

If a relationship between the sedimentation rate and the organic carbon concentration is established, on the contrary, it may have more practical application. This is particularly true in undrilled areas where the prime data is a seismic record. I think that the estimation of sedimentation rate from the seismic data is easier than that of the depositional environment. The average rate of sedimentation of a given geologic interval in an undrilled area may be estimated from the thickness of the interval and the corresponding geologic time, both of which can be estimated using seismic data from the regional correlation.

The results described in this chapter suggest that this method may only be used to predict possible maximum concentrations of organic carbon in undrilled areas—the true values can be significantly lower. Because any decision on petroleum exploration must first be made by a positive idea on a prospect, this approach may provide useful information on the best potential of concentration of organic carbon. Comparison of the estimated best potentials of different prospects may allow a petroleum geologist to rank these prospects. Because the information on synclinal areas, from where most petroleum is believed to have derived, is seldom available owing to the lack of well data there, such estimates of the best potentials of the organic carbon concentration in these synclinal areas can be quite valuable.

POSSIBLE RELATIONSHIP BETWEEN SEDIMENTATION RATE AND ORGANIC CARBON CONCENTRATION

Dow (1978, p. 1588) states that, 'If the supply of organic matter is constant, its concentration in sediments should be inversely related to the depositional rate of mineral particles. Therefore, areas of high sedimentation rates, such as deltas, should contain sediments with relatively low organic-carbon concentrations. If the sedimentation rate is too slow, however, much of the organic matter reaching the bottom may be consumed by heterotrophic organisms before it can be protected by burial. Intermediate sedimentation rates which minimize the effects of both dilution and consumption commonly result in the most organic-rich sediments'. He then summarized (p. 1588) that 'the most organic-rich sediments are deposited in areas of high organic productivity, where the supply of bottom oxygen is minimal, the water is reasonably quiet, and the sedimentation rate of mineral particles is intermediate'.

Figure 2.3 shows the maximum organic carbon % and the rate of

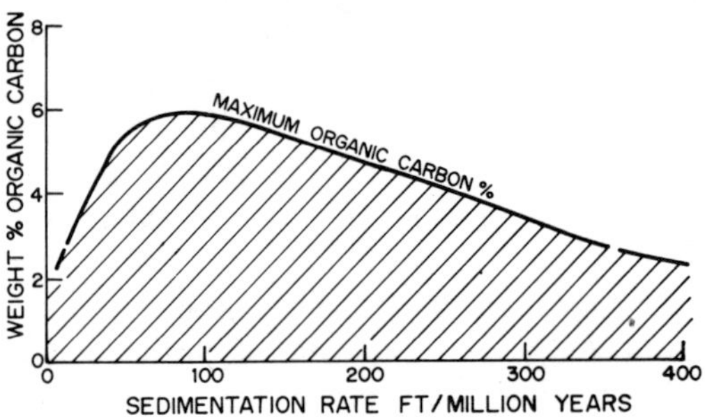

FIG. 2.3. Maximum weight % organic carbon versus average rate of sedimentation for the Canadian Arctic region. About 300 data points are plotted in the shaded area. The data are derived from Powell (1978).

sedimentation for the Canadian Arctic region. Approximately 300 samples from Powell (1978) were plotted to establish the possible maximum organic carbon % (curved line in Fig. 2.3). The results suggest that the organic carbon % is relatively high for the rate of sedimentation or burial, around 80 ft/million years (ft/MY). The maximum organic carbon percentage decreases as the rate of sedimentation increases or decreases from this value.

There are many points plotted below the curved line in Fig. 2.3, suggesting that, even if the rate of sedimentation were optimum, the organic matter was not always available in the areas of deposition. The relationship shown by this curved line suggests only the possible maximum organic carbon percentage—the real percentage could be lower. There is no compaction correction made for the estimation of the rate of sedimentation in Fig. 2.3 (the effect of compaction correction will be discussed in the next section).

Figure 2.4 is a plot of the organic carbon percentage versus the sedimentation rate for the Pennsylvanian and Permian rocks of the Palo Duro and Midland Basins of Texas. The data are derived from Dutton (1980). It shows that the highest organic carbon percentage corresponds to 80–90 ft/MY (rate of sedimentation).

Another example from the Japanese Tertiary formations is shown in

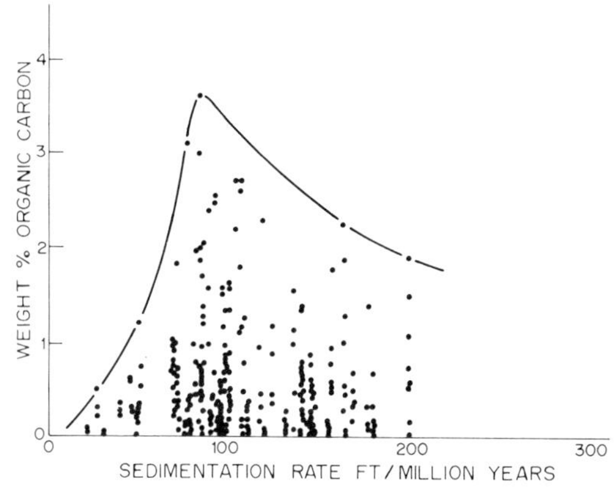

FIG. 2.4. Plot of weight % organic carbon versus sedimentation rate for Pennsylvanian and Permian rocks of Palo Duro and Midland Basins in Texas. The data are derived from Dutton (1980).

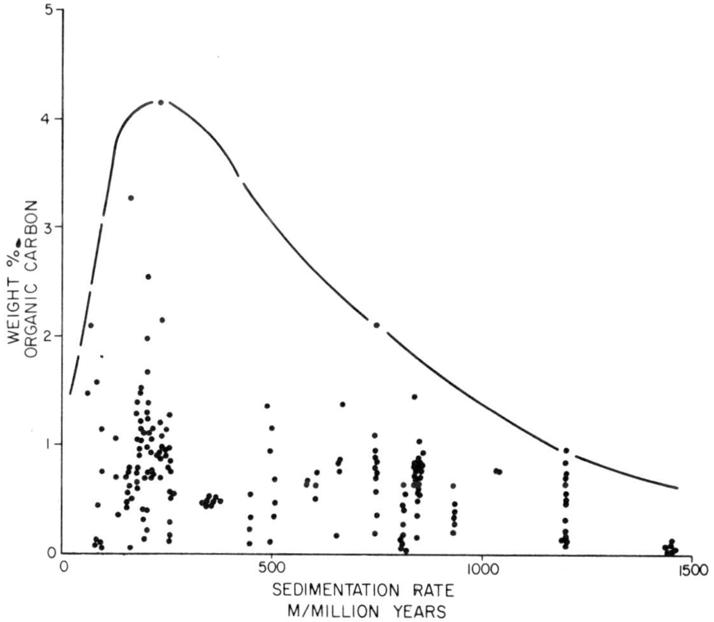

FIG. 2.5. Plot of weight % organic carbon versus sedimentation rate for Tertiary sediments of Japanese basins. Data are derived from Japanese Association of Petroleum Technologists (1973).

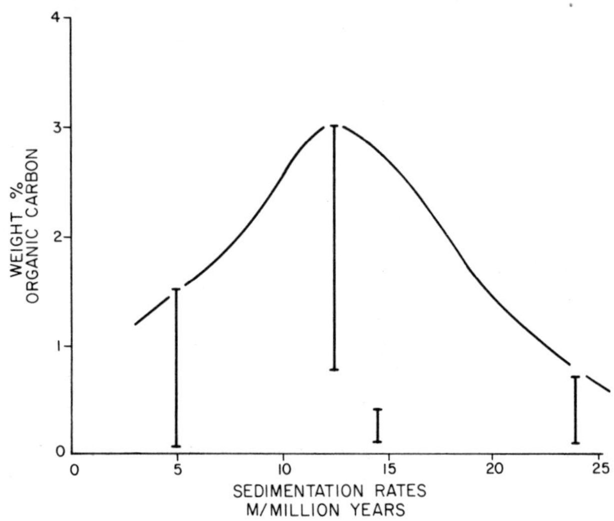

FIG. 2.6. Plot of weight % organic carbon versus sedimentation rate for Lower Cretaceous–Jurassic sections of Peregrina Canyon, Mexico. (Data from Scholle and Arthur, 1980.)

Fig. 2.5 (data compiled by Japanese Association of Petroleum Technologists, 1973). The optimum sedimentation rate for the organic carbon concentration in this case is about 250 m/MY, which is significantly higher than those of the previous two examples (Figs. 2.3 and 2.4). Figure 2.6 is a similar plot of the Lower Cretaceous–Jurassic sections of the Peregina Canyon, Mexico (Scholle and Arthur, 1980).

Figure 2.7 shows a plot of the total organic carbon (TOC) percentage and sedimentation rate (m/MY) on logarithmic paper, constructed by Ibach (1982) using the Deep Sea Drilling Project (DSDP) data. Depositional environment and geologic age are also depicted in this figure. Lithologic zonation of the same data is shown in Fig. 2.8, in which black shale has the highest percentage of organic carbon followed by siliceous shale. Figure 2.9 represents summary plots of the linear regression analyses for the data of different lithologies (see Table 2.1 for the statistical data), showing that the rate of sedimentation corresponding to the peak TOC percentage ranges from about 14 to 41 m/MY (or from about 46 to 134 ft/MY).

All the examples shown by Figs. 2.3–2.9 seem to support Dow's (1978) concept that there is an optimum rate of sedimentation for the maximum concentration of organic carbon in sediments.

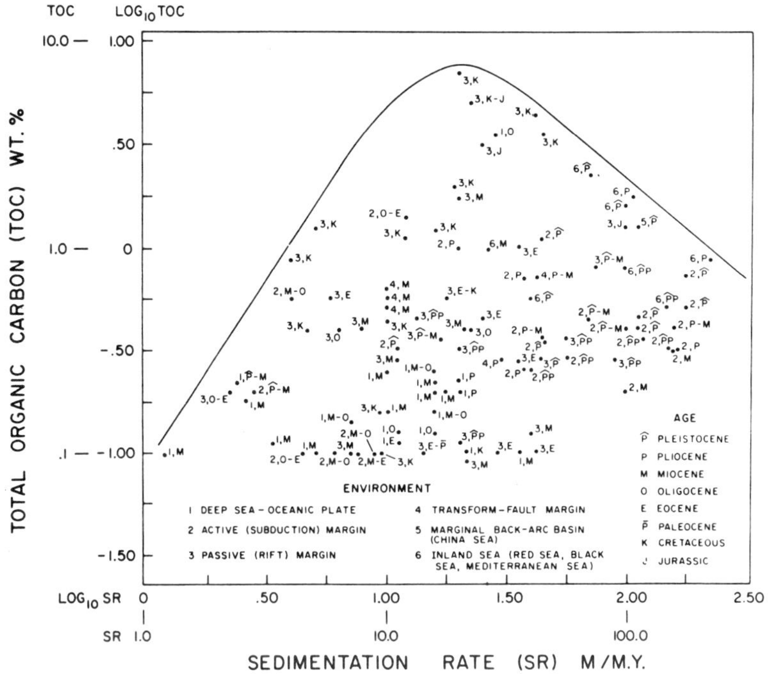

FIG. 2.7. Relationship between total organic carbon percentage and sedimentation rate from Deep Sea Drilling Project data. Depositional environment and geologic age are also shown. (From Ibach, 1982.) (Reproduced by permission of the American Association of Petroleum Geologists.)

It may be argued that, under a certain restricted and reducing environment, even relatively slow rates of sedimentation may accumulate significant amounts of organic carbon. In such a case, the peak concentration of organic carbon would correspond to a relatively slow sedimentation rate. Even in this case, if the sedimentation rate was extremely slow or nearly nil, there must have been some destructive effects on organic matter. On the other hand, if the sedimentation rate was relatively fast, there would have been some dilution effects on organic carbon concentration.

In conclusion, I believe that Dow's (1978) basic concept can be applied to almost any geologic environment as the principal mechanism. However, the optimum rate of sedimentation for the maximum organic concentration is not always constant, but would vary from basin to

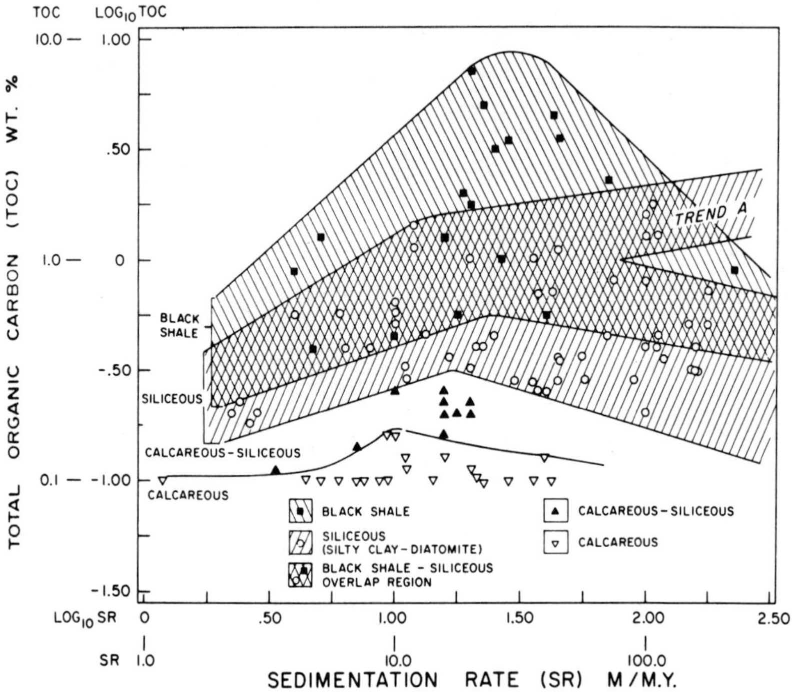

FIG. 2.8. Relationship between total organic carbon percentage and sedimentation rate from Deep Sea Drilling Project data. Lithologic field zonation is also shown. (From Ibach, 1982.) (Reproduced by permission of the American Association of Petroleum Geologists.)

basin, or from area to area within a basin, depending upon its depositional environment. This further suggests that the optimum sedimentation rate must be empirically obtained within each basin or area, and the result may only be applied within the same basin or area.

Figure 2.10 shows a schematic cross section of the Louisiana Gulf Coast (Dow, 1978). The shaded areas in Fig. 2.10 show productive intervals, and the interval between the vitrinite reflectance values of $0{\cdot}6\,R_0$ and $1{\cdot}35\,R_0\%$ indicates the estimated oil generation zone. The thickness of the Miocene section is depicted at the top of this diagram; its greatest thickness is in areas of paleo continental slope, and it thins in the northward and southward directions. Since the sediments shown in this cross section were deposited within a given geologic time interval

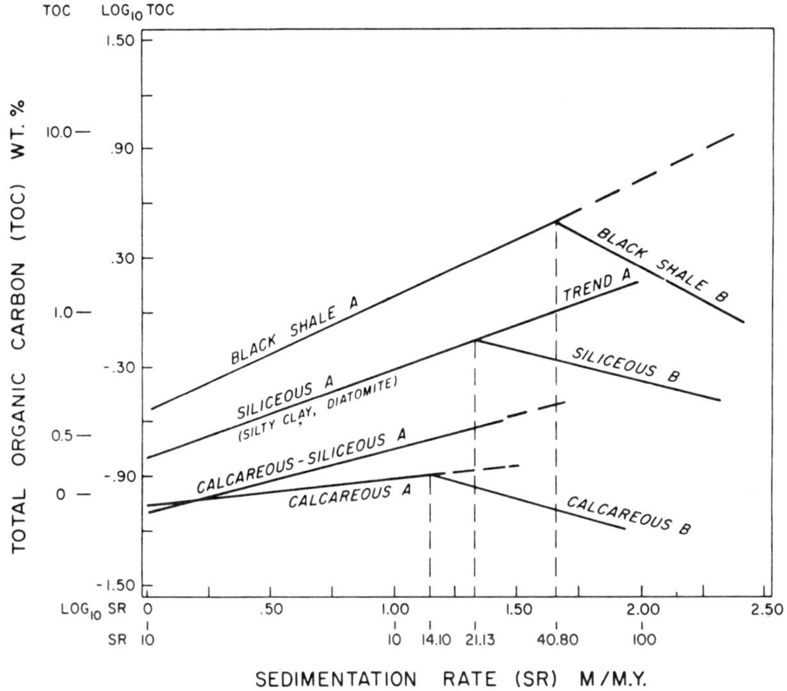

FIG. 2.9. Relationship between total organic carbon percentage and sedimentation rate from Deep Sea Drilling Project data. Summary of lithologic field zonation is shown. (From Ibach, 1982.) (Reproduced by permission of the American Association of Petroleum Geologists.)

(Miocene), the average rate of sedimentation during the Miocene time can also be shown schematically by the diagram at the top of Fig. 2.10.

The position of the maximum sediment thickness or of the maximum sedimentation rate will move southward for younger geologic intervals (Pliocene and Pleistocene).

If the supply of organic matter or carbon is uniform in this region, we expect profiles of organic carbon percentage to be either A or B (Fig. 2.11). Profile A may be expected if the average sedimentation rate was relatively fast; the organic carbon percentage is depressed in the area of the maximum rate of sedimentation owing to the dilution effect as discussed by Dow (1978). The organic carbon percentage increases towards the areas of slower sedimentation rates within the slope region.

TABLE 2.1
Total organic carbon versus sedimentation rate (mathematical expression and statistical correlation for each lithology)

Lithology	*Relationship A* *incr. TOC with sed. rate*	*R*	*Relationship B* *decr. TOC with sed. rate*	*R*	*Critical sed. rate*
Black shale	Log(TOC) = 0·64 log (SR) − 0·55	0·52	Log(TOC) = −0·73 log (SR) + 1·72	0·91	40·8 log = 1·66
Siliceous (silty clay, diatomite)	Log(TOC) = 0·51 log (SR) − 0·80	0·88	Log(TOC) = 0·34 log (SR) + 0·30	0·45	21·1 log = 1·325
Calcareous siliceous	Log(TOC) = 0·37 log (SR) − 1·10	0·75			
Calcareous	Log(TOC) = 0·16 log (SR) − 1·07	0·67	Log(TOC) = −0·35 log (SR) − 0·48	0·88	14·0 log = 1·15

R = Pearson's coefficient of correlation (0·0–1·0).
TOC = Total organic carbon (wt.%).
SR = Sedimentation rate (m/M.Y.).

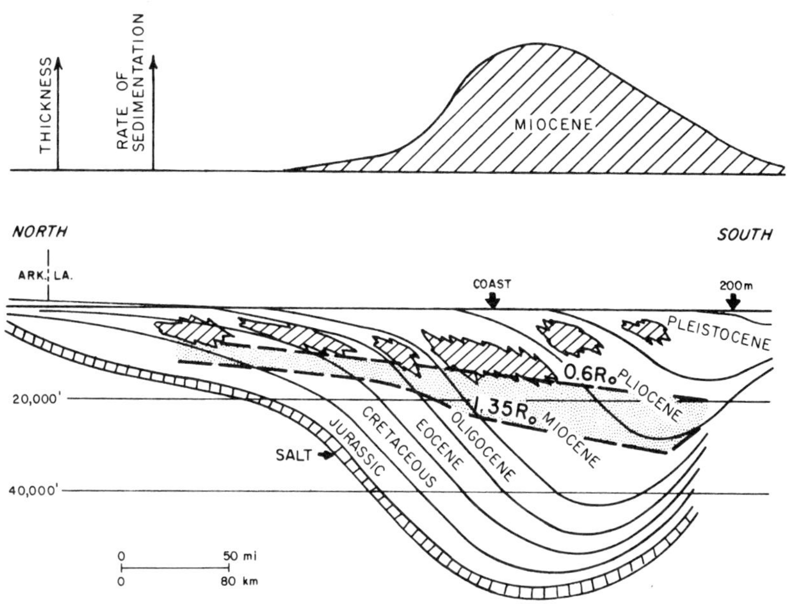

FIG. 2.10. Cross section showing distribution of most productive intervals of oil in Gulf Coast by Dow (1978). Estimated vitrinite reflectance values of 0·6 and 1·35% R_0 are superimposed to show possible oil generation zone. Top diagram shows thickness and sedimentation rate of Miocene section. (Reproduced by permission of the American Association of Petroleum Geologists.)

Because the effect of the organic matter destruction by oxygen seems to become significant in the areas of slow sedimentation rates, the organic carbon percentage may decline towards the shelf and rise regions. Closs *et al.* (1974) reported a significant oxidizing effect in the continental rise as well as in the shelf (Fig. 2.12).

If the average rate of sedimentation was moderate, the profile B may be expected (Fig. 2.11). In this case, the optimum rate of sedimentation occurred in the middle of the continental slope region. If the average sedimentation rate was slow, we may expect an organic carbon profile similar to that of Fig. 2.11B, but the value would be further reduced because of the poor preservation condition there.

The concept described above may be documented by the actual organic carbon percentage of the Tertiary sediments of the Louisiana Gulf Coast, shown in Fig. 2.13 (Dow and Pearson, 1975). There seems to be a dilution effect in the lower bathyal zone (0·59%), whereas the values

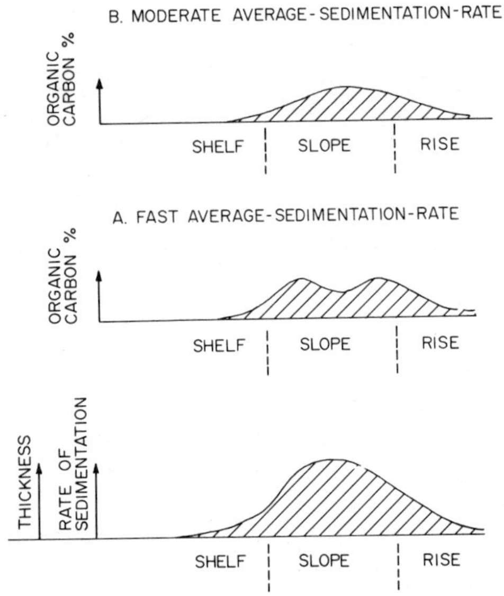

FIG. 2.11. Schematic diagram showing possible profiles of organic carbon percentage for fast sedimentation rate (A) and moderate sedimentation rate (B) throughout continental shelf–slope–rise region.

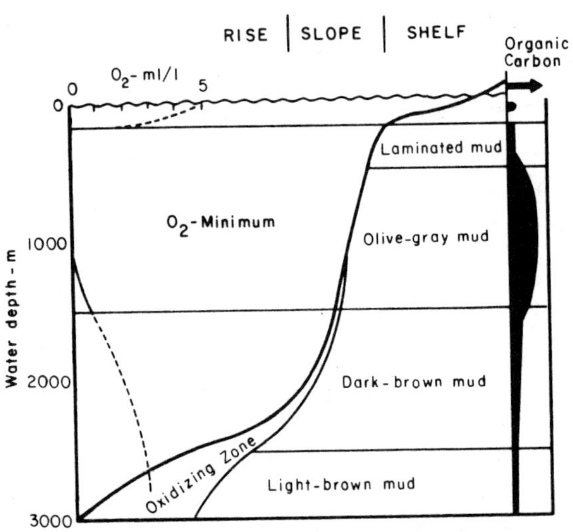

FIG. 2.12. Schematic section across India–Pakistan continental margin showing oxygen contents in sea water and quantitative contents of selected components of surface sediments. Organic carbon is most abundant in upper slope oxygen-minimum zone. (From Closs *et al.*, 1974.) (Reproduced by permission of Springer-Verlag, Federal Republic of Germany.)

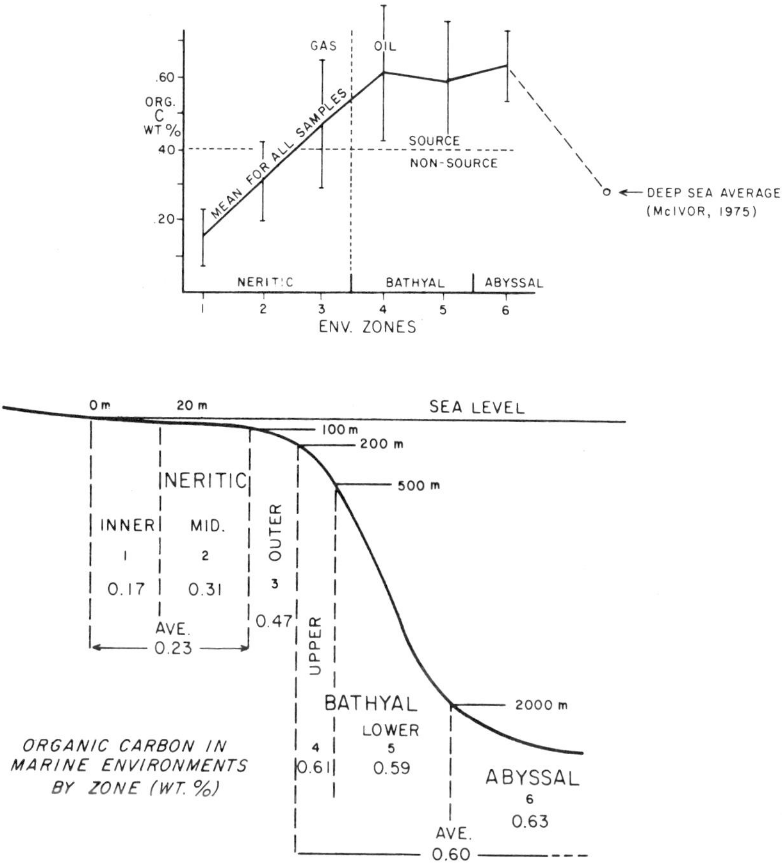

FIG. 2.13. Mean organic carbon content by environmental depth zones in Louisiana Gulf Coast Tertiary. Bars in top diagram represent standard deviation of mean and approximate range in values. Average organic carbon percentage of deep sea sediments (McIvor, 1975) is added to top diagram. Figure adapted from Dow and Pearson (1975). (Reproduced by permission of the authors.)

in the upper bathyal (0·61%) and abyssal (0·63%) zones are slightly higher. The average value (0·3%) of the organic carbon percentage from deep sea (McIvor, 1975) is added to the top diagram of Fig. 2.13. This top diagram is similar in shape to the schematic one shown in Fig. 2.11A.

Figure 2.14A shows a schematic relationship of the sedimentation rate and organic carbon percentage. Where the average sedimentation rate was relatively fast, there would have been some dilution effects; the trend

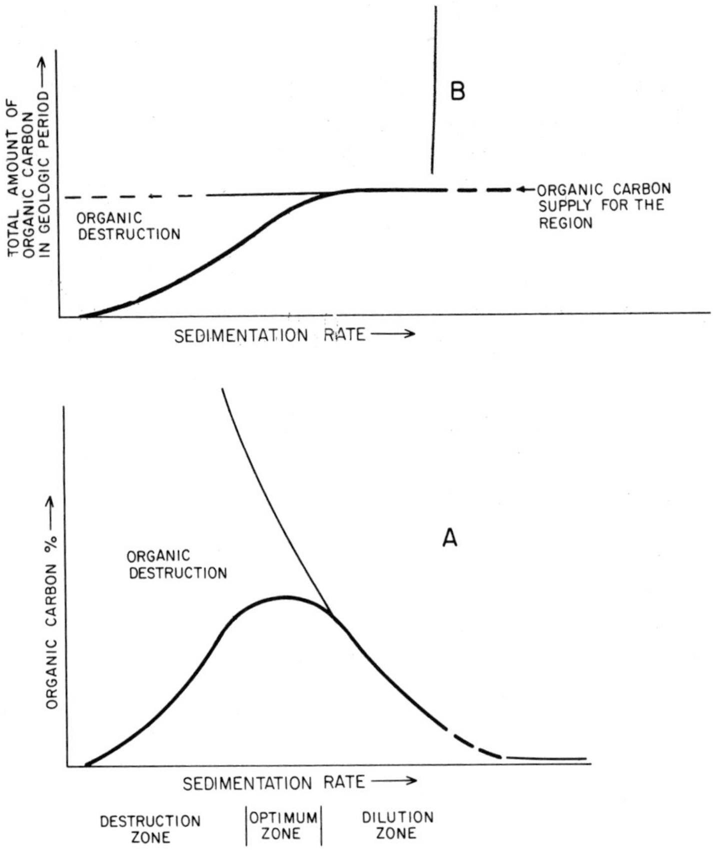

FIG. 2.14. Schematic diagrams showing relationships between sedimentation rate and organic carbon percentage (A), and total amount of organic carbon deposited within a given geologic period (B, thick lines). Thin lines show relationships for a uniform organic carbon supply without organic destruction.

in this area or zone may be approximated by a hyperbola, suggesting that, as the sedimentation rate is doubled, the organic carbon percentage is halved (see a thin curved line in Fig. 2.14A). Even in this area or zone, the total amount of organic carbon that was deposited within a given geologic time interval would be relatively constant, as shown in Fig. 2.14B.

If the rate of sedimentation was relatively slow, there would have been some destructive effects on the organic carbon, as shown in Fig. 2.14A.

The total amount of the organic carbon deposited within a given geologic period would also be reduced in this case (Fig. 2.14B). The maximum organic carbon percentage may be expected in the areas of moderate sedimentation rates (Fig. 2.14A), but the total amount may not be the largest there (Fig. 2.14B).

Figure 2.15 shows similar relationships for two cases of effective (A) and reduced (B) organic carbon supply. If the amount of organic carbon supply is reduced by the decrease of production and/or by the ineffective

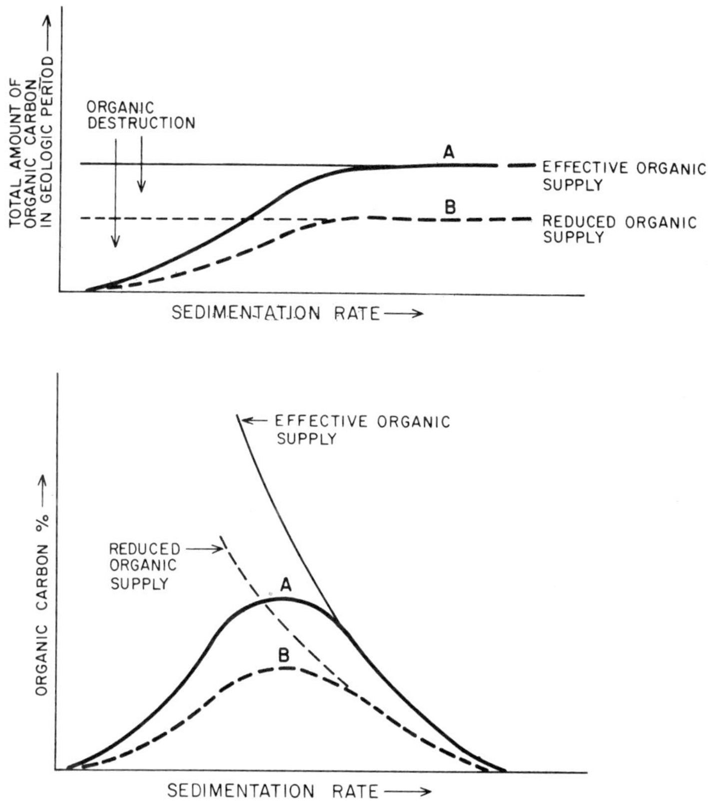

FIG. 2.15. Schematic diagrams showing relationships between sedimentation rate and organic carbon percentage (bottom), and total amount of organic carbon deposited within a given geologic period (top). Relationships for two cases of organic carbon supply, effective (A) and reduced (B), are shown by solid and dashed heavy lines, respectively. Thin lines show relationships for two cases of organic carbon supply without organic destruction.

transportation, the organic carbon percentage for a given sedimentation rate will be reduced (two curves labelled 'B' in Fig. 2.15). The effect of the organic destruction will further reduce the organic carbon percentage if the rate of sedimentation is relatively slow. Many points plotted below the maximum organic carbon percentage line in Figs. 2.3–2.7 may be explained by the combination of the reduced organic supply and of the extensive organic destruction.

EFFECT OF SEDIMENT COMPACTION AND EROSION, AND ORGANIC MATURATION

The plots of Figs. 2.3–2.9 were made on the basis of the rate of sedimentation calculated from the present or compacted sediment thickness. If a proper correction for compaction is applied, the points will move to the right in the plot (Fig. 2.16). Refer to Chapter 6 of this book for the compaction correction. This effect can be significant if a plot is made for sedimentary rocks of variable levels of compaction; point for more compacted rock must move to the right relative to a less compacted rock in the plot (Fig. 2.16).

If there was some erosional effect in the geological past, the average

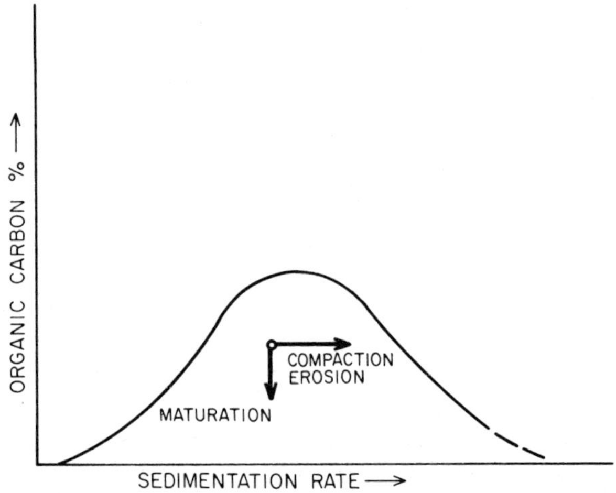

FIG. 2.16. Schematic diagram showing effects of compaction and erosion (horizontal arrow), and of organic maturation (vertical arrow) on plot of sedimentation rate versus organic carbon percentage.

rate of sedimentation would be underestimated because of the eroded sediment's thickness. In this case, the true point must move to the right, as well (Fig. 2.16).

The organic carbon percentage is likely to decrease as the organic maturation advances because of the production of hydrocarbons. This effect tends to move the point downward for a more matured rock, as shown in Figure 2.16. The correction may be made if the level of maturation and the amount of carbon lost by maturation are evaluated.

The effect of maturation on the organic carbon percentage may influence the commonly accepted guideline of the minimum requirement of the organic matter concentration (0·5%) as described above. Tissot (1978) stated that 'in the deeper parts of sedimentary basins, where maturation of the organic matter has reached the dry gas stage, a residual 0·5% organic carbon may be witness of an original 1% or more. Thus, it may have been a potential source rock'.

CHANGE OF ORGANIC CARBON SUPPLY WITH GEOLOGIC TIME

Using the concept described above, one may be able to calculate the total organic carbon supply within formations of a given geologic age. Figure 2.17, showing the significant variation with time, is the result of such calculation based on the data of Fig. 2.18 (Canadian Arctic region). The calculation is made on the basis of the maximum organic carbon % in the relatively undestructed zones. Therefore, the figure shows the variation of the possible *maximum* organic carbon supply with geologic age. The unit used here is grams of organic carbon per cm^2(area) per one million years. In order to estimate the total maximum organic carbon supply for a given formation within a region, one may simply multiply the value in Fig. 2.17 by the corresponding geologic time period and the total area of distribution. The scale on the right-hand side of Fig. 2.17 may be used to estimate the maximum organic carbon supply in million tons per square mile per million years.

Within each geologic period, certain areas may have received a smaller quantity of organic carbon owing to transportation and production problems. Note that Fig. 2.17, which is developed on the basis of the maximum organic carbon percentage shown in Fig. 2.18, illustrates the estimated maximum organic carbon supplies. If significant petroleum maturation occurred, a large quantity of organic carbon would have

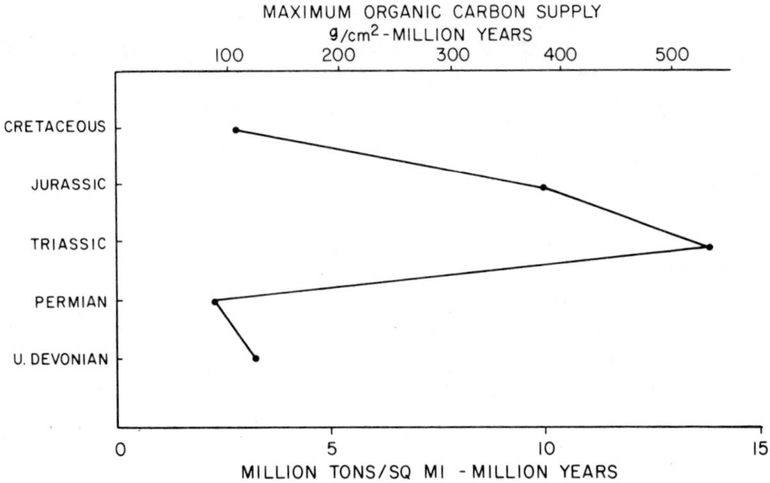

FIG. 2.17. Plot showing change of maximum organic carbon supply with geologic age in Canadian Arctic region, estimated from data of Fig. 2.18.

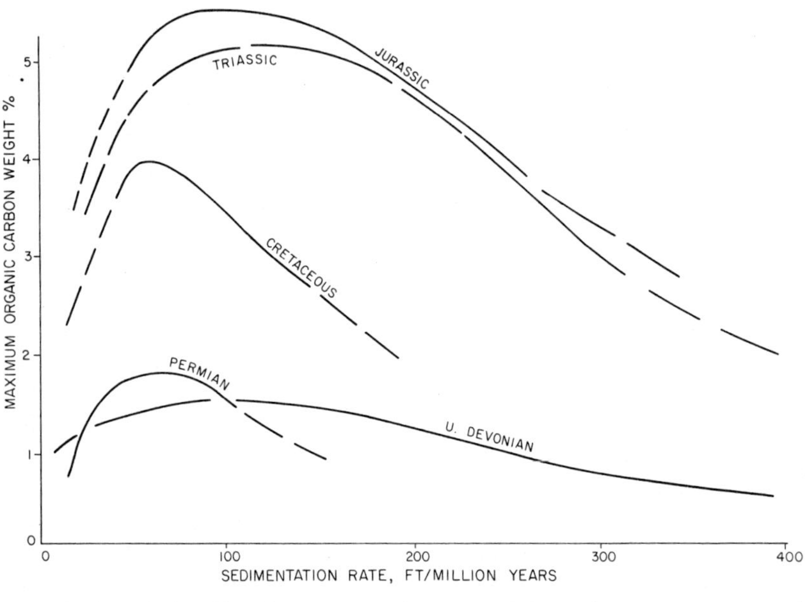

FIG. 2.18. Relationships between maximum organic carbon percentage and sedimentation rate for formations of different geologic ages in Canadian Arctic region. Data derived from Powell (1978).

been converted to the form of hydrocarbons and may have been lost from the source rocks. The relatively low values for both the Permian and Upper Devonian rocks in this region may be due partly to this maturation effect.

REFERENCES

Barker, C. (1977) Aqueous solubility of petroleum as applied to its origin and primary migration: discussion. *A.A.P.G.*, **61**, 2146–9.

Bordovskiy, O. K. (1965) Accumulation and transformation of organic substances in marine sediments. *Marine Geology*, **3**, 3–114.

Closs, H., Narain, H. and Garde, S. C. (1974). Continental margins of India. In *The Geology of Continental Margins*, New York, Springer-Verlag, pp. 629–39.

Dow, W. G. (1978) Petroleum source beds on continental slopes and rises. *A.A.P.G.*, **62**, 1584–1606.

Dow, W. G. and Pearson, D. B. (1975) Organic matter in Gulf Coast sediments, 7th Offshore Tech. Conf. Preprints, Paper OTC 2343.

Dutton, S. P. (1980) Petroleum source rock potential and thermal maturity, Palo Duro Basin, Texas, University of Texas at Austin, Bureau of Economic Geology, Geological Circular.

Hunt, J. M. (1979) *Petroleum Geochemistry and Geology*, W. H. Freeman, San Francisco, 617 p.

Ibach, L. E. J. (1982) Relationship between sedimentation rate and total organic carbon content in ancient marine sediments, *A.A.P.G.*, **66**, 170–88.

Japanese Association of Petroleum Technologists (1973) Nihon no Sekiyu Kogyo to Gijitsu: Petroleum Mining and Technology of Japan, 430 p (in Japanese).

Magara, K. (1978) *Compaction and Fluid Migration—Practical Petroleum Geology*, Elsevier, Amsterdam, 319p.

McIvor, R. D. (1975) Hydrocarbon occurrences from JOIDES Deep Sea Drilling Project, *9th World Petroleum Congress Proc.*, Vol. 2, pp. 269–80.

Momper, J. A. (1978) Oil migration limitations suggested by geological and geochemical considerations. In Physical Constraints on Petroleum Migration, *A.A.P.G.*, **1**, B1–B60.

Powell, T. G. (1978) An assessment of the hydrocarbon source rock potential of the Canadian Arctic Islands, Geological Survey of Canada, Paper 78-12, 82 p.

Ronov, A. B. (1958) Organic carbon in sedimentary rocks (in relation to the presence of petroleum). *Geochemistry*, **5**, 510–36.

Scholle, P. A. and Arthur, M. A. (1980) Carbon isotope fluctuations in Cretaceous pelagic limestones: potential stratigraphic and petroleum exploration tool. *A.A.P.G.*, **64**, 67–87.

Tissot, B. (1978) The application of the results of organic geochemical studies in oil and gas exploration. In Hobson, G. D. (ed.), *Developments in Petroleum Geology – 1*, Chap 2, Applied Science Publishers, London, pp. 53–82.

Tissot, B. and Welte, D. H. (1978) *Petroleum Formation and Occurrence: a New Approach to Oil and Gas Exploration*, Springer-Verlag, New York, 538 p.

Chapter 3

Models of Petroleum Generation and Primary Migration

INTRODUCTION

As discussed in Chapter 1, timings of the formation or destruction of the main geologic factors for a petroleum accumulation are important in assessing a prospect. Such an estimation of timing could be made in terms of absolute geologic age of each factor, if sufficient data are available. If not, one would be able to estimate such merely in terms of burial depth.

In this chapter, three primary migration models for oil and gas based on the following assumptions are considered: (1) migration is a function of petroleum generation; (2) migration occurs in aqueous solution; (3) migration occurs as a separate hydrocarbon phase. Among these conceptual models, the first one ignores several physical, chemical, and mineralogical factors (e.g. compaction, aquathermal effect, clay mineral conversion, capillary pressure, etc.) involved in the primary hydrocarbon migration. It simply assumes that a certain proportion of the generated hydrocarbons moves at the time of primary migration (concept often used by geochemists). Because such an assumption is not always valid, its application may be limited. However, this model requires the least amount of data for applications.

The second model is believed to simulate most natural conditions for gas migration because the aqueous solubility of gas is relatively high, whereas the third model simulates those for oil migration since its aqueous solubility is extremely low. Optimum burial depth ranges for both the oil and gas *migration* with five selected shale-compaction profiles were estimated and are described in this chapter.

Depth ranges for the oil and gas *generation* can also be estimated by relating the shale-compaction profiles to possible geothermal gradients.

If the *generation*-depth range coincides with the *migration*-depth range, the most effective petroleum migration can be expected; otherwise, chances of migrating petroleum would be quite small.

Estimation of the time of petroleum generation and maturation in terms of absolute geologic age will be described in Chapter 5.

MODEL 1: PETROLEUM MIGRATION BEING A FUNCTION OF ITS GENERATION

In the first model; the primary oil and gas migration is assumed to be primarily controlled by the amount of oil and gas already generated in source rocks. A certain proportion of the generated petroleum is expelled from the source rocks in this model, so that more generation would directly cause more expulsion or migration of petroleum. In other words, possible influences by other physical, chemical and mineralogical factors on migration are not considered in this model.

There may be two options in choosing this migration model as follows:

1. Quantity of the migrated petroleum within a given geologic period or a given burial-depth interval is a function of the petroleum generated within the same period or interval.
2. Quantity of the migrated petroleum within a given geologic period or a given burial-depth interval is a function of the total or cumulative amount of petroleum which has been accumulated in the source rock up to this geologic period or burial-depth interval.

The significance of choosing one of these options can be demonstrated by using the schematic diagrams of oil and thermal gas generation shown in Fig. 3.1. Figure 3.1A shows a schematic of amounts of both oil and thermally generated gas with increasing burial depth, and Fig. 3.1B is a plot of the thermal gas derived from Fig. 3.1A. Critical vitrinite reflectance values, R_0, which are important in assessing the levels of petroleum generation and maturation, are also shown. Refer to Chapter 5 for the vitrinite reflectance.

Based on the first assumption shown above, the most important stage of oil migration is interpreted to be at $R_0 = 0{\cdot}6–0{\cdot}7\%$, or at the peak *rate* of oil generation (Fig. 3.1A, see also Magara, 1978*c*). For the thermal gas, it can be shown at the stage between $R_0 = 1{\cdot}0$ and $1{\cdot}35\%$ (Fig. 3.1B).

If the second assumption is chosen, on the contrary, the most significant oil migration stage is interpreted to be at $R_0 = 1{\cdot}0\%$ or at the peak oil

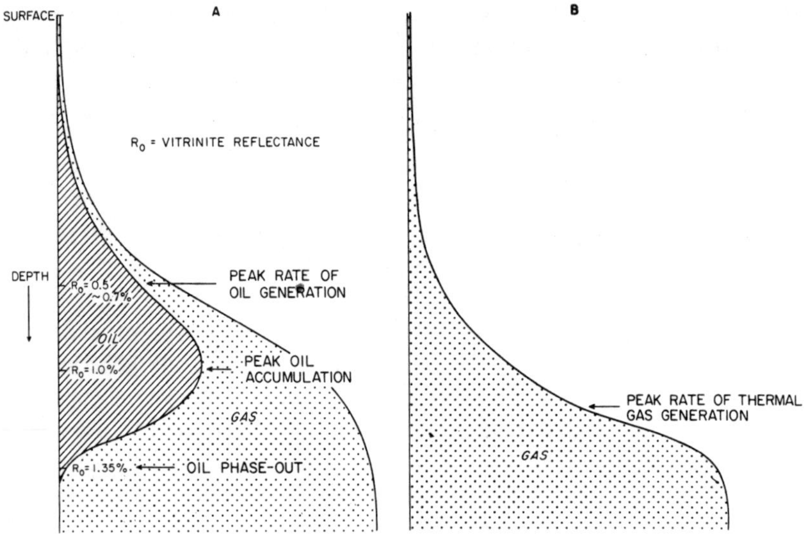

FIG. 3.1. Schematic diagrams showing amounts of oil and gas generated due to thermal effect with burial (A), and amount of gas (B). Critical vitrinite reflectance values, R_0, are also shown.

accumulation in the source rock (Fig. 3.1A), and the peak gas migration can be expected below the level of $R_0 = 1{\cdot}35\%$ or at the peak gas *accumulation* (Fig. 3.1B).

In conclusion, the first assumption was chosen for the migration model because of the following reasons:

1. Let us assume an incremental burial increase and associated petroleum generation from a given source bed. Within a given short geologic period or burial-depth interval, this source bed generated a certain quantity of oil, and a certain proportion of the generated oil was flushed out, leaving behind some unflushed oil in the bed. In the next short geologic period, more oil will be generated, and a part of it will be expelled. However, the chances of flushing the previously unflushed oil would be quite small, even though it may not be impossible. In any situation, the amount to be flushed at this later stage from the previously unflushed oil would be negligibly small.
2. On the basis of the second option, the depth, temperature and

geologic time required for significant petroleum migration would be quite deep, hot and long (or late). However, as pointed out by Wilson (1975), this concept contradicts many subsurface evidences of relatively early migration reported in many petroleum-producing regions. To provide a possible solution for this problem, Magara (1978*c*) proposed a concept similar to the first assumption mentioned above.

3. If the total amount of oil accumulated up to a certain geologic stage controls the amount of migrating oil, then most oil shales, which contain a large quantity of unflushed oil, must have been excellent source rocks. However, I believe that this contradicts the subsurface evidence in many regions.

On the basis of these reasons, the first option is included in the model being discussed in this chapter, in that the quantity of the flushed oil in a given geologic period is primarily controlled by the amount generated within the same period.

A. Oil Migration

The relationship between the total oil generated (Q), oil remaining in source rocks (Q_R), oil thermally destroyed (Q_T), and oil expelled from source rocks (Q_E) can be expressed as follows,

$$Q = Q_R + Q_T + Q_E \tag{3.1}$$

If the amount of oil generated but not thermally destroyed is termed Q_O, which is given as,

$$Q_O = Q - Q_T \tag{3.2}$$

then we obtain

$$Q_O = Q_R + Q_E \tag{3.3}$$

The shaded area in Fig. 3.1 shows schematically the changing quantity Q_O with burial.

If the incremental change of the above quantities with burial or geologic time is considered, the first derivatives of eqn. (3.3) must be obtained as

$$\frac{dQ_O}{dt} = \frac{dQ_R}{dt} + \frac{dQ_E}{dt} \tag{3.4}$$

Based on the first assumption described above, the amount expelled within a short geologic period is proportional to the amount generated within the same period. Hence we obtain

$$\frac{dQ_E}{dt} = a \cdot \frac{dQ_O}{dt} \tag{3.5}$$

where a is a factor ranging from 0 (no expulsion) to 1 (perfect expulsion).

As shown in Fig. 3.1, the total depth range of the oil generation is from the surface to the level of $R_0 = 1{\cdot}0\%$, and below the $R_0 = 1{\cdot}0\%$ level, oil is being destroyed thermally. In the range between the surface and the $R_0 = 1{\cdot}0\%$ level, the rate of the thermally undestroyed oil generation (dQ_O/dt) is positive or zero. In addition, the rate of oil expulsion (dQ_E/dt) is also considered to be positive or zero. In this model a negative oil expulsion is denied because this means that the oil is returning from the reservoirs to the source rocks.

At the surface, the rate (dQ_O/dt) is almost zero because there is virtually no petroleum generation there, so that both (dQ_R/dt) and (dQ_E/dt) are also nearly zero, because neither (dQ_R/dt) nor (dQ_E/dt) can be negative. At the level of $R_0 = 1{\cdot}0\%$, again the rate (dQ_O/dt) is zero, so that both (dQ_R/dt) and (dQ_E/dt) are zero. This means that at both the surface and the $R_0 = 1{\cdot}0\%$ level, the rates of increases of both the oil expulsion and remaining oil are almost nil.

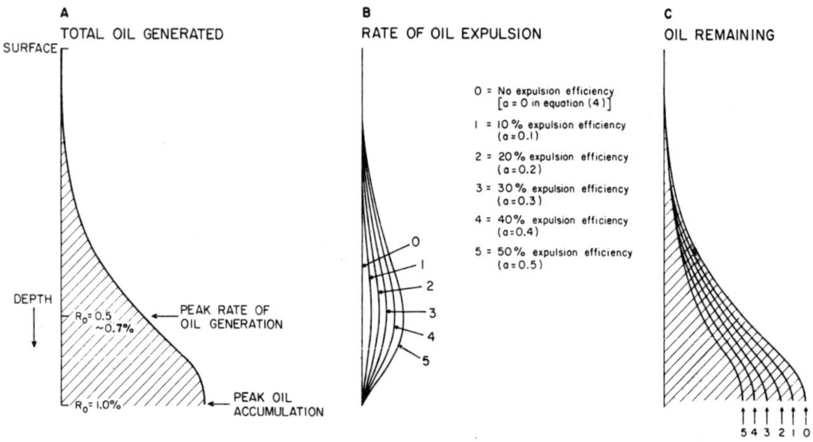

FIG. 3.2. Schematic diagrams showing total oil generated (A) in oil generation zone ($R_0 = 0{\cdot}5$–$1{\cdot}0\%$), rate of oil expulsion (B), and remaining oil with increasing burial depth (C), based on geologic model 1 described in text.

At an intermediate burial stage where R_0 is between 0·6 and 0·7% (steepest curve of oil in Fig. 3.1A), the rate (dQ_O/dt) is highest, so that both (dQ_R/dt) and (dQ_E/dt) can be highest, according to the conditions given above (see eqn. (3.5)). The most effective oil expulsion can be expected at this burial stage.

Figure 3.2 shows schematically the total oil generated (A), which is a copy of part of Fig. 3.1A, the rate of oil expulsion (B) based on the expulsion efficiency factors of 0 (no expulsion, 0·1 (10% expulsion), 0·2 (20%), 0·3 (30%), 0·4 (40%), and 0·5 (50%), respectively (see factor *a* in eqn. (3.5)), and the oil remaining in source rocks (C). Note that the peak rate of oil expulsion is reached at the level of $R_0 = 0{\cdot}6\text{–}0{\cdot}7\%$.

B. Thermal Gas Generation

The relationship between the amount of the total thermal gas generated from liquid petroleum (Q_G), the gas remaining in source rocks (Q_{RG}), and the gas expelled (Q_{EG}) can be similarly shown as,

$$Q_G = Q_{RG} + Q_{EG} \tag{3.6}$$

For the rates, we obtain,

$$\frac{dQ_G}{dt} = \frac{dQ_{RG}}{dt} + \frac{dQ_{EG}}{dt} \tag{3.7}$$

Using the curves shown in Fig. 3.1B, the rate (dQ_G/dt) is nearly zero at both the surface and the deepest depth point. The maximum thermal gas generation rate can be obtained at about the mid-point (peak rate of thermal gas generation in Fig. 3.1B) between $R_0 = 1.0$ and 1.35%, where the peak gas migration also occurs.

As mentioned previously, the above model is based on the assumption that the petroleum expulsion is a function of its generation, apart from the other possible chemical, physical and mineralogical factors which have influenced or even controlled petroleum migration. More realistic subsurface models will be discussed in the following sections.

MODEL 2: MIGRATION AS AQUEOUS SOLUTION

In this second model, petroleum is assumed to migrate in solution in water which moves from the source rocks to the reservoirs during burial. In this case, both the amount of compaction water expelled and the solubility of petroleum in the water are the most important factors.

Migration of petroleum by diffusion is not considered here.

The empirical function of shale porosity and depth at the compaction equilibrium condition is given by the following general form (Hubbert and Rubey, 1959),

$$\phi = \phi_0 \exp(-cZ) \tag{3.8}$$

where ϕ is shale porosity at depth Z, ϕ_0 shale porosity at surface or at $Z = 0$ (zero), and c a coefficient which determines the rate of porosity reduction with depth.

If depth Z is measured in ft, the unit of c is ft^{-1}. Figure 3.3 shows six porosity–depth trends on semi-log paper, for $c = 0$ (no compaction), $-0{\cdot}0001$, $-0{\cdot}0002$, $-0{\cdot}0003$, $-0{\cdot}0004$, and $-0{\cdot}0005\,\mathrm{ft}^{-1}$, respectively. The surface porosity ϕ_0 is 0·8 (or 80%). Note that since eqn. (3.8) is an

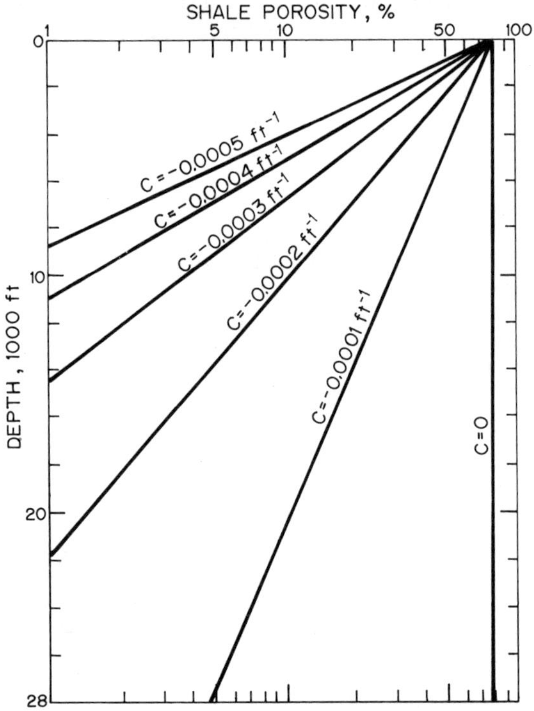

FIG. 3.3. Six shale porosity–depth relationships, when value c ranges from 0 to $-0{\cdot}0005\,\mathrm{ft}^{-1}$ in $\phi = \phi_0 \cdot \exp(-cZ)$ where ϕ is shale porosity at depth Z, ϕ_0 surface shale porosity (80%).

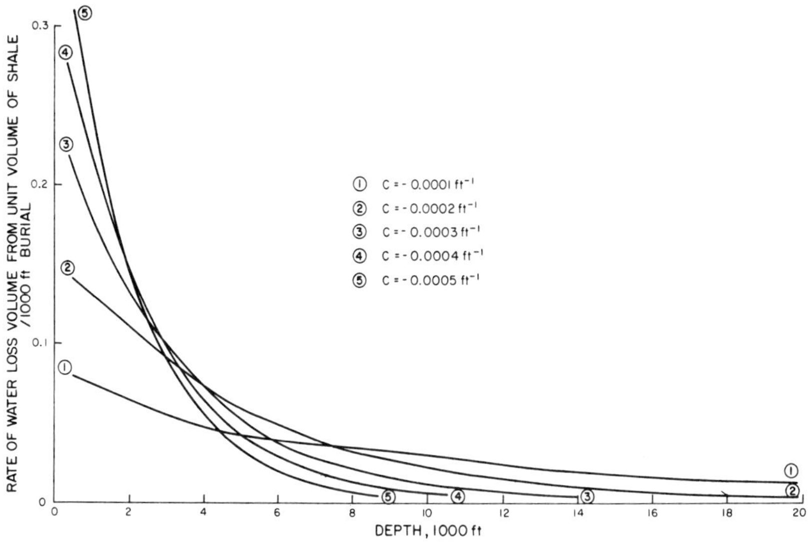

FIG. 3.4. Plots of rate of water loss volume from unit volume of shale per 1000 ft burial vs depth for five compaction rates ranging from $c = -0{\cdot}0001\,\text{ft}^{-1}$ to $c = -0{\cdot}0005\,\text{ft}^{-1}$. See text for explanation.

exponential function it can be shown as a straight line for a given c value on semi-log paper. The value c of the normal compaction trend of the Gulf Coast area based on Dickinson's (1953) porosity–depth curve was estimated to be nearly $-0{\cdot}0001\,\text{ft}^{-1}$ and that for the Paleozoic shales in Oklahoma based on Athy's curve (1930) is about $-0{\cdot}000433\,\text{ft}^{-1}$ (Hubbert and Rubey, 1959).

Figure 3.4 shows the plots of the rate of water loss volume per unit volume of the original shale per one thousand ft of burial vs depth for the five compaction cases from $c = -0{\cdot}0001$ to $-0{\cdot}0005\,\text{ft}^{-1}$. As expected from Fig. 3.3, the rate of water loss associated with shale compaction for each c value continuously decreases with burial. When $c = -0{\cdot}0001\,\text{ft}^{-1}$ or at the slowest compaction case among the five cases mentioned above, for example, the rate of water loss is relatively low at shallow depths but stays reasonably high at greater depth (Fig. 3.4). On the contrary, when $c = -0{\cdot}0005\,\text{ft}^{-1}$ or at the fastest compaction case, the rate of water loss is very high at shallower depths but diminishes relatively quickly with depth (Fig. 3.4).

Similar shale compaction trends on the basis of the sonic transit time data have been established by different workers (Hottman and Johnson, 1965; Reynolds, 1973; Fertl, 1974; and Magara, 1978*b*). By combining eqn. (3.8) with Wyllie *et al.*'s (1956) time average equation, we obtain,

$$\frac{\Delta t - \Delta t_m}{\Delta t_w - \Delta t_m} = \frac{\Delta t_0 - \Delta t_m}{\Delta t_w - \Delta t_m} \cdot \exp(-cZ) \tag{3.9}$$

or

$$\Delta t - \Delta t_m = (\Delta t_0 - \Delta t_m) \exp(-cZ) \tag{3.10}$$

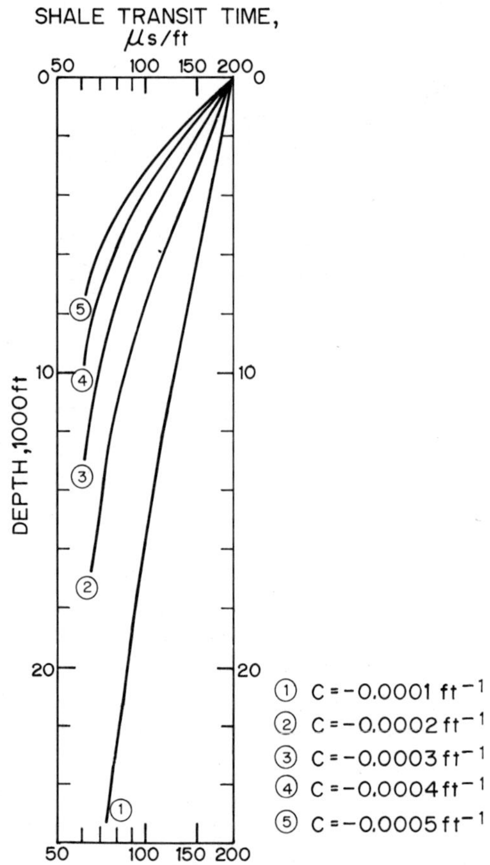

FIG. 3.5. Normal compaction trends shown as transit time vs depth plots on semi-log paper, when value c ranges from $-0{\cdot}0001$ to $-0{\cdot}0005\,\text{ft}^{-1}$. Refer to text for explanation.

where Δt is shale transit time (μs/ft) at depth Z, Δt_m matrix (or grain) transit time (μs/ft), Δt_w transit time (μs/ft) of pore-water, and Δt_0 shale transit time (μs/ft) at surface or at $Z=0$ (zero).

Equation (3.10) suggests that if the difference between the measured transit time and the matrix transit time $(\Delta t - \Delta t_m)$ instead of Δt is plotted for a given constant lithology (Δt_m is thus constant), the normal compaction trend can be shown as a straight line on semi-log paper. However, if a plot of the transit time (Δt) vs depth is made, the normal trend cannot be shown by a straight line. Figure 3.5 shows five curved lines representing $c = -0{\cdot}0001$, $-0{\cdot}0002$, $-0{\cdot}0003$, $-0{\cdot}0004$, and $-0{\cdot}0005\ \text{ft}^{-1}$, respectively, when $\Delta t_m = 60$ μs/ft (approximate shale matrix transit time).

If only the relatively shallow portion of each curve in Fig. 3.5 is considered, the trend can be approximated by a straight line for practical purposes. Therefore, most workers used these approximate straight lines as normal compaction trends, but, theoretically speaking, the trends are curves so that extrapolating the straight line trends to depth can cause erroneous interpretations.

A. Oil Migration

Price (1976) studied aqueous solubility of oil with increasing laboratory temperature as shown in Fig. 3.6. Effect of pressure on solubility is not considered in his experiments. Using the highest solubilities from the Farmer's whole oil sample, the subsurface solubility was estimated and is shown in Fig. 3.7. Surface temperature and geothermal gradient are assumed to be 70°F and 1·4°F/100 ft, respectively. This figure shows continuous increase of aqueous oil solubility with burial, but the amount transported with compaction water at depth can be relatively small because the rate of compaction-water loss decreases continuously (Fig. 3.4).

In order to estimate the efficiency of oil transportation with burial by aqueous solution Fig. 3.8, which shows the plot of the product of the rate of water loss (Fig. 3.4) and the change of the aqueous oil solubility (Fig. 3.7) against depth, was produced. Although the transportation efficiency increases with burial when $c = -0{\cdot}0001\ \text{ft}^{-1}$, which is close to the Gulf Coast case, all the other cases show continuous decrease of the efficiency with burial. This suggests that, unless the rate of shale compaction is relatively slow, aqueous solubility cannot be the principal mechanism for transporting oil. Even in this case, if severe undercompaction occurs at depth, the transport efficiency would drop significantly because of difficulty of expelling fluid there.

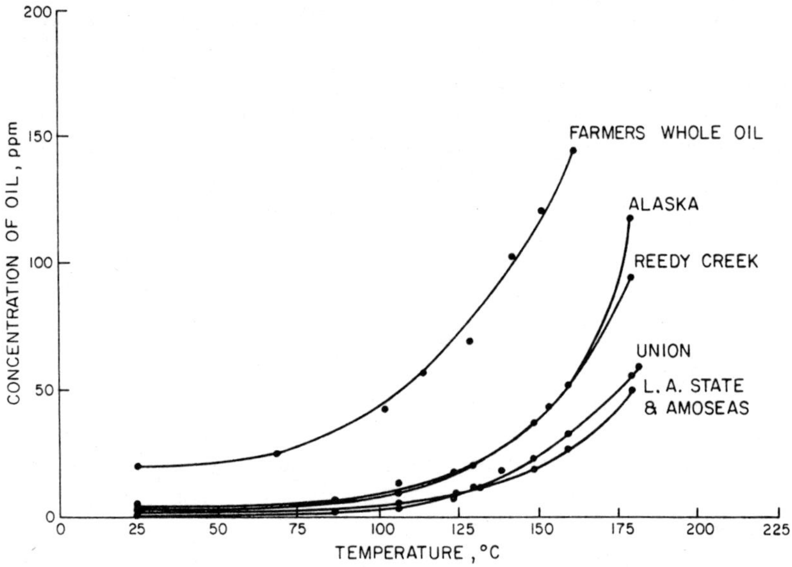

FIG. 3.6. Aqueous solubilities (ppm) of two whole oils (Wyoming Farmers and Louisiana State) and four topped oils (Amoseas Lake, Reedy Creek, Alaska, and Union Moonie) as functions of temperature. Topping temperature is 200°C (from Price, 1976). (Reproduced by permission of the American Association of Petroleum Geologists.)

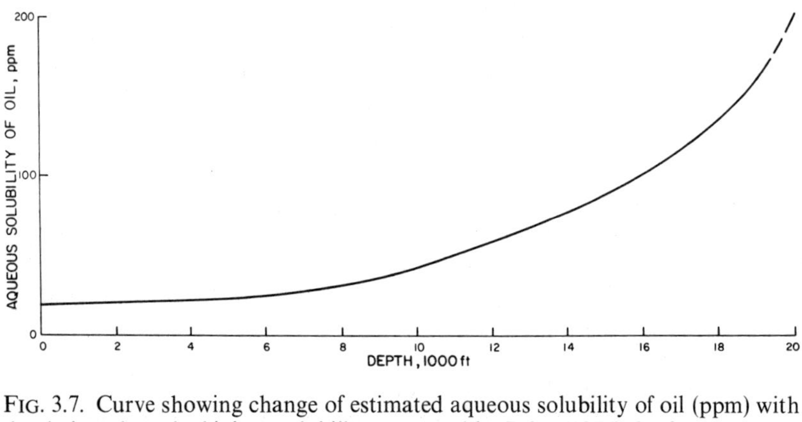

FIG. 3.7. Curve showing change of estimated aqueous solubility of oil (ppm) with depth, based on the highest solubility measured by Price (1976). Surface temperature and geothermal gradient are assumed to be 70°F and 1·4°F/100 ft, respectively.

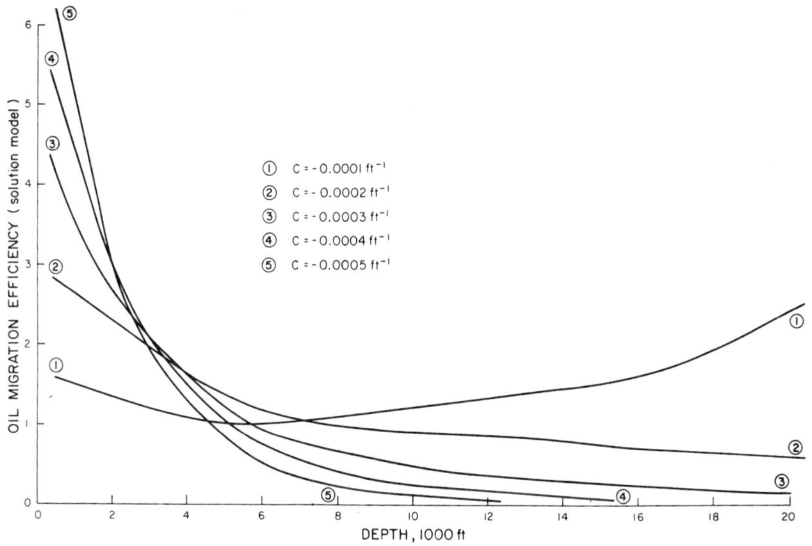

FIG. 3.8. Oil migration efficiency vs depth plots for five compaction rates from $c = -0{\cdot}0001$ to $-0{\cdot}0005\,\mathrm{ft}^{-1}$, based on solution model. Refer to text for explanation.

B. Gas Migration

Solubility of gas in formation water has been studied by several workers (Dodson and Standing, 1944; Bonham, 1978 and Price, 1979). Figure 3.9 shows the chart made by Dodson and Standing in which the solubility increases with increasing water pressure. The effect of temperature is relatively small. Figure 3.10 was constructed in that the estimated gas solubility (ft^3/bbl of water) is plotted against burial depth. The temperature values are the same as previously assumed (Fig. 3.7), and the pressure is assumed to increase at 0·465 psi/ft of burial (hydrostatic pressure gradient).

The transport efficiency for gas in solution in compaction water was estimated and is shown in Fig. 3.11. The efficiency is high at relatively shallow depth ranges for $c = -0{\cdot}0003$ to $-0{\cdot}0005\,\mathrm{ft}^{-1}$, and diminishes very rapidly with depth. When the rate of compaction is relatively slow ($c = -0{\cdot}0001$ and $-0{\cdot}0002\,\mathrm{ft}^{-1}$), the efficiency stays relatively high at depths where significant thermal gas generation is expected.

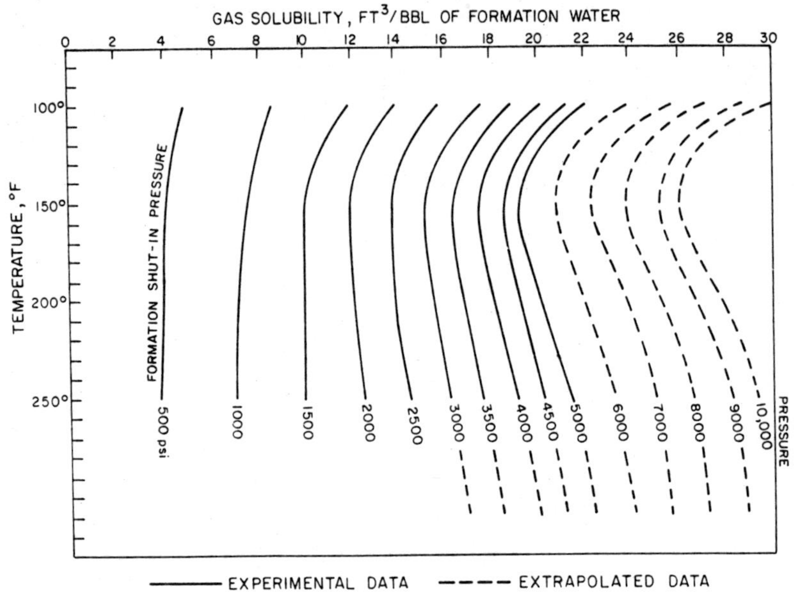

FIG. 3.9. Solubility of natural gas in formation water (from Dodson and Standing, 1944). (Reproduced by permission of the American Petroleum Institute.)

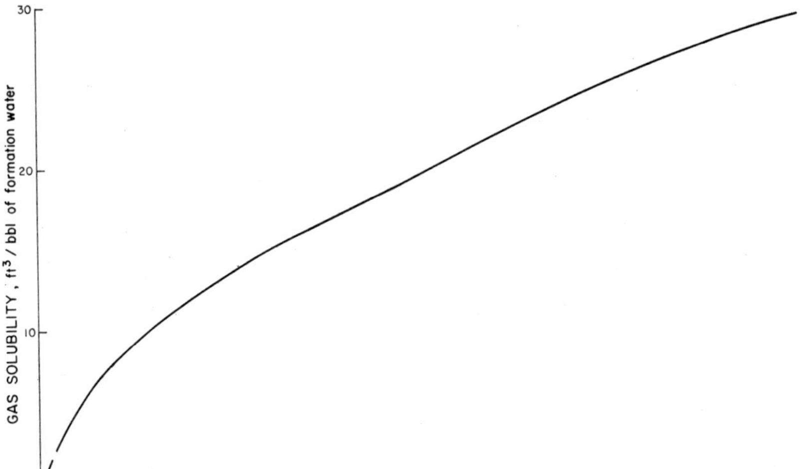

FIG. 3.10. Curve showing change of gas solubility with depth based on Dodson and Standing's (1944) chart (Fig. 3.9). Assumed surface temperature and geothermal gradient are 70°F and 1·4°F/100 ft, respectively. Fluid pressure gradient is assumed to be 0·465 psi/ft (hydrostatic).

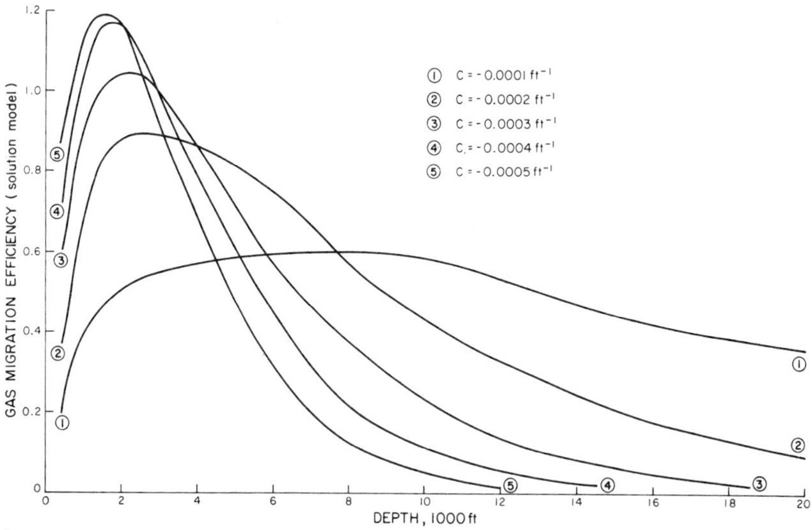

FIG. 3.11. Gas migration efficiency vs depth plots for five compaction rates from $c = -0{\cdot}0001$ to $-0{\cdot}0005\,\mathrm{ft}^{-1}$ based on solution model. Refer to text for detailed explanation.

MODEL 3: MIGRATION IN SEPARATE HYDROCARBON PHASE

Dickey (1975) pointed out the difficulty of migrating a significant amount of liquid petroleum in aqueous solution and proposed a mechanism of migrating oil in its own separate phase in the presence of structured or semi-solid water in shales which is more viscous than the oil. Magara (1978*a*, 1979, and 1981) further provided subsurface and laboratory evidence to support Dickey's hypothesis.

A. Oil Migration

Magara's (1978*a*) model is shown in Fig. 3.12. The top diagram shows a schematic of relative permeability vs degree of shale compaction. As the shale compacts, the relative permeability to water decreases and that to oil increases because of the continuous increase of viscosity of the remaining pore-water. Nevertheless, even with this increased relative permeability to oil, the absolute permeability of the shale will continually decrease as the shale becomes more compacted (middle diagram in Fig. 3.12). Assuming that the pressure gradient in the shale remains relatively unchanged during compaction, oil migration in the oil phase will reach

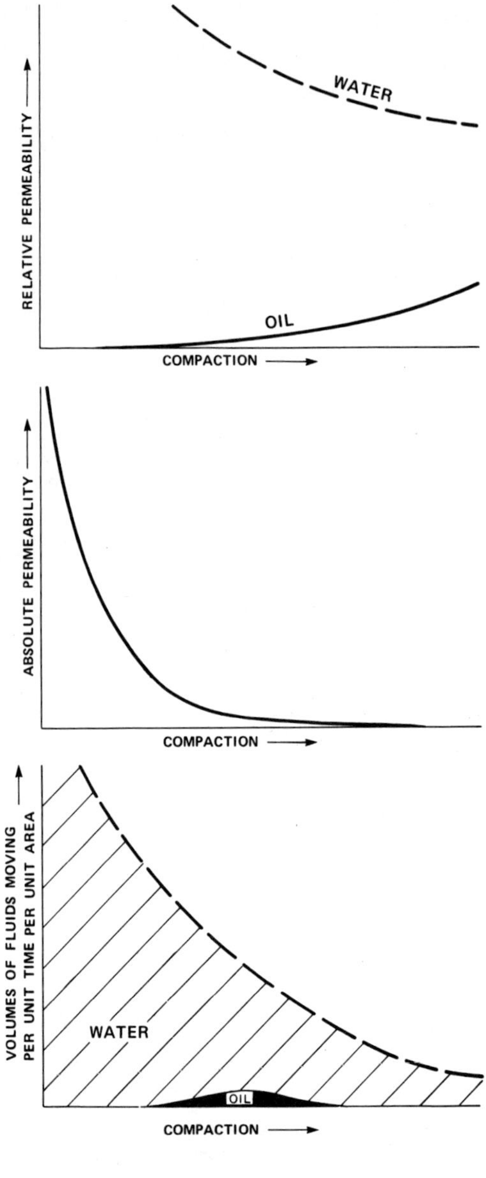

FIG. 3.12. Hypothetical relationships of relative permeability, and fluid movement to degree of shale compaction (Magara, 1978*a*). (Reproduced by permission of the Canadian Society of Petroleum Geologists.)

its maximum at an intermediate, not a late, compaction stage. After the intermediate compaction stage, the oil phase migration will decline as the absolute permeability of the shale decreases (bottom diagram, Fig. 3.12).

Although the data on relative permeability of oil over the viscous water in clay or shale are unavailable, the laboratory-measured viscosity of pore-water in montmorillonite is available (Low, 1976). Based on the comparative study of the subsurface shale porosity in the Beaufort Basin, Canada, and Low's water viscosity data, it is possible to relate the shale porosity to the viscosity of the remaining water in shales (Magara, 1979).

Figure 3.13 shows such estimated viscosity values with increasing depth, using the porosity–depth relationships of five of the compaction cases in Fig. 3.3. If the type of clay mineral varies, the viscosity value at a given compaction level will vary. The curves in Fig. 3.13 show possible changing viscosity with burial, if the shales contain mainly montmorillonite. They depict continuously increasing tendencies of viscosity, which may be interpreted as the general tendencies of increasing relative

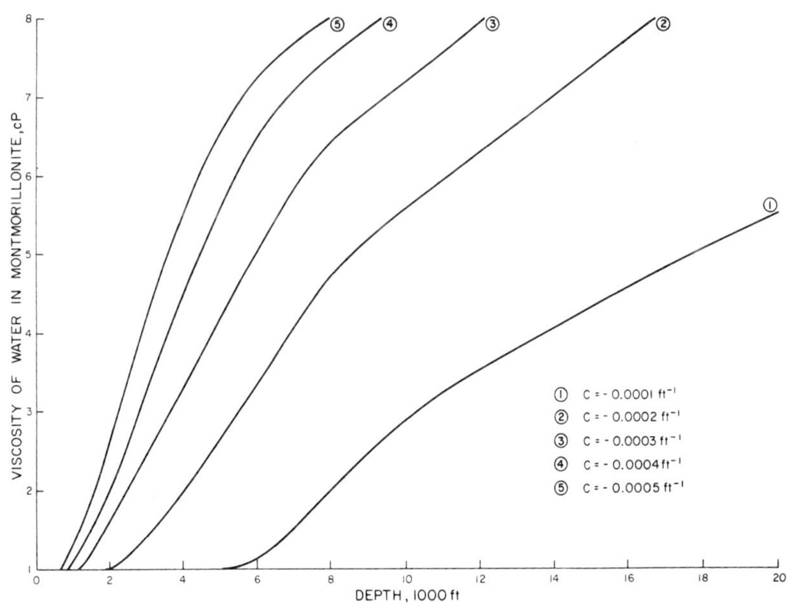

FIG. 3.13. Curves showing changing viscosity of water in montmorillonite with depth for five compaction rates from $c = -0{\cdot}0001$ to $-0{\cdot}0005\ \text{ft}^{-1}$, based on Low's (1976) data.

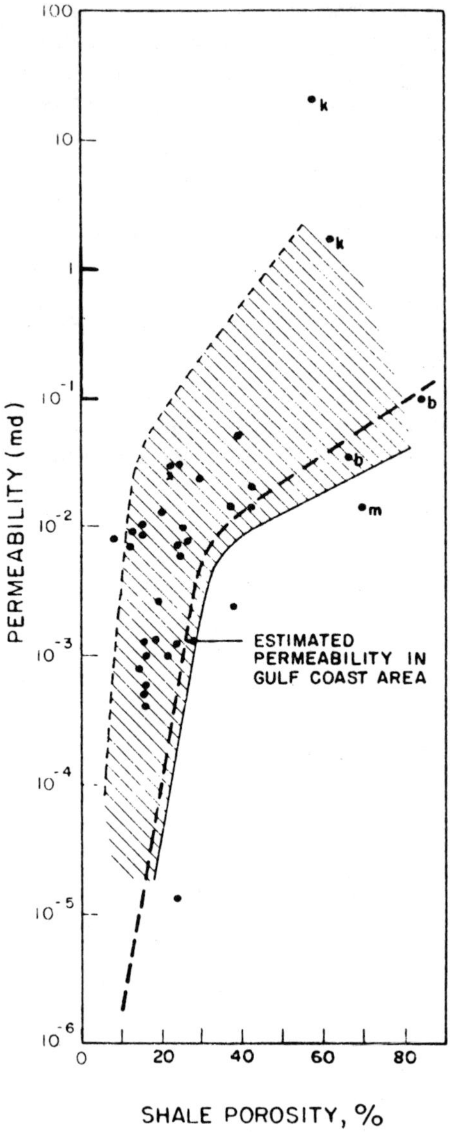

FIG. 3.14. Range of shale porosity and permeability measured in laboratory and estimated porosity–permeability relationship in the subsurface of Gulf Coast area (from Magara, 1971). (Reproduced by permission of the Society of Petroleum Engineers, Texas.)

permeability for oil, since the mobility of the viscous water continuously decreases with burial.

The absolute permeability values of shales were compiled, and the theoretical aspects were studied by Magara (1969, 1971). Using the estimated porosity–permeability relationship in the Gulf Coast shown in Fig. 3.14, the absolute permeability of shales for the five compaction cases as discussed above were estimated and are plotted against depth in Fig. 3.15. They show continuous decreasing trends with burial. The scale showing arbitrary permeability factors is also depicted on the right-hand side of Fig. 3.15.

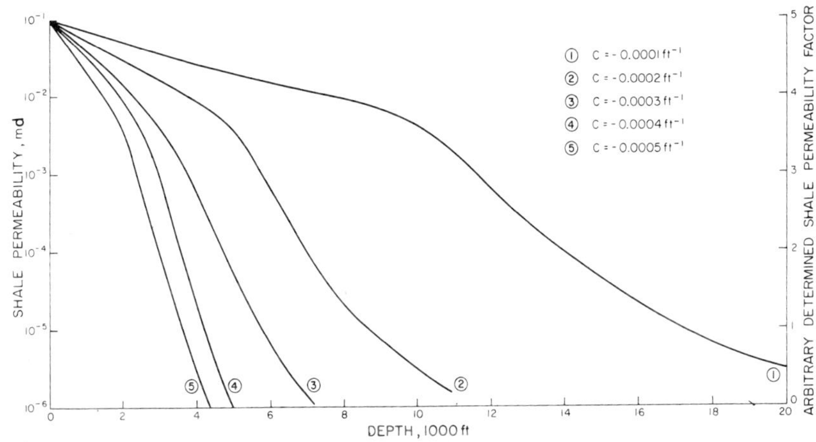

FIG. 3.15. Curves showing changes of shale permeability with depth for five compaction rates from $c = -0{\cdot}0001$ to $-0{\cdot}0005\,\mathrm{ft}^{-1}$, based on Magara's porosity–permeability relationship (see Fig. 3.14).

Since, with the currently available data, it is not possible to evaluate the migration efficiency of oil in its own phase in a quantitative way such as in the solution model, I made a product plot of the remaining water viscosity (Fig. 3.13) and of the permeability factor (Fig. 3.15). This product may be interpreted as a qualitative measure of the migration efficiency of oil in its own phase and is shown in Fig. 3.16. As suggested by Magara (1978, Fig. 12), the oil migration efficiency in the oil phase reaches its peak at an intermediate depth range. For a slower compaction rate, it falls in a relatively deep interval, but for a faster compaction rate, it occurs in a relatively shallow interval.

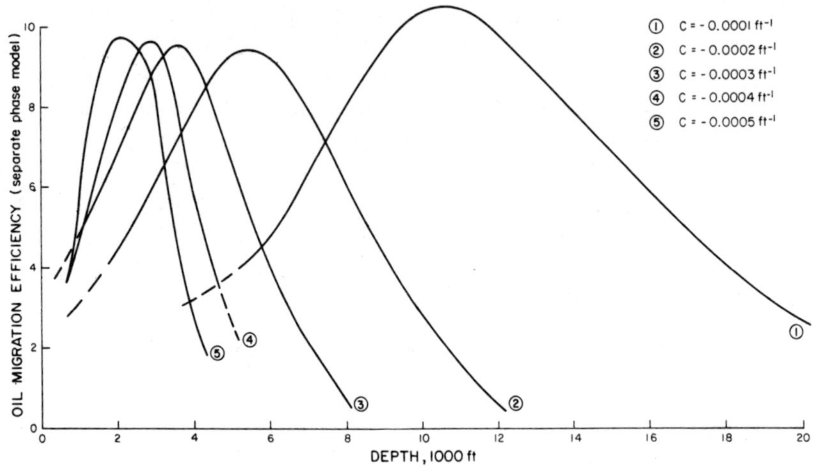

FIG. 3.16. Oil migration efficiency vs depth plots for five compaction rates from $c = -0{\cdot}0001$ to $-0{\cdot}0005\,\mathrm{ft}^{-1}$, based on separate phase model. Refer to text for explanation.

B. Gas Migration

As shown in Fig. 3.10, solubility of natural gas in water is relatively high. Therefore, it is assumed here that most gas migrates in aqueous solution (see *Gas migration* in Model 2).

COMBINED GENERATION AND MIGRATION WINDOWS FOR OIL AND GAS

In geologic model 1, the time of most active migration falls within the range of the most active generation of petroleum. Therefore, if this model is applied, migration and generation are interpreted to have occurred concurrently. As stated previously, this model ignores many other physical, chemical and mineralogical constraints involved in petroleum migration.

However, in model 2 or 3, the time of most effective migration does not necessarily coincide with the time of most active generation of petroleum. Comparison of these two timings or burial depth ranges would be worthwhile in assessing the timeliness of these two important events. For gas, model 2, or the aqueous solution model, seems to simulate most subsurface conditions because of its relatively high solubility in water, whereas, for oil, model 3, or the separate phase model, seems to be the more important one to use because of its low solubility in water.

Petroleum generation is controlled by both geologic time and temperature (Connan, 1974). However, in the above geologic models, the effect of the geologic time on a normal compaction trend is not considered. Magara (1978*b*) examined this possibility using the examples from Canada and concluded that the rate of sedimentation is probably more important in causing the slope change of a normal compaction trend than is the total geologic age of the sedimentary rocks. This concept is also supported by Terzaghi's concept of subsurface stress balance, in that, if the compaction equilibrium is once reached in the normal compaction zone, shales cannot be compacted any further, even if a significant geologic time has elapsed after the establishment of the equilibrium compaction. Possible effect of cementation in shales is ignored in this concept.

The changing normal compaction trend, however, may be related to changing geothermal gradient. Reynolds (1973) reported the empirical relation between the slope of the normal trend of sonic transit time and geothermal gradient in the Gulf Coast. Reynolds' relation shows that the geothermal gradient is higher for a faster shale compaction with depth and vice versa.

Because the thermal conductivities of most rock materials are significantly higher than that of water, the average thermal conductivity of porous media increases with compaction (Lewis and Rose, 1970). Therefore, the overall average thermal conductivity of more compacted rocks is higher than that of less compacted rocks. The subsurface heat flow usually occurs more effectively in the areas of higher thermal conductivity than in those of lower conductivity, so that higher geothermal gradient may be expected in the former areas than in the latter. Possible presence of undercompacted sections (low thermal conductivity) in the deep subsurface of rapidly deposited basins (where the value of c is a relatively low negative figure or close to $-0{\cdot}0001\,\mathrm{ft}^{-1}$) may further retard the effective heat flow there. This effect may also cause relatively low geothermal gradients when the value of c is close to $-0{\cdot}0001\,\mathrm{ft}^{-1}$.

As mentioned above, Reynolds' (1973) empirical relationship is based on the slope of shale transit time. A comparable but approximate relationship was established for the shale porosity as follows,

$c = -0{\cdot}0001\ \text{ft}^{-1}$: geothermal gradient $\simeq 1{\cdot}0°\text{F}/100\ \text{ft}$
$c = -0{\cdot}0002\ \text{ft}^{-1}$: geothermal gradient $\simeq 1{\cdot}2°\text{F}/100\ \text{ft}$
$c = -0{\cdot}0003\ \text{ft}^{-1}$: geothermal gradient $\simeq 1{\cdot}5°\text{F}/100\ \text{ft}$
$c = -0{\cdot}0004\ \text{ft}^{-1}$: geothermal gradient $\simeq 1{\cdot}8°\text{F}/100\ \text{ft}$
$c = -0{\cdot}0005\ \text{ft}^{-1}$: geothermal gradient $\simeq 2{\cdot}2°\text{F}/100\ \text{ft}$.

Pursey (1973) established the oil window between 150°F and 300°F on the basis of the 262 known giant oil fields compiled by Halbouty (1970), ranging from 1 to approximately 200 million years of geologic age. For the thermal gas generation, it may be assumed that the temperature range between 300°F and 350°F is most important. These temperature ranges are plotted vs depth for selected geothermal gradients from 1 to 2·2°F/100 ft in Fig. 3.17. Possible oil and gas generation windows can be estimated for a given geothermal gradient and may be related to a selected normal compaction trend. For example, if the slope c of a normal compaction trend is about $-0{\cdot}0001\ \text{ft}^{-1}$, which corresponds to about 1°F/100 ft geothermal gradient, the oil window can be expected in the interval between about 8000 ft and 23 000 ft (Fig. 3.17).

Figure 3.18A shows a cross plot of the estimated oil generation

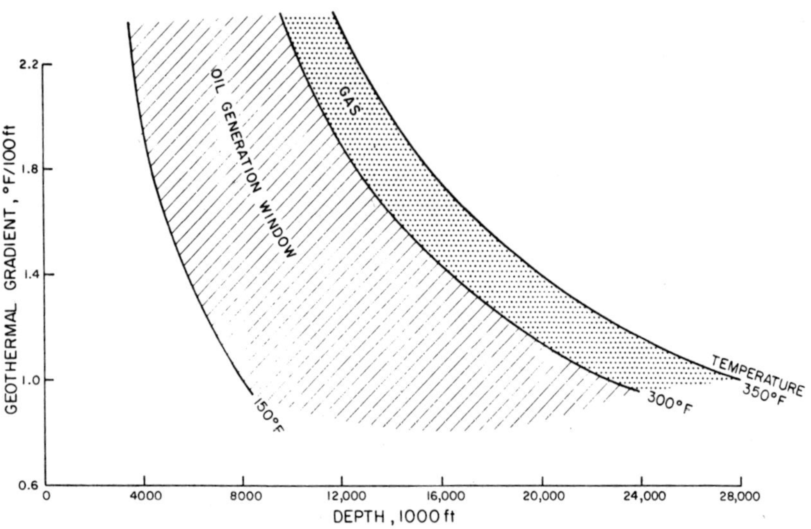

FIG. 3.17. Relationships between geothermal gradient and depth for oil (150–300°F) and gas (300–350°F) generation windows. This oil generation window is based on Pursey's (1973) data of giant oil fields in the world.

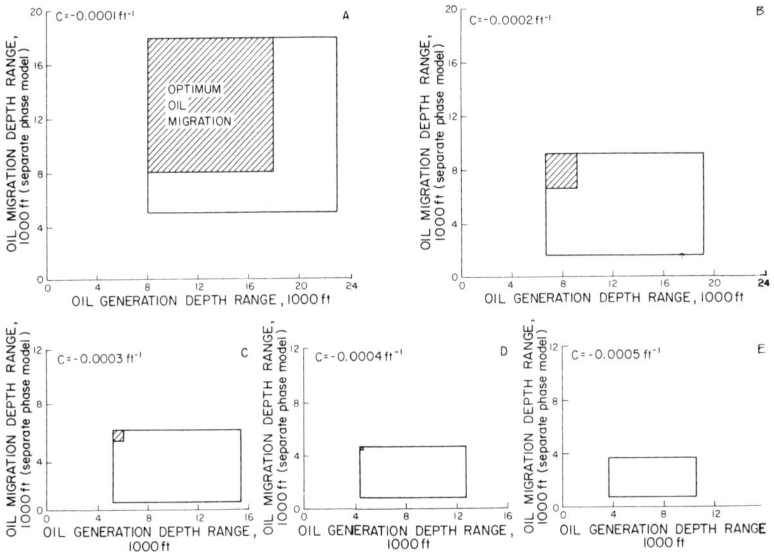

FIG. 3.18. Open blocks show depth ranges of oil generation and migration, based on separate phase migration model, for five compaction rates (A: $c = -0{\cdot}0001\ \mathrm{ft}^{-1}$, B: $-0{\cdot}0002$, C: $-0{\cdot}0003$, D: $-0{\cdot}0004$, E: $-0{\cdot}0005$, respectively). Shaded block in each diagram shows optimum depth range for oil migration. Refer to text for detailed explanation.

window (X axis) for $c = -0{\cdot}0001\ \mathrm{ft}^{-1}$ from Fig. 3.17 and the estimated peak oil migration window (Y axis) from Fig. 3.16 (the product value or oil migration efficiency greater than 4). An open block shows these estimated depth ranges for oil generation and migration. To cause an effective oil migration, however, the intervals for both generation and migration must fall in the same depth range, which is shown by a shaded block in Fig. 3.18A. It is interpreted that the most effective oil migration can occur in the depth range between about 8000 ft and 18 000 ft in this case. Note that this migration window can exist if the shales are normally compacted to the bottom depth (18 000 ft). However, if shales are severely undercompacted in the deeper section, suggesting ineffective fluid expulsion, the bottom of the migration window would become shallower than 18 000 ft. It is interesting to note that Galloway *et al.* (1981) reported the subsurface data of effective oil migration from the Gulf Coast shales between about 8000 and 14 000 ft. Below 14 000 ft, severely undercompacted shales are encountered.

Figures 3.18B–3.18E show similar plots when $c = -0{\cdot}0002$ to $-0{\cdot}0005\ \text{ft}^{-1}$, respectively. The most effective oil migration window becomes smaller as the rate of compaction increases, suggesting that the chances of effective oil migration declines with increasing rate of compaction.

For the thermal gas generation from liquid petroleum, the temperature between 300°F and 350°F can be important. However, gas can also be generated directly from source rock, so that the true gas generation window can be very wide. In the following discussion, the temperature range for gas generation is thus assumed to be from 150°F to 350°F, which covers the oil generation window as well.

Similar cross plots for gas are shown in Fig. 3.19. The best gas migration was estimated using the gas solubility–compaction water product plots shown in Fig. 3.11 (best migration zone is the zone where the product value or gas migration efficiency exceeds 0·4). Figure 3.19A shows that, for $c = -0{\cdot}0001\ \text{ft}^{-1}$, the possible gas generation zone can exist between 8000 and 28 000 ft, but the effective migration condition

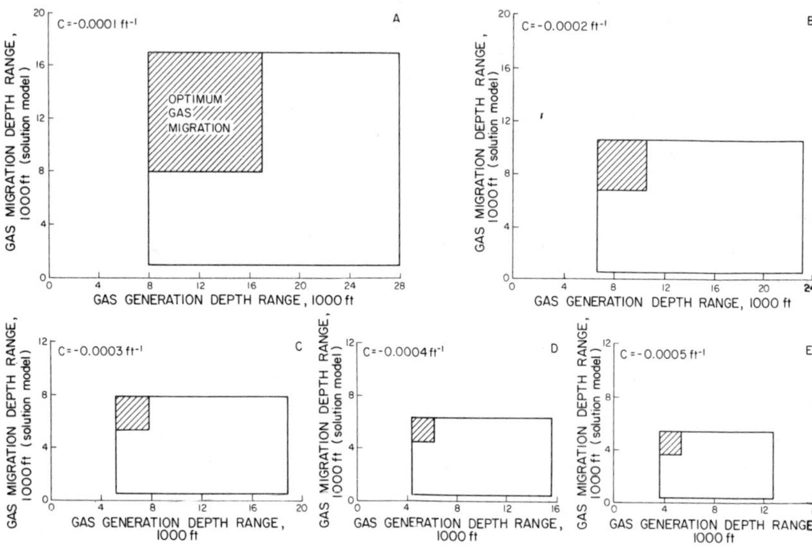

FIG. 3.19. Open blocks show depth ranges of gas generation and migration, based on solution model, for five compaction rates (A: $c = -0{\cdot}0001\ \text{ft}^{-1}$, B: $-0{\cdot}0002$, C: $-0{\cdot}0003$, D: $-0{\cdot}0004$, and E: $-0{\cdot}0005$, respectively). Shaded block in each diagram shows optimum depth range for gas migration. Refer to text for detailed explanation.

exists between 1000 and 17 000 ft (open block in Fig. 3.19A). Hence, the optimum depth interval for both generation and migration can be between 8000 and 17 000 ft (shaded block in Fig. 3.19A). Similar cross plots for $c = -0{\cdot}0002$ to $-0{\cdot}0005\,\text{ft}^{-1}$ are shown in Figs. 3.19B–3.19E, respectively. As for oil migration (Fig. 3.18), the optimum gas migration window becomes shallower and smaller from Fig. 3.19A ($c = -0{\cdot}0001\,\text{ft}^{-1}$) to Fig. 3.19E ($c = -0{\cdot}0005\,\text{ft}^{-1}$), suggesting that the chances of effective gas migration decrease with increase of the rate of compaction.

REFERENCES

Athy, L. F. (1930) Density, porosity and compaction of sedimentary rocks. *A.A.P.G.*, **14**, 1–24.

Bonham, L. C. (1978) Solubility of methane in water at elevated temperatures and pressures. *A.A.P.G.*, **62**, 2478–81.

Connan, J. (1974) Time–temperature relations in oil genesis. *A.A.P.G.*, **58**, 2516–21.

Dickey, P. A. (1975) Possible primary migration of oil from source rocks in oil phase. *A.A.P.G.*, **59**, 337–45.

Dickinson, G. (1953) Geological aspects of abnormal reservoir pressures in Gulf Coast Louisiana. *A.A.P.G.*, **37**, 410–32.

Dodson, C. R. and Standing, M. B. (1944) Pressure–volume–temperature and solubility relations for natural gas–water mixtures. In *Drilling and Production Practice*, American Petroleum Institute, pp. 173–8.

Dow, W. G. (1978) Petroleum source beds on continental slopes and rises. *A.A.P.G.*, **62**, 1584–1606.

Fertl, W. H. (1976) *Abnormal Formation Pressures*, Elsevier, Amsterdam, 382p.

Galloway, W. E., Hobday, D. K. and Magara, K. (1981) Frio Formation of the Texas Gulf Coastal Plain—Depositional systems, structural framework, and hydrocarbon origin, migration, distribution and exploration potential, The University of Texas at Austin, Bureau of Economic Geology, Report No. 122, 78p.

Halbouty, M. T. (1970) *Geology of a Giant Petroleum Field*, A.A.P.G., Tulsa, Ok., 575 p.

Hottman, C. E. and Johnson, R. K. (1965) Estimation of formation pressures from log-derived shale properties. *Petrol. Technol. J.*, **17**, 717–22.

Hubbert, M. K. and Rubey, W. W. (1959) Role of fluid pressure in mechanics of overthrust faulting. *Geol. Soc. Amer. Bull.*, **70**, 115–206.

Lewis, C. R. and Rose, S. C. (1970) A theory relating high temperatures and overpressures. *Petrol. Technol. J.*, **22**, 11–16.

Low, P. F. (1976) Viscosity of interlayer water in montmorillonite. *Soil Sci. Soc. Amer. Proc.*, **40**, 500–5.

Magara, K. (1969) Porosity–permeability relationship of shale. *Canadian Well Logging Soc. J.*, **2**, 47–93.

Magara, K. (1971) Permeability considerations on generation of abnormal pressures. *Soc. Petrol. Engrs. J.*, **11**, 236–42.

Magara, K. (1978*a*) The significance of the expulsion of water in oil-phase primary migration. *Canadian Petrol. Geology*, **26**, 123–31.

Magara, K. (1978*b*) *Compaction and Fluid Migration—Practical Petroleum Geology*, Elsevier, Amsterdam, 319 p.

Magara, K. (1978*c*) Time of oil generation and migration and oil window. *Canadian Petrol. Geology Bull.*, **27**, 87–93.

Magara, K. (1979) Structured water and its significance in primary oil migration. *Canadian Petrol. Geology Bull.*, **27**, 87–93.

Magara, K. (1981) Possible primary migration of oil globules. *Petrol. Geology J.*, **3**, 325–31.

Price, L. C. (1976) Aqueous solubility of petroleum as applied to its origin and primary migration. *A.A.P.G.*, **60**, 213–44.

Price, L. C. (1979) Aqueous solubility of methane at elevated pressures and temperatures. *A.A.P.G.*, **63**, 1527–33.

Pursey, W. C. (1973) How to evaluate potential gas and oil source rocks. *World Oil*, **176**, no. 5, April.

Reynolds, E. B. (1973) The application of seismic techniques to drilling techniques, SPE 4643, 48th AIME Fall Meeting, Las Vegas, Nevada.

Wilson, H. H. (1975) Time of hydrocarbon expulsion, paradox for geologists and geochemists. *A.A.P.G.*, **59**, 69–84.

Wyllie, M. R. J., Gregory, A. R. and Gardner, L. W. (1956) Elastic wave velocities in heterogeneous and porous media. *Geophysics*, **21**, 41–70.

Chapter 4

Evidence of Primary Oil Migration

INTRODUCTION

Conventional oil and gas accumulations are usually found in pore spaces and fracture openings of reservoir rocks, such as sandstones, limestones, and dolomites. A certain proportion of this oil and gas may have generated from various types of organic matter that were deposited in the reservoirs themselves. These hydrocarbons did not migrate but accumulated in situ.

The proportion of such oil and gas in the total accumulation is usually considered to be relatively small, compared with oil and gas that were generated in, and migrated from, the surrounding rocks, particularly from shales. This is because the concentration of organic carbon in average shales is higher than that in average sandstone and carbonate rocks. Figure 2.1 in Chapter 2 shows the difference in the Russian Platform, based on 25 742 samples. A similar plot of organic carbon percentage vs grain size of sediments in the Bering Sea is shown in Fig. 2.2. The highest concentrations of organic carbon and bitumen are observed in the silt-clay mud.

Volumes or thicknesses of shales in most oil-producing sedimentary basins are usually three to four times greater than reservoir volumes. In addition, shales usually contain more fluid (70–80% at the time of deposition) than do sandstones (35–40%), and such fluid is considered to have been an important agent for hydrocarbon migration at later stages. Variations of fluid potential in shales caused by continuous burial are usually much greater than those in sandstones. A strong fluid-potential difference (or gradient) would be necessary for significant primary hydrocarbon migration.

Although significant volumes of this fluid may have been expelled from shales prior to petroleum generation, remaining fluid volumes at onset of petroleum generation may still have been significant. Dickey (1975) estimated that about 11% of porosity loss occurred during burial between 5000 and 15 000 ft, corresponding to the temperature range of petroleum generation.

Another reason that shale may be more important than sandstone as a petroleum source rock is related to the quality of organic matter. Although some sandstones contain exceptionally large amounts of organic matter, this material is most likely to be primarily land-derived. Marine and amorphous organic matter, both considered better petroleum sources, seem more likely to occur in shale. Argillaceous limestones can also be principal source rocks for petroleum.

Another example of the in-situ or unmigrated petroleum is 'recycled petroleum', usually considered rare. Levorsen (1967) states that, '... petroleum hydrocarbons and related hydrocarbon compounds eroded from the land may conceivably be carried by the rivers to the ocean and re-deposited along with recycled sedimentary particles in both reservoir and nonreservoir sediments. Eroded petroleum may come from two sources: (1) the oil disseminated through the nonreservoir shales and carbonates and (2) the oil contained in any fields that have undergone erosion since the beginning of Paleozoic time'. And he continues, '... presumably the more stable, heavier hydrocarbon fractions and the asphaltic fractions eroded from the nonreservoir rocks and from the oil pools would be the most susceptible to preservation and redeposition'.

What is the evidence for the belief held by most petroleum geologists that oil moves from source rocks to a reservoir by primary migration? Such evidence may be statistical and indirect (based on a large amount of data), or it may be geochemical and more direct.

STATISTICAL AND INDIRECT EVIDENCE

Curtis *et al.* (1960) gave a statistical relationship between the total ultimate recoverable oil in a sandstone and its average thickness, based on data of 7241 sandstone reservoirs in the United States. This relationship indicates that the total recoverable oil is proportional to the square of the average sandstone thickness. The gas in the reservoirs was converted to an oil equivalent (Curtis *et al.*, 1960). Smith *et al.* (1971) showed this relationship in graph form (Fig. 4.1). In this figure, Curtis *et*

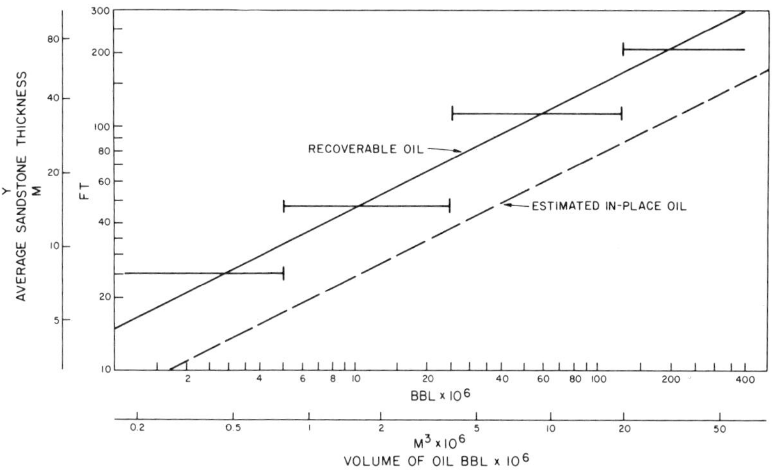

FIG. 4.1. Graph showing relationship between average sandstone thickness and volume of recoverable oil or in-place oil, based on 7241 sandstone reservoirs in USA (revised from Smith *et al.*, 1971).

al.'s (1960) statistical data are indicated by four horizontal lines and the average relationship is shown by a diagonal solid line (*recoverable oil*).

The total in-place oil, I_0, in sandstone reservoirs was given as follows,

$$I_0 = C \cdot Y^2 \tag{4.1}$$

where, according to Smith *et al.* (1971), Y is the average gross thickness of the sandstones and C is constant. When the average recovery factor of 27% is used, the value of C is $2{\cdot}85 \times 10^4$ m^3 oil/m^2 (Smith *et al.*, 1971). The relationship when the recovery factor is 27% is shown graphically by a broken straight line (see the 'estimated in-place' line, in Fig. 4.1).

As mentioned previously, the 'recoverable oil' line of Fig. 4.1 is drawn through the average of the statistical data. In other words, the actual statistical data are plotted in a wider range. Similarly, the in-place oil must be shown in a range rather than as a straight line as in Fig. 4.1. The value of C in eqn. (4.1), therefore, is not constant but changes within a range. The range of C obtained from the statistical data by Curtis *et al.* (1960) is from $1{\cdot}3 \times 10^4$ to about $6{\cdot}3 \times 10^4$ m^3 oil/m^2. The value of C may vary because of such factors as sandstone body types (e.g. narrow stringer sand, sheet sand), reservoir quality, source-rock properties, etc. However, it is interesting to note that the range of C is not large, and that the most important factor controlling the volume of recoverable in-place oil is sandstone thickness. Smith *et al.* (1971) stated that the dimensional dependence of eqn. (4.1) (their eqn. (8)) suggests that the area of contact between sandstones and surrounding lithologies could be the factor controlling the amount of oil migrating into a sandstone.

The observation by Smith *et al.* (1971) is quite interesting. Let us assume that there is a sandstone model, the width and length of which become doubled when its thickness is doubled. The volume of the sandstone body and possibly the total pore space in this sandstone will increase in this case by a factor of 8. If total pore space is the most important factor controlling the amount of accumulated oil in this model, the reserve must increase by a factor of 8 when the thickness is doubled. If, for the sake of argument, the organic matter in the reservoir section is assumed to control the amount of accumulated oil, the reserve must still increase by a factor of 8 when the reservoir thickness is doubled.

The statistical data indicate that the reserve increases by a factor of 4 when the thickness is doubled, suggesting that the most important factor controlling reserve volume is the contact area between the sandstone body and the surrounding shales, as Smith *et al.* (1971) have stated. This fact becomes readily understandable if the main hydrocarbon source were in the surrounding shales and the hydrocarbons were squeezed from them during compaction. The efficiency of the fluid expulsion generally increases as the contact area increases.

If a different model, the width and length of which do not become doubled when the thickness is doubled, is assumed, the previous logic cannot apply. If there were such a sandstone type in the subsurface, its presence might be inferred from the range of accumulated oil for a certain average thickness of sandstone, as indicated in Fig. 4.1. A certain sandstone type in terms of the thickness–width or thickness–length ratio could contain greater or smaller amounts of oil than could the other type for the same thickness. The average sandstone thickness for the 7241 reservoir sandstones of the American oil fields is 39 ft, according to Curtis *et al.* (1960).

GEOCHEMICAL AND DIRECT EVIDENCE

Figure 4.2 shows plots of shale transit time vs depth, hydrocarbon and total extract vs depth, and percentage organic carbon vs depth for the North Sabine H49 well in the Canadian Arctic Islands (Magara, 1980*a*).

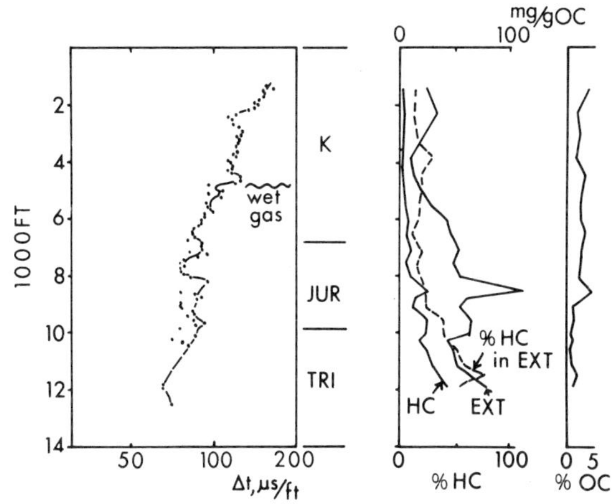

FIG. 4.2. Plots of shale transit time (Δt), hydrocarbon extract (HC) and total extract (EXT), and percentage of organic carbon (OC), against depth of Canadian Arctic Island well (from Magara, 1980*b*). Dashed line in central plot shows percentage of hydrocarbons in total extract. Geochemical data derived from Powell (1978). (Reproduced by permission of the American Association of Petroleum Geologists.)

Below about 4800 ft, the gas found in cuttings is more than 60% wet, indicating a geochemically mature facies. The shales between 4400 and 4800 ft are slightly undercompacted. This type of shale is called a 'pressure seal' (Evans *et al.*, 1975) and is believed to prevent the upward leakage of the subsurface fluids (see Chapter 6 of this book).

The amount of the total extract increases with depth between 4800 and 7000 ft and stays relatively constant between 7000 and 9000 ft, except for an anomalously high value at 8500 ft. In these intervals, the amount of hydrocarbons remains less than those below 9000 ft.

There may be two possible explanations for the relatively small quantity of hydrocarbons in the total extract from the above intervals: (1) these rocks represent a relatively early genetic stage in which hydrocarbons are still being generated; or (2) a part of the generated hydrocarbons has been expelled from the shales. Probably the plot represents the combined results of both factors.

As the transit-time plot indicates, the shales in these intervals are relatively well compacted, indicating an effective paleofluid drainage condition. If a preferential expulsion of hydrocarbons over asphaltenes, such as observed in the Devonian shales of Algeria (Fig. 1.7 in Chapter 1) had taken place, the amount of hydrocarbons remaining in the shales would be relatively small.

Below 9000 ft, where the shales are significantly undercompacted, the amount of hydrocarbons remaining in the shales increases rapidly with depth. In this deeper interval, the generated hydrocarbons were probably not flushed effectively from the shales because of the relatively poor drainage environment.

Figure 4.3 is a similar plot for the Helicopter J12 well in the Canadian Arctic Islands (Magara, 1980*a*). The percentage hydrocarbons in the total extract (dashed line in the central column) increases sharply at about 5000 ft, at which significant undercompaction begins. Geochemically mature facies from the cuttings-gas analysis begin at about 3300 ft. Therefore, it is quite probable that the hydrocarbons generated in the mature section between 3300 and 5000 ft were preferentially expelled because of the effective drainage conditions there.

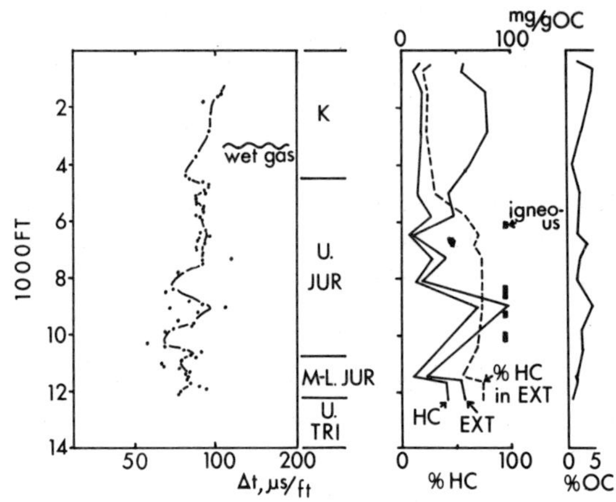

FIG. 4.3. Plots of shale transit time (Δt), hydrocarbon extract (HC) and total extract (EXT), and % of organic carbon (OC) against depth of Canadian Arctic Island well (from Magara, 1980*b*). Dashed line in central plot shows percentage of hydrocarbons in total extract. Geochemical data derived from Powell (1978). (Reproduced by permission of the American Association of Petroleum Geologists.)

Another example from the Tertiary shales in the Niger Delta studied by Fujita (1977) is shown in Fig. 4.4. The left-hand column in this figure shows the plot of shale porosity vs depth, and the central plot is the ratio of weight (mg) of hydrocarbons over grams of organic carbon. Fujita

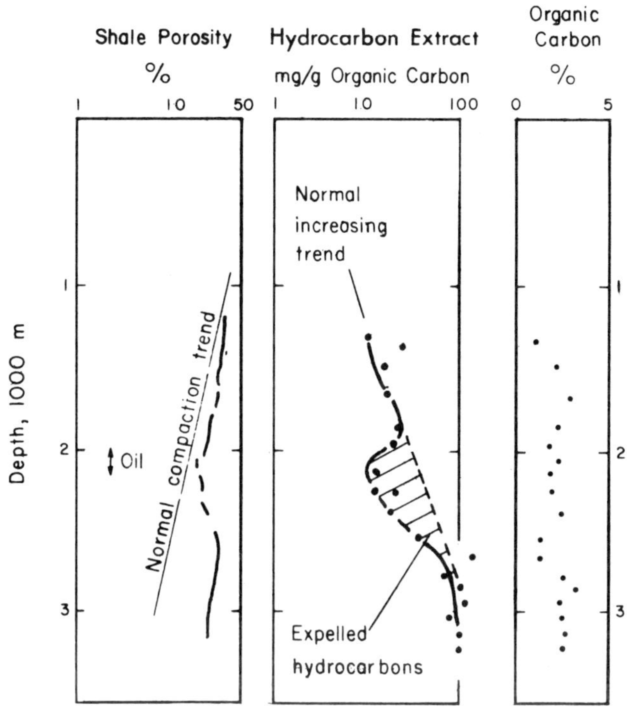

FIG. 4.4. Plots of shale porosity, hydrocarbon extract, and percentage organic carbon against depth of Niger Delta well, Nigeria (revised from Fujita, 1977).

assumed that this ratio would increase continuously with maturation or burial depth if most of the generated hydrocarbons were to remain in the shales. The result is shown by the curved line labelled 'normal increasing trend'.

The actual ratio declines toward the zone around 2100 m, where the shales have reached the level of maximum compaction. There is good reservoir development in this section, and the reservoirs sometimes contain accumulated hydrocarbons. Therefore, the declining trend is interpreted to show primary migration of hydrocarbons from the shales into the interbedded reservoir sections. Figure 4.5 is a similar plot for a well in Niigata, Japan (Fujita, 1977). The best drainage section exists between 3200 and 4100 m.

Figure 4.6 is an example from northern Japan, in which the best drainage section is shown between 2000 and 3000 m. Another example from the same general area is shown in Fig. 4.7. The most effective

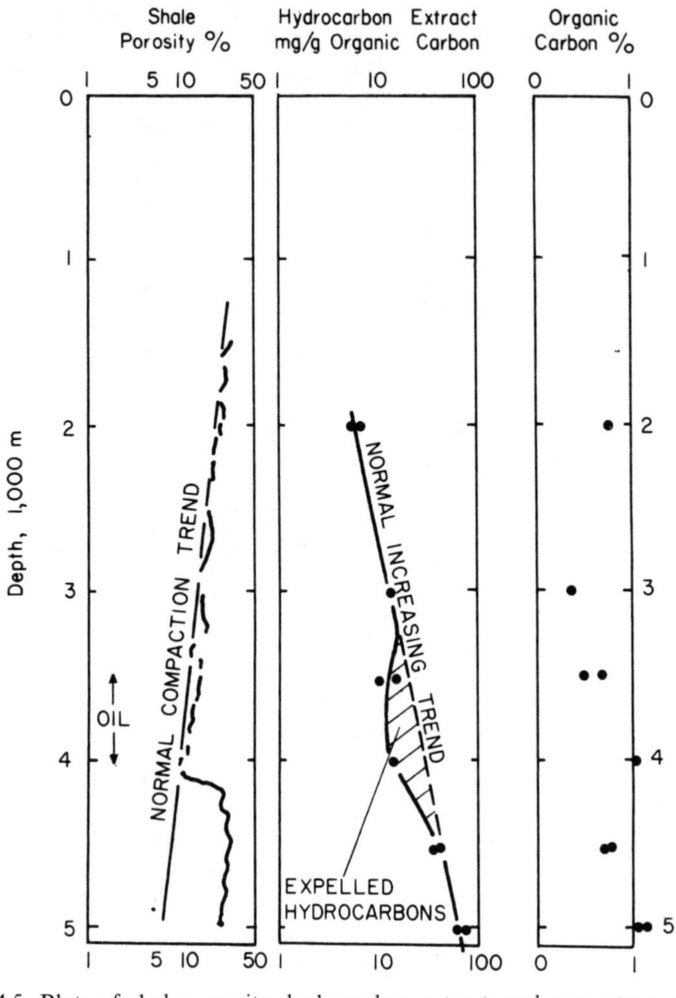

FIG. 4.5. Plots of shale porosity, hydrocarbon extract, and percentage organic carbon against depth of MITI-Yoshida well, Japan (revised from Fujita, 1977).

expulsion of hydrocarbons is indicated in the zone between 1800 and 3100 m. The hydrocarbon extract data at 3200 m suggest a relatively small amount of expulsion, while the shale compaction there is still normal. This may suggest that at such deep burial (3200 m), the expulsion efficiency declines significantly even though the shales are normally compacted.

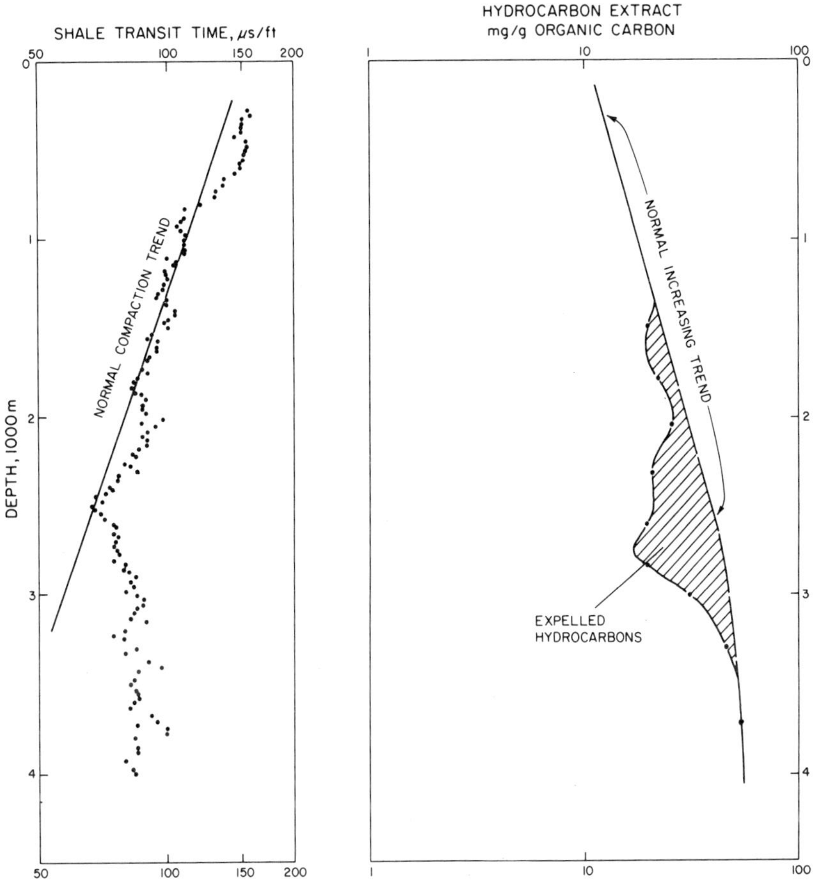

FIG. 4.6. Plots of shale transit time and hydrocarbon extract against depth of a well in northern Japan (from Magara, 1980*b*). (Reproduced by permission of the American Association of Petroleum Geologists.)

The direction of fluid migration seems to have been primarily vertical (or across the shale's bedding planes). For such vertical hydrocarbon migration to be possible, the shales must have had strong vertical potential differences, which are commonly observed in the intermediate depth ranges (3000–8000 ft) of relatively young sedimentary basins in the world. In the older sedimentary rocks, such vertical potential differences may have been dissipated, but shale porosity profiles sometimes indicate

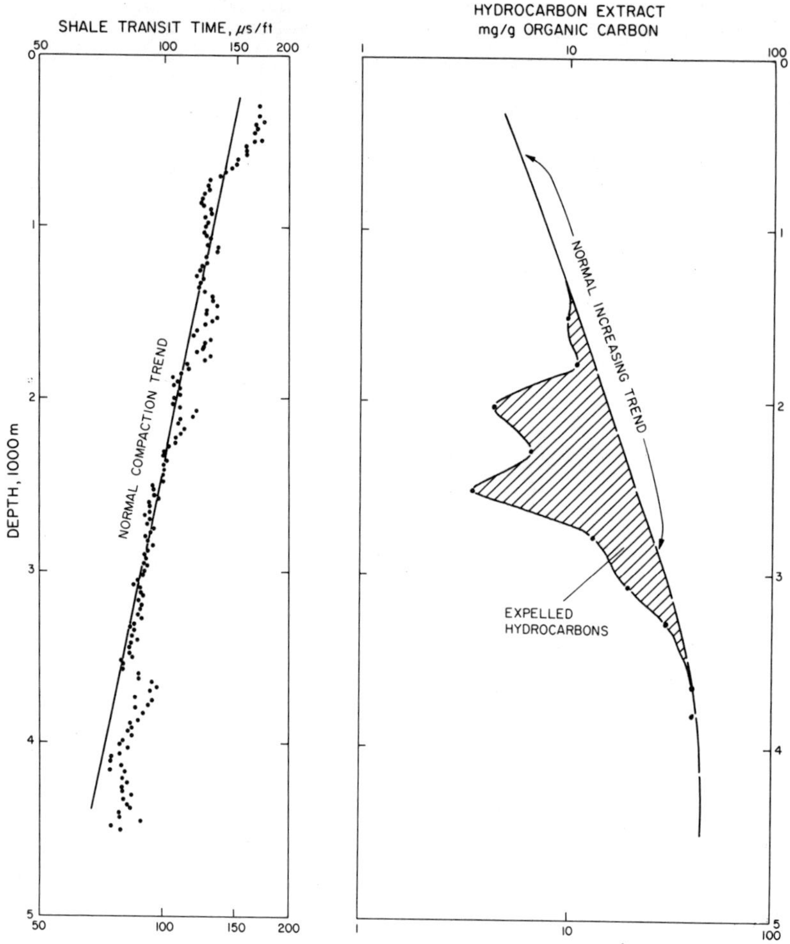

FIG. 4.7. Plots of shale transit time and hydrocarbon extract against depth of a well in northern Japan (from Magara, 1980*b*). (Reproduced by permission of the American Association of Petroleum Geologists.)

fossil conditions of undercompaction (see Chapter 5 of Magara, 1978).

There must have been a portion of hydrocarbons which migrated horizontally along the bedding planes. However, the significant differences of remaining hydrocarbons in the vertical section of source rocks suggest that the net vertical migration must have occurred.

EFFICIENCY OF HYDROCARBON EXPULSION

On the basis of the data shown in Figs. 4.2–4.7 and Fig. 1.7, we may be able to conclude that up to 30–70% of the total hydrocarbons generated could be squeezed from source rocks in zones of effective fluid expulsion. Based on the study of four major sedimentary basins (Los Angeles, Permian, Arabian Gulf, and West Siberian), McDowell (1975) concluded that hydrocarbons expelled from source rocks can be as high as 50% of the total generated.

The summary plots of hydrocarbons ($C15^+$) extract vs maximum burial depth for ten wells drilled in the Canadian Arctic region (Fig. 4.8) may provide additional evidence of hydrocarbon expulsion from source rocks. The geochemical data are derived from Powell (1978). The maximum burial depth is estimated using the method described in Chapter 7 of this book. Some shales in this region have been severely metamorphosed by the effect of volcanic intrusions, which occurred in the geological past; the data from these shales are not included in this plot.

Because the chemical kinetic function of hydrocarbon generation can be expressed by the Arrhenius equation (exponential function, Connan (1974)), it may be shown as a straight line on semi-log paper such as used in Fig. 4.8. This function is shown by a thick line 'hydrocarbon generation trend' in this figure. It would bend at depth as the effective source material is exhausted by production of a significant quantity of hydrocarbons, as shown in the deeper part of Fig. 4.8. This hydrocarbon generation line is drawn through the highest measured values, on the assumption that they represent the least expulsion cases.

Many points are plotted below the generation trend, suggesting the possibility of hydrocarbon expulsion. The percentage hydrocarbons expelled in the total generated can be estimated, using the four dashed lines representing 20, 40, 60, and 80% expelled hydrocarbons, respectively.

There is a chance that some source rocks at a given burial depth produced smaller quantities of hydrocarbons owing to their poor source quality. The values of these poor source rocks would be plotted below the generation trend which corresponds to the maximum generation cases. However, I believe that many, if not most, points in Fig. 4.8 still show the effect of significant hydrocarbon expulsion, on the basis of previously mentioned observation and interpretations.

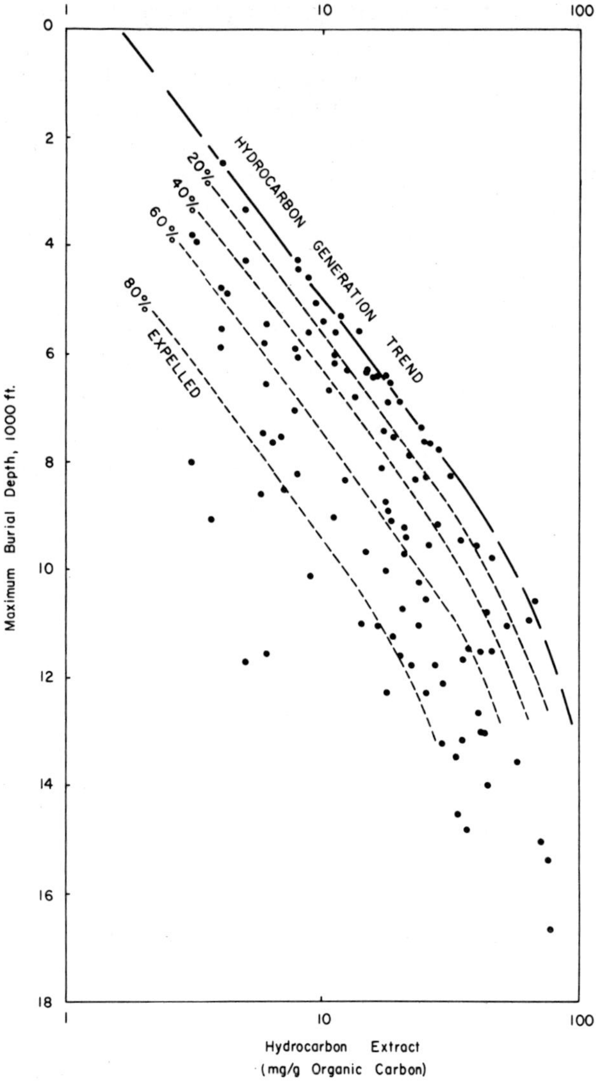

FIG. 4.8. Plot of hydrocarbon extract versus maximum burial depth in Canadian Arctic Island. Geochemical data are derived from Powell (1978). Thick line shows possible hydrocarbon generation trend and four dashed lines represent hydrocarbon expulsion of 20, 40, 60, and 80%, respectively, of total generated (from Magara, 1980*b*). (Reproduced by permission of the American Association of Petroleum Geologists.)

EVIDENCE FROM THE WORLD'S MAJOR OIL-PRODUCING BASINS

After determining source-rock oil-generating capacity and the amounts of accumulated oil in the several oil-producing sedimentary basins, McDowell (1975) concluded that the amount of oil expelled (Q_E) from the source rock is between 0 and 50% of the total (Q) generated ($Q_E = 0$–50% of Q—see Fig. 4.9). McDowell considered that the amount lost to the surface (Q_L) is no more than 10% of Q_E ($Q_L = 0$–10% of Q_E—see Fig. 4.9). The amount in reservoirs (Q_R) was considered to be less than 50% of Q_E or less than 25% of Q ($Q_R = 0$–50% of $Q_E = 0$–25% of Q—see Fig. 4.9). These figures suggest that as much as 75–100% of the total oil generated

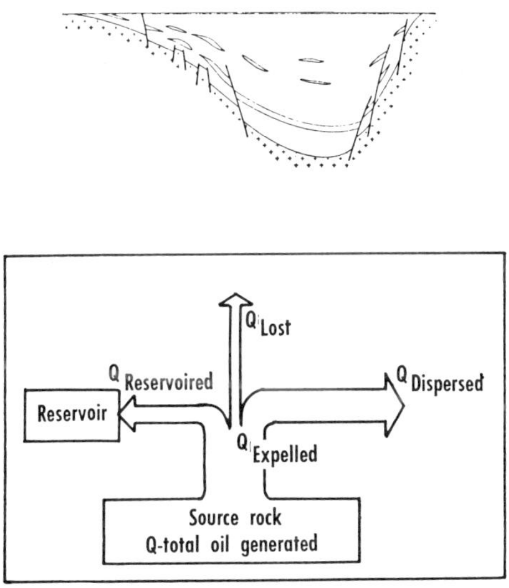

$Q_E = 0$ to 50+% of Q

$Q_L = 0$ to 10% of Q_E

$Q_R = 0$ to 50% of $Q_E = 0$ to 25+% of Q

FIG. 4.9. McDowell (1975) model of oil generated, expelled, dispersed, lost, and reservoired (from McDowell, 1975). (Reproduced by permission of the *Oil and Gas Journal*, Oklahoma.)

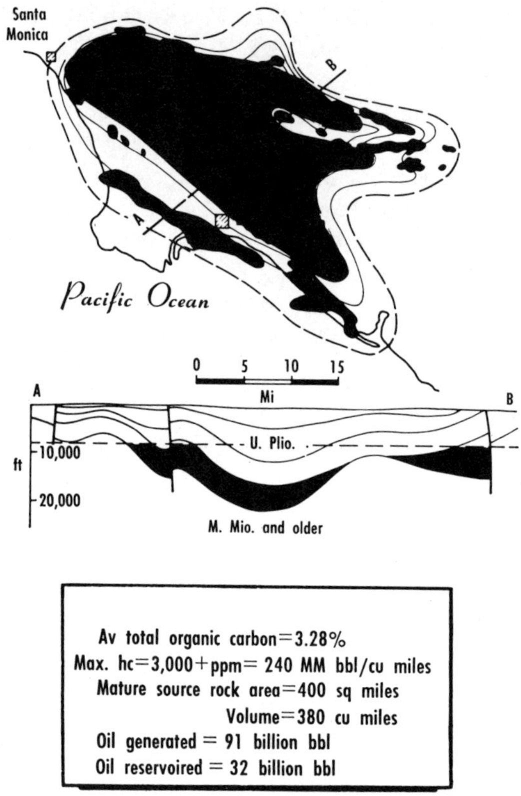

FIG. 4.10. Estimated oil generated and reservoired in Los Angeles Basin (from McDowell, 1975). (Reproduced by permission of the *Oil and Gas Journal*, Oklahoma.)

can be disseminated or dispersed in the sedimentary rocks, some of which may be called 'low-quality reservoirs'.

It would thus be a mistake to assume that the amount of oil (Q_E) expelled from the source rock equals the amount in reservoirs (Q_R). Because the amounts accumulated in good-quality reservoirs are usually only a small proportion of the total oil generated, one might thence conclude that the oil-expulsion mechanism in source rock is very ineffective. But this conclusion is unsound, because the premise ignores the significant amount of oil dispersed in the sedimentary rocks and lost to the surface.

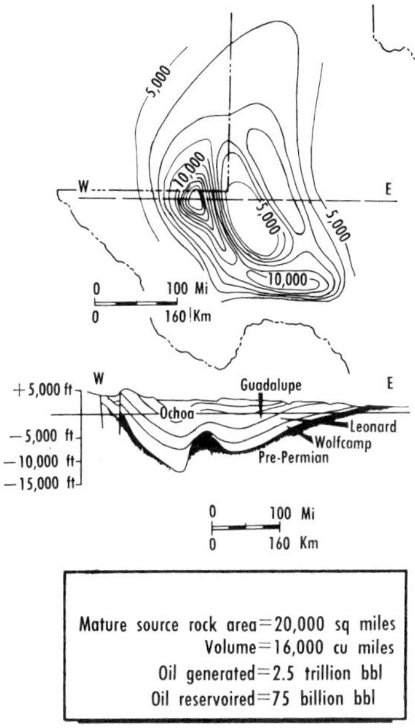

FIG. 4.11. Estimated oil generated and reservoired in Permian Basin (from McDowell, 1975). (Reproduced by permission of the *Oil and Gas Journal*, Oklahoma.)

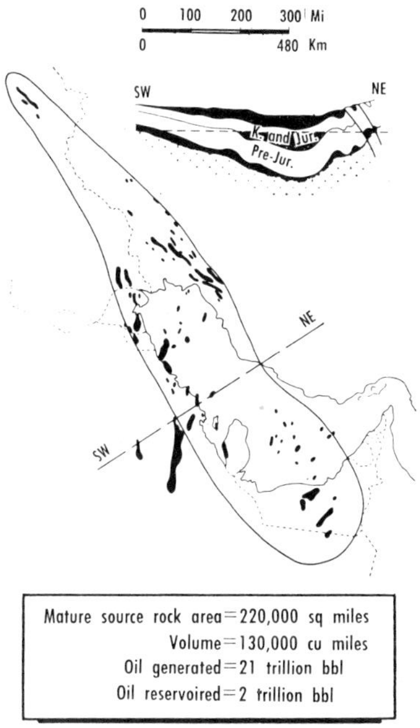

FIG. 4.12. Estimated oil generated and reservoired in Arabian Gulf Basin (from McDowell, 1975). (Reproduced by permission of the *Oil and Gas Journal*, Oklahoma.)

McDowell (1975) studied the four major sedimentary basins in the world (Los Angeles, Permian, Arabian Gulf and West Siberian), and made estimates of the oils generated and accumulated, as shown in Figs. 4.10, 4.11, 4.12 and 4.13.

REFERENCES

Connan, J. (1974) Time–temperature relation in oil genesis. *A.A.P.G.*, **58**, 2516–21.

Curtis, B. F., and Sandstone Reservoir Committee (1960). Characteristics of sandstone reservoirs in United States. In Peterson, J. A. and Osmond, L. C. (eds.), *Geometry of Sandstone Bodies*, A.A.P.G., pp. 208–19.

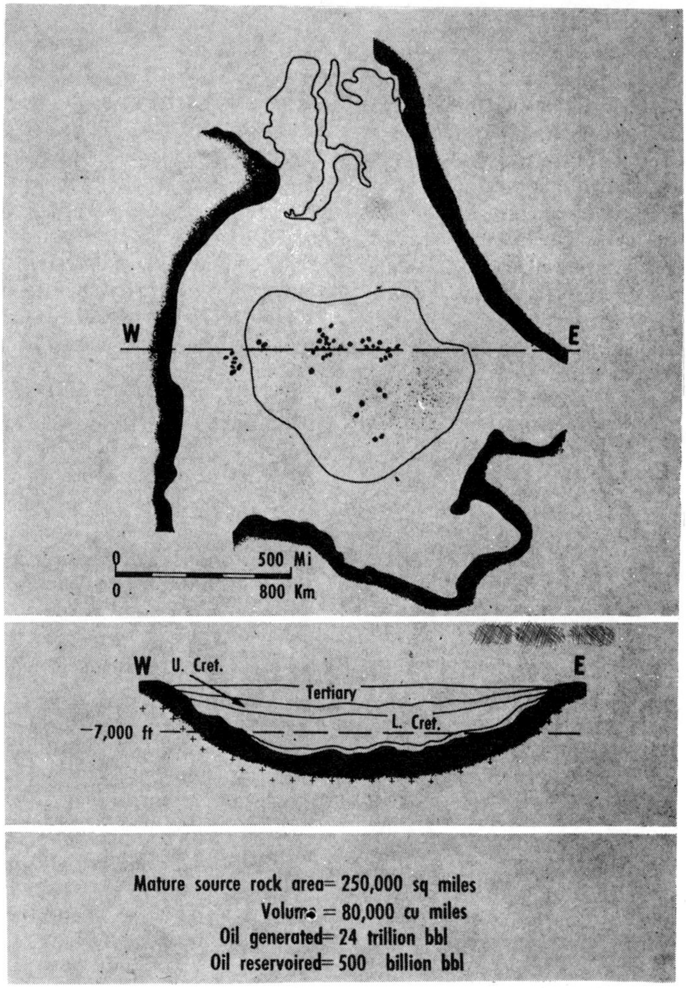

FIG. 4.13. Estimated oil generated and reservoired in West Siberian Basin (from McDowell, 1975). (Reproduced by permission of the *Oil and Gas Journal*, Oklahoma.)

Dickey, P. A. (1975) Possible primary migration of oil from source rock in oil phase. *A.A.P.G.*, **59**, 337–45.

Evans, C. R., McIvor, D. K. and Magara, K. (1975) Organic matter, compaction history, and hydrocarbon occurrence—Mackenzie Delta, Canada, *Proc. 9th World Petroleum Congress*, **3**, 149–57.

Fujita, Y. (1977) The role of shale porosity anomaly in hydrocarbon exploration. *Japan. Assoc. Petrol. Technol. J.*, **42**, 107–16.

Levorsen, A. I. (1967) *Geology of Petroleum*, 2nd edn, W. H. Freeman, San Francisco, 724 p.

Magara, K. (1978) *Compaction and Fluid Migration—Practical Petroleum Geology*, Elsevier, Amsterdam, 319 p.

Magara, K. (1980*a*) Primary migration of oil and gas. In Miall, A. (ed.), *Facts and Principles of the World's Petroleum Occurrences*, Canadian Soc. Petrol. Geologists Memoir 6.

Magara, K. (1980*b*) Evidences of primary oil migration. *A.A.P.G.*, **64**, 2108–17.

McDowell, A. N. (1975) What are the problems in estimating the oil potential of a basin? *Oil Gas, J.*, June 9, 85–90.

Powell, T. G. (1978) An assessment of the hydrocarbon source rock potential of the Canadian Arctic Islands, Canada Geol. Survey Paper 78–12, 82 p.

Smith, J. E., Erdman, J. G. and Morris, D. A. (1971) Migration, accumulation, and retention of petroleum in the earth. *8th World Petroleum Cong. Proc.*, **2**, 13–26.

Tissot, B. and Pelet, R. (1971) Nouvelles données sur les mechanismes de génése et de migration du petrole simulation mathematique et application á la prospection. *8th World Petroleum Cong. Proc.*, 35–46.

Chapter 5

Estimation of Vitrinite Reflectance from Burial History Plots

INTRODUCTION

Vitrinite reflectance has been widely used as an index of the level of organic maturation, which is important in the assessment of the status of petroleum generation in source rocks. Figure 5.1 summarizes several maturation indices including vitrinite reflectance in relation to oil and gas generation (Dow, 1977).

Because vitrinite reflectance values are measured using rock samples, a

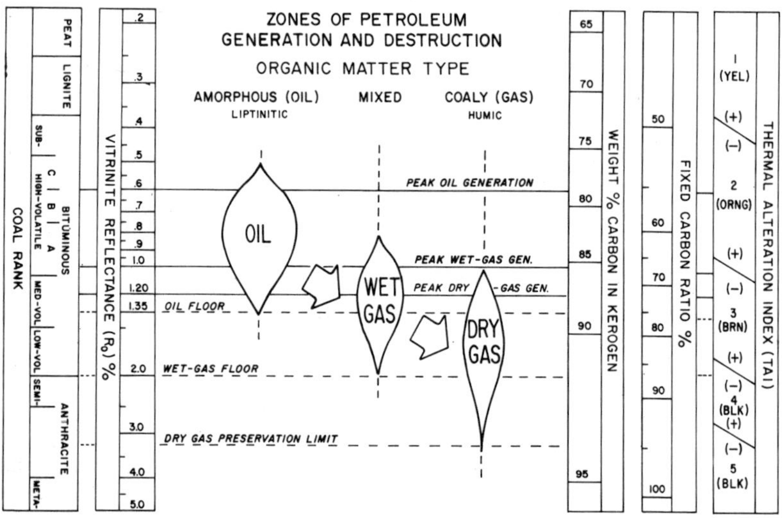

FIG. 5.1. Correlation of coal-rank scale with various maturation indices and zones of petroleum generation and destruction (from Dow, 1977). (Reproduced by permission of Elsevier Science Publishers, Amsterdam.)

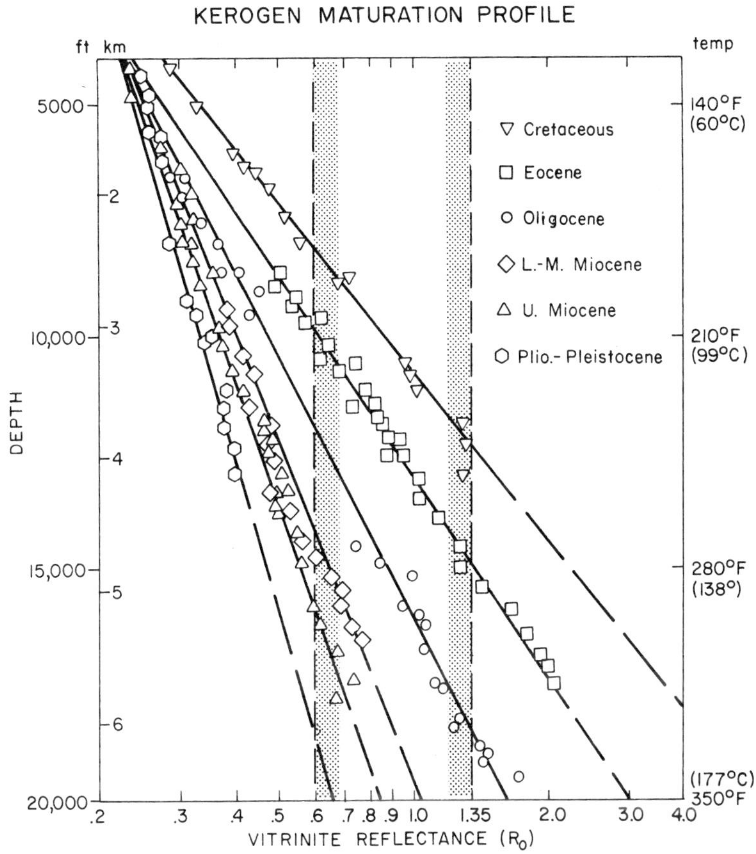

FIG. 5.2. Composite vitrinite reflectance (R_0) profiles for two wells in Gulf Coast. All wells have uniform geothermal gradients close to 1·4°F/100 ft. Zone between 0·6% R_0 and 1·35% R_0 is considered to be liquid generating zone. Note that higher temperatures are required to achieve equivalent maturation in younger rocks than in older ones (from Dow, 1978). (Reproduced by permission of the American Association of Petroleum Geologists.)

proper estimate of organic maturation in the subsurface can be made only after drilling a well. However, in most exploration plays it is desirable to have such an estimate before drilling. This chapter describes a method of estimating reflectance values from a burial history plot which can be constructed from geologically interpreted seismic data.

Both organic maturation and vitrinite reflectance are primarily controlled by temperature and duration of time (Connan, 1974). Therefore, it is possible to estimate the reflectance if a relationship between the reflectance, geologic age, and depth of burial (or temperature) in a study area is known. Figure 5.2 shows such an empirical relationship for the Gulf Coast sedimentary rocks (Dow, 1978). Where rates of sediment burial were relatively uniform throughout the entire geologic period, a

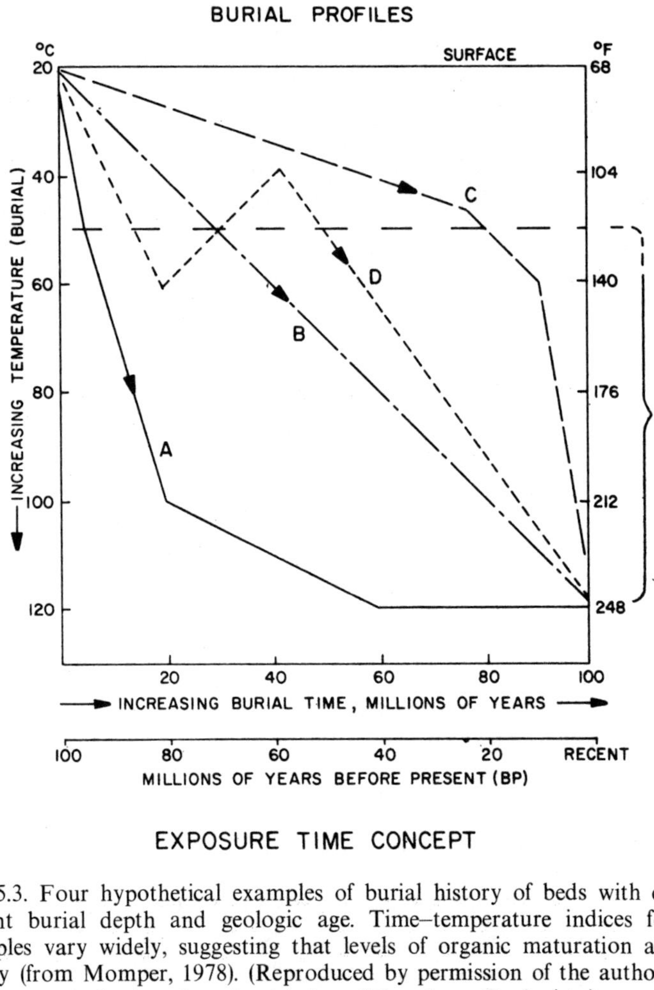

FIG. 5.3. Four hypothetical examples of burial history of beds with common present burial depth and geologic age. Time–temperature indices for these examples vary widely, suggesting that levels of organic maturation also vary greatly (from Momper, 1978). (Reproduced by permission of the author and of the American Association of Petroleum Geologists.)

plot of depth of a bed of known geologic age provides a reliable estimate of vitrinite reflectance, such as in Fig. 5.2. However, if the rate of burial significantly changed in the geological past, or if major erosional periods were involved, a reliable estimate cannot be made by simply plotting the depth or temperature value on a chart such as that of Fig. 5.2.

Momper (1978) presented hypothetical examples of pairing of the present burial depth and geologic age with variable burial histories (cases A, B, C, and D in Fig. 5.3). In these four cases of variable burial histories, the time spent by sediments at different depths differ, suggesting that the levels of organic maturation and of vitrinite reflectance are also different. To make a proper estimate of vitrinite reflectance, the burial history and the time–temperature index of a given sedimentary rock must be estimated (Waples, 1980).

BURIAL HISTORY PLOT

Let us plot the burial history for a reservoir at the base of the Triassic formations shown in Fig. 5.4. Depth in ft is shown on the left and geologic age in millions of years on the right of the column in Figure 5.4. Figure 5.5 shows the depths of the reservoir at five different stages of burial.

Stage 1 is when the deposition of the reservoir section had just been completed. At stage 2, which is the end of the Triassic age, the reservoir had been buried to 5000 ft, because that is the thickness of the Triassic formations above it (14 500 − 9500 = 5000 ft, Fig. 5.4). The effect of decompaction of the sedimentary sequence is not considered in this case. That is, a 5000 ft section today must have been thicker than 5000 ft when it was less deeply buried (or at the end of the Triassic). Therefore, the reservoir must have been buried deeper than 5000 ft at that time. Corrections for decompaction effect can be made if the porosity–depth relationship of the area is known (see Chapter 6 of Magara's book, 1978). The age of the reservoir at stage 2 was 45 million years since deposition (225 − 180 = 45 million years, Fig. 5.4). The burial history from stage 1 to stage 2 is plotted in Fig. 5.6 (solid line connecting numbers 1 and 2).

At stage 3, or at the end of Jurassic age, the reservoir reached a depth of 8000 ft as shown in Fig. 5.5 (14 500–6500 = 8000 ft, Fig. 5.4). The geologic age of the reservoir at that time was 90 million years (225 − 135 = 90 million years, Fig. 5.4). The burial of the reservoir at stage 3 is shown by number 3 on Fig. 5.6.

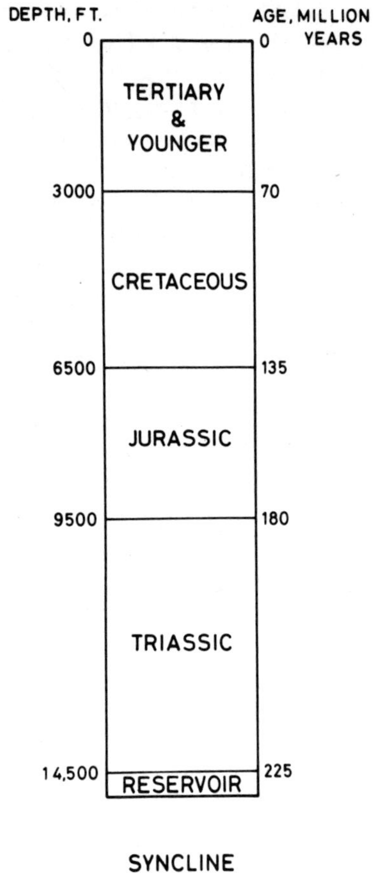

FIG. 5.4. Schematic geologic column of a syncline location.

The burial history for the later stages is illustrated by columns 4 and 5 in Fig. 5.5 and points 4 and 5 in Fig. 5.6.

If the correction for the effect of decompaction is properly applied, we would expect the burial history plotted as 1, 2′, 3′, 4′ and 5 in Fig. 5.6. In other words, at intermediate stages of burial, the points are plotted deeper than in the previous uncorrected cases, but the two end points of the burial history (stage 1, initial, and stage 5, present) will be the same in both cases. Refer also to Van Hinte (1978) for the method of compaction or decompaction correction.

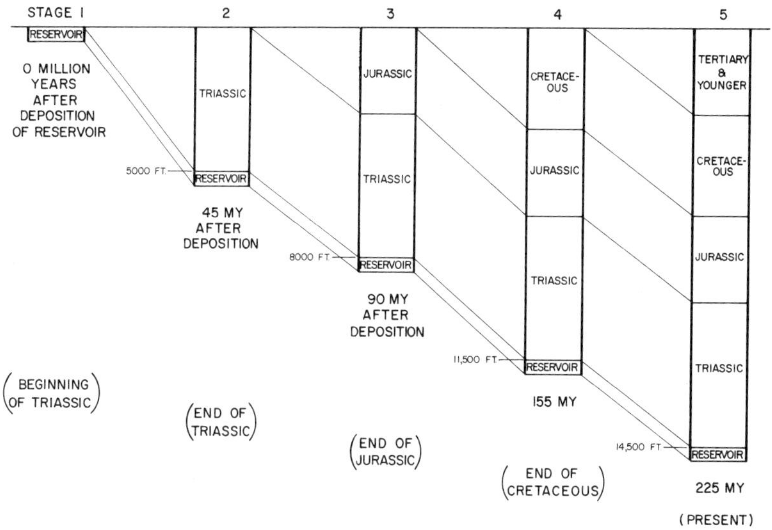

FIG. 5.5. Diagram showing successive change of burial depth of reservoir (shown in Fig. 5.4) with geologic time.

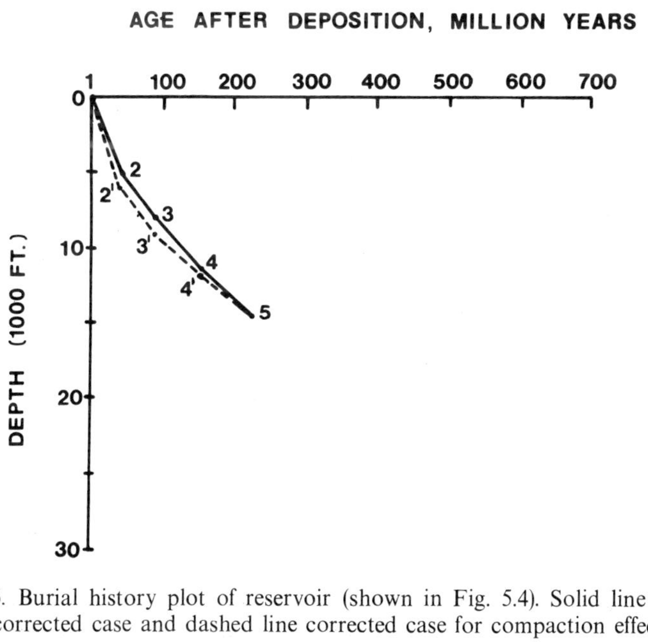

FIG. 5.6. Burial history plot of reservoir (shown in Fig. 5.4). Solid line shows uncorrected case and dashed line corrected case for compaction effect.

TIME–TEMPERATURE INDEX AND VITRINITE REFLECTANCE

Based on Lopatin's (1971) fundamental concept, Waples (1980) stated that the time–temperature index (TTI) or the maturity indicator of a given sediment is defined as the sum of the maturities acquired in each time–temperature interval, or

$$\mathrm{TTI} = \sum_{n\,\min}^{n\,\max} \Delta t_n \cdot r^n \tag{5.1}$$

where Δt_n is the length of time spent by the sediment in the nth temperature interval, r^n the temperature factor of the same interval shown in Table 5.1, and n max and n min are the n-values of the highest and lowest temperature intervals encountered.

TABLE 5.1
List of temperature intervals, index values, n, and temperature factors, r^n, for calculation of time–temperature integrals (TTI). (From Waples, 1980)

Temperature interval (°C)	*Index value n*	*Temperature factor r^n*
20–30	−8	2^{-8}
30–40	−7	2^{-7}
40–50	−6	2^{-6}
50–60	−5	2^{-5}
60–70	−4	2^{-4}
70–80	−3	2^{-3}
80–90	−2	2^{-2}
90–100	−1	2^{-1}
100–110	0	1
110–120	1	2
120–130	2	2^2
130–140	3	2^3
	m	2^m

Once the TTI value is calculated for a given sample, it can be converted to vitrinite reflectance, R_0, using Fig. 5.7, which is a plot of Waples' correlation between the TTI and R_0 derived from 402 worldwide samples.

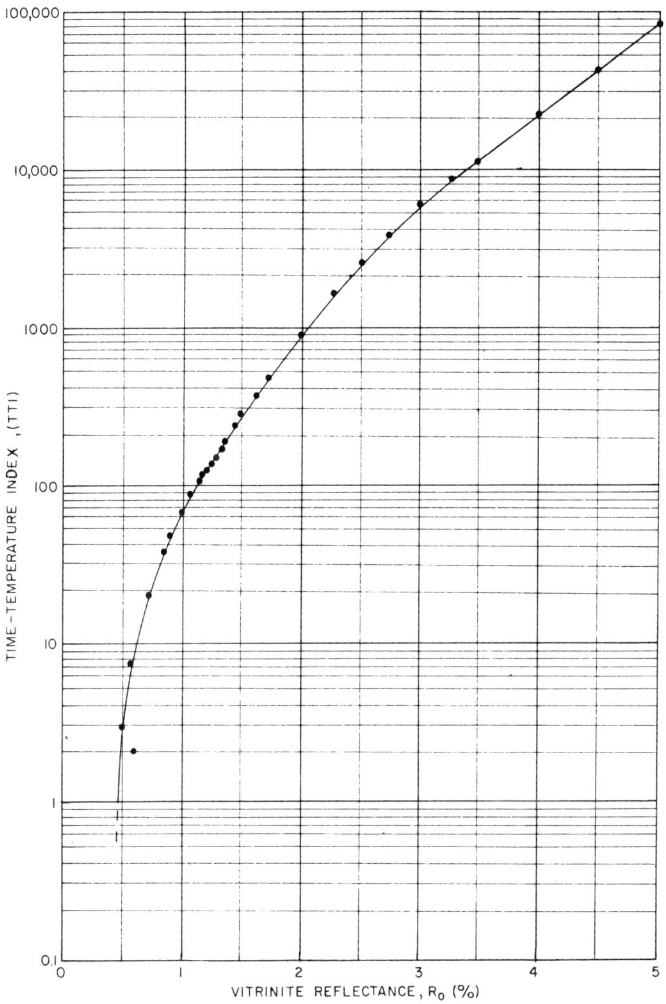

FIG. 5.7. Empirical relationship between time–temperature index (TTI) and vitrinite reflectance (R_0) derived from 402 samples studied by Waples (1980).

EXAMPLES OF CALCULATION IN TEXAS

Burial history plots for the Oligocene beds at 12 500 ft (30 million years old) and at 9000 ft (21 million years old) of the General Crude-DOE

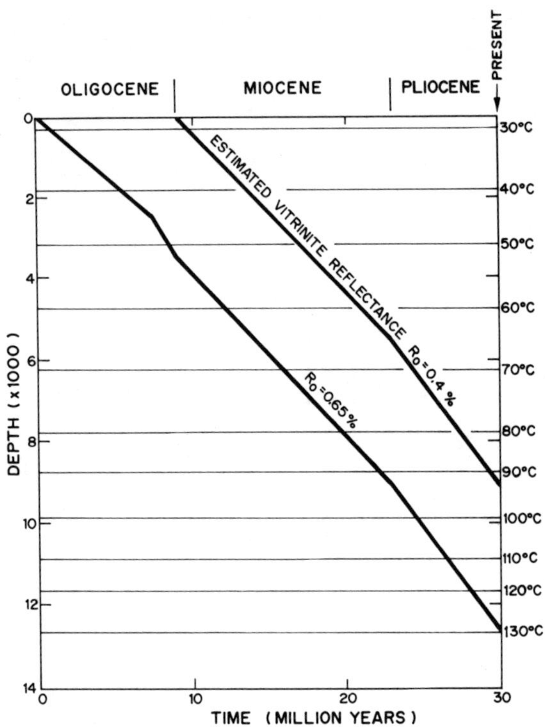

FIG. 5.8. Burial history plots of two Oligocene beds at 12 500 and 9000 ft at General Crude-DOE Pleasant Bayou No. 1 well in Texas Gulf Coast. Estimated vitrinite reflectance values (R_0) are also shown. Refer to Table 5.2 and text for detailed calculation of R_0.

Pleasant Bayou No. 1 well are shown in Fig. 5.8. Both cases represent relatively uniform rates of burial. The paleo-geothermal gradient during the burial stages is assumed to be the same as the present geothermal gradient determined by Gregory and Backus (1979). Details of calculation of the TTI and vitrinite reflectance, R_0, are shown in Table 5.2. Estimated R_0 values are about 0·65% for 12 500 ft and about 0·4% for 9000 ft, respectively, which are comparable with the actually measured values shown in Fig. 5.9.

Figure 5.10 shows the burial history plots of the beds at the top (4000 ft) and bottom (7000 ft) of the Pennsylvanian formations (281–325 million years old) of the Baria Werner No. 1 Mayes well in the Palo

TABLE 5.2
Temperature interval r^n, Δt_n, $\Delta t_n \cdot r^n$, and TTI of two Oligocene beds at 12 500 and 9000 ft (present burial depths) of General Crude-DOE Pleasant Bayou No. 1 well in Gulf Coast, Texas

Temperature interval (°C)	r^n	Δt_n *(m.y.)*	$\Delta t_n \cdot r^n$	*TTI*
Bed at 12 500 ft (30 million years old)				
30–40	2^{-7}	5·5	0·04	
40–50	2^{-6}	3·0	0·05	
50–60	2^{-5}	3·5	0·11	
60–70	2^{-4}	4·0	0·25	
70–80	2^{-3}	3·8	0·47	
80–90	2^{-2}	2·5	0·62	
90–100	2^{-1}	2·2	1·10	
100–110	1	2·0	2·00	
110–120	2	1·6	3·20	
120–130	2^2	1·9	7·60	15·44 ($R_0 \doteqdot 0{\cdot}65\%$)
Bed at 9000 ft (21 million years old)				
30–40	2^{-7}	4·5	0·04	
40–50	2^{-6}	3·5	0·06	
50–60	2^{-5}	4·0	0·12	
60–70	2^{-4}	3·5	0·22	
70–80	2^{-3}	3·0	0·37	
80–90	2^{-2}	2·0	0·50	
90–100	2^{-1}	0·5	0·25	1·56 ($R_0 \doteqdot 0{\cdot}4\%$)

Duro Basin, Texas. Based on the study of available sonic logs in this basin, at least a 1000 ft of section is estimated to have been removed by erosion during the Tertiary period (refer to Chapter 7 of this book for this method). This erosional event is indicated by the wavy lines in Fig. 5.10. There must have been more erosional events before the Tertiary, but they are not shown because the amount and duration of erosion cannot be estimated at the present time. The geothermal gradient is assumed to have been 1·2°F/100 ft, which is the present average figure for this region. The estimated R_0 values are about 0·53% for the top and 0·72% for the bottom of the Pennsylvanian formations. The details of calculation of the TTI and R_0 are included in Table 5.3.

According to Dutton (1980), the average vitrinite reflectance, R_0, for

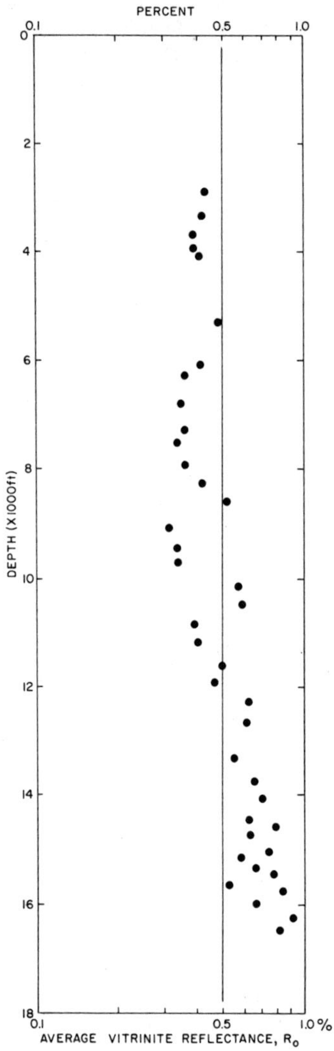

FIG. 5.9. Plot showing average vitrinite reflectance (R_0) of rocks at General Crude-DOE Pleasant Bayou Nos. 1 and 2 wells in Texas Gulf Coast.

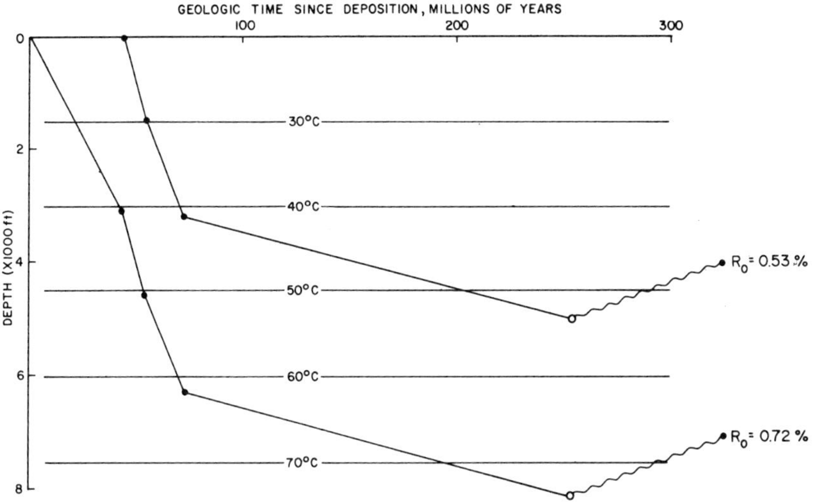

FIG. 5.10. Burial history plots of two Pennsylvanian beds at 7100 and 4000 ft at Baria Werner No. 1 Mayes well in Palo Duro Basin, Texas. Estimated vitrinite reflectance values (R_0) are also shown. Refer to Table 5.3 and text for detailed calculation of R_0.

TABLE 5.3
Temperature interval r^n, Δt_n, $\Delta t_n \cdot r^n$, and TTI of two Pennsylvanian beds at 7100 and 4000 ft (present burial depths) of Baria Werner No. 1 Mayes well in Palo Duro Basin, Texas

Temperature interval (°C)	r^n	Δt_n*(m.y.)*	$\Delta t_n \cdot r^n$	*TTI*
Bed at 7100 ft (325 million years old)				
20–30	2^{-8}	22	0·1	
30–40	2^{-7}	22	0·2	
40–50	2^{-6}	10	0·2	
50–60	2^{-5}	15	0·5	
60–70	2^{-4}	121	7·6	
70–80	2^{-3}	103	12·9	
70–60	2^{-4}	32	2·0	23·5 ($R_0 \doteqdot$ 0·72%)
Bed at 4000 ft (281 million years old)				
20–30	2^{-8}	11	0·1	
30–40	2^{-7}	16	0·1	
40–50	2^{-6}	122	1·9	
50–60	2^{-5}	100	3·1	
50–40	2^{-6}	32	0·5	5·7 ($R_0 \doteqdot$ 0·53%)

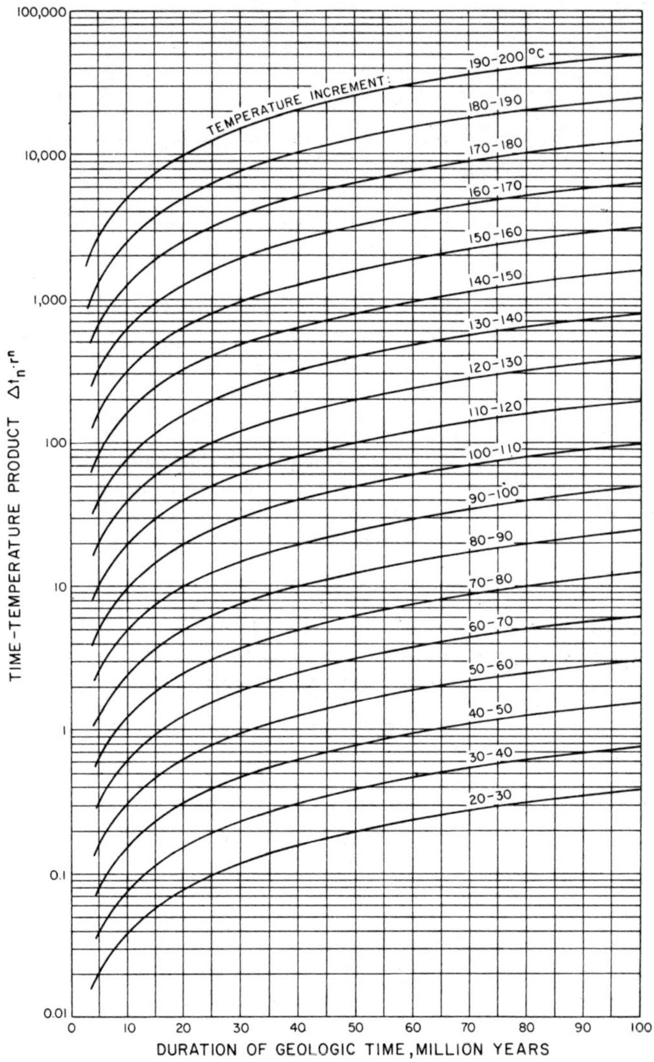

FIG. 5.11. Chart relating time–temperature product ($\Delta t_n \cdot r^n$), temperature increment, and duration of geologic time for 0–100 million years.

the Pennsylvanian source rocks in this basin is 0·52% and the highest R_0 is about 0·7%. The estimated values of R_0 (Table 5.3) are comparable with these actual values, although the actual burial histories of this basin were probably much more complicated than those assumed in Fig. 5.10.

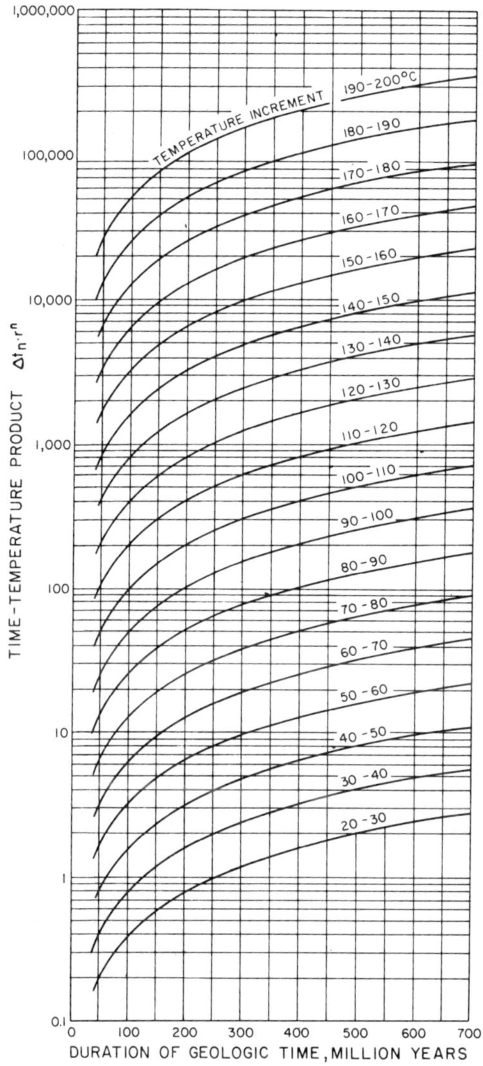

FIG. 5.12. Chart relating time–temperature product ($\Delta t_n \cdot r^n$), temperature increment, and duration of geologic time for 100–700 million years.

In the above two examples (Figs. 5.8 and 5.10), there is no correction made for the changing thickness of formations during burial and compaction. Note that the thickness of a formation at a shallower burial stage or at a less compacted stage was greater than that at the present burial stage. The burial history plot shown in these figures is, however, made on the basis of the present and compacted thickness of formations.

ESTIMATION OF TTI AND R_0 USING CHARTS

On a burial history plot such as that shown in Fig. 5.10, a series of horizontal lines corresponding to temperature increments of 10°C are drawn, and the duration of geologic time (million years) within each temperature increment is estimated.

Using either Fig. 5.11 (0–100 million years) or Fig. 5.12 (100–700 million years), the time–temperature product, $\Delta t_n \cdot r^n$ (see eqn. (5.1)), for each temperature increment can be estimated graphically. The sum of these values equals the TTI, which may be converted to the R_0, using Fig. 5.7.

REFERENCES

Connan, J. (1974) Time–temperature relations in oil genesis. *A.A.P.G.*, **58**, 2516–21.

Dow, W. G. (1977) Kerogen studies and geological interpretations. *Geochem. Exploration J.*, **7**, 79–99.

Dow, W. G. (1978) Petroleum source beds on continental slopes and rises. *A.A.P.G.*, **62**, 1584–1606.

Dutton, S. P. (1980) Petroleum source rock and thermal maturity, Palo Duro Basin, Texas, Geol. Circular, The University of Texas, Bureau of Economic Geology, 48 p.

Gregory, A. R. and Backus, M. M. (1979) Geopressured formation parameters, geothermal well, Brazoria County, Texas, *4th Geopressured Geothermal Energy Conf. Proc.*, The University of Texas at Austin, pp. 235–311.

Lopatin, N. V. (1971) Temperature and geologic time as factors in coalification (in Russian), *Akad Nauk SSSR Ixv. Ser. Geol.*, no. 3, pp. 95–106.

Magara, K. (1978) *Compaction and Fluid Migration—Practical Petroleum Geology*, Elsevier, Amsterdam, 319 p.

Momper, J. A. (1978) Oil migration limitations suggested by geological and geochemical considerations. In *A.A.P.G. Short Course, Physical and Chemical Constraints on Petroleum Migration*, W. H. Roberts, III and R. J. Cordell (Eds), vol. 1, pp. B1–B60.

Van Hinte, J. E. (1978) Geohistory analysis—application of micro-paleontology in exploration geology. *A.A.P.G.*, **62**, 210–22.

Waples, D. W. (1980) Time and temperature in petroleum formation: Application of Lopatin's method to petroleum exploration. *A.A.P.G.*, **64**, 916–26.

Chapter 6

Burial History Models

INTRODUCTION

As described in Chapter 5, the level of petroleum generation and maturation expressed by vitrinite reflectance can be estimated from a burial history plot. This chapter discusses other applications of the burial history plot for estimating times of trap and seal formations as well as the estimation of the organic carbon concentration.

There are three basic types of traps; structural, stratigraphical, and hydrodynamic. Another trap type called 'diagenetic', which is a result of the diagenetic processes in sedimentary rocks, was proposed by Wilson (1977). A burial history plot may be used for interpreting the time of the structural trap (anticline, dome or fault) formation.

Two types of seals are known; capillary seal and pressure seal. The timing analysis using a burial history plot is possible only for the pressure seal.

Possible maximum organic carbon percentage may be estimated from the slope of a burial history plot, because the slope is determined by the average rate of burial or sedimentation. The relationships between sedimentation rate and organic carbon percentage have been developed in many sedimentary basins (Chapter 2).

TRAP FORMATION

Let us now estimate the time of trap development by using geologic columns from both synclinal and anticlinal locations as shown in Fig. 6.1. The anticlinal location experienced significant erosion (2000 ft) after Tertiary deposition. Refer to Chapter 7 for the methods of estimating

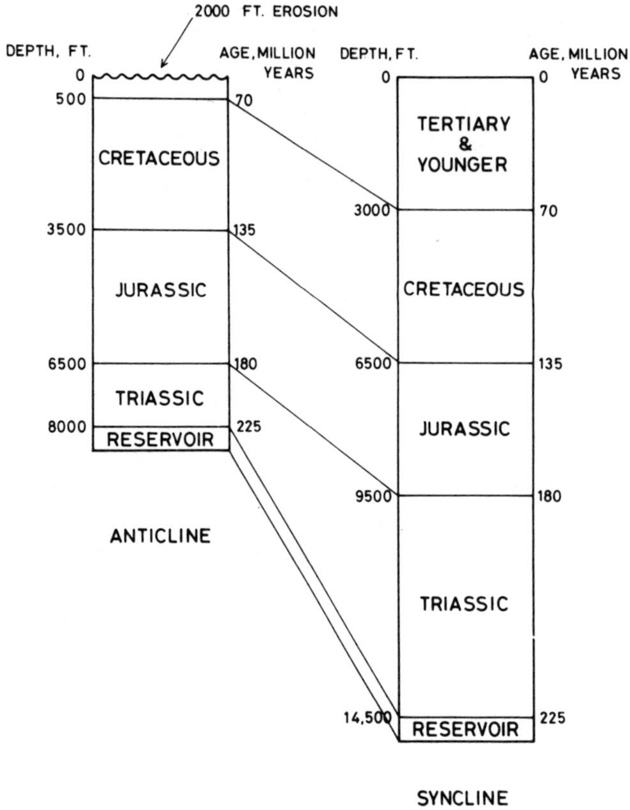

FIG. 6.1. Schematic geologic columns of both synclinal and anticlinal locations. Latter column is identical to one shown in Fig. 5.4.

thickness of erosion. Burial histories for both synclinal and anticlinal locations are shown in Fig. 6.2. Although the synclinal location experienced continuous burial from stage 1 to stage 5, the anticlinal location had two periods of slow burial or significant uplift relative to the synclinal location—one between stages 1 and 2, and one several million years before stage 5 (erosional period shown by a wavy line in Fig. 6.2). These two periods of trap development can be shown by the tendency of the two burial history lines from the synclinal and anticlinal locations to separate from each other (Fig. 6.2). No compaction correction has been made in the plot of Fig. 6.2.

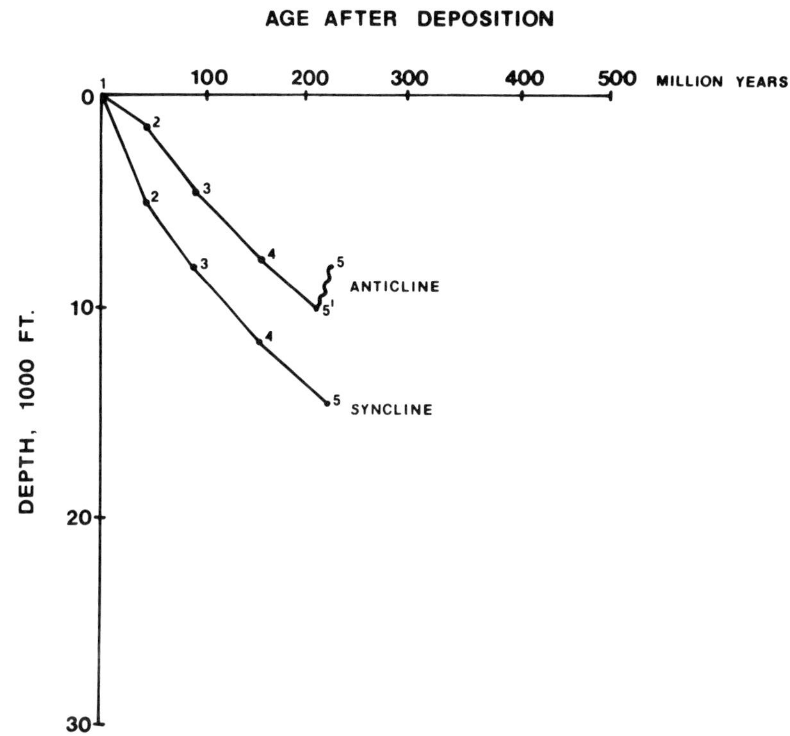

FIG. 6.2. Burial history plots of both synclinal and anticlinal locations shown in Fig. 6.1. Refer to text for explanation of timing of trap formation.

Between these two periods, the two burial history lines are nearly parallel, indicating that both the synclinal and anticlinal locations subsided at the same rate.

CONSIDERATION OF WATER-DEPTH CHANGE

In the preceding example, the depositional surface is assumed to be at sea level. However, this assumption may not be warranted. Changing depth of the depositional surface can be included in the burial history plot, if reliable paleobathymetry data are available.

Figure 6.3 shows examples of analyses in the offshore Louisiana area made by Van Hinte (1978). These plots were based on paleontologic

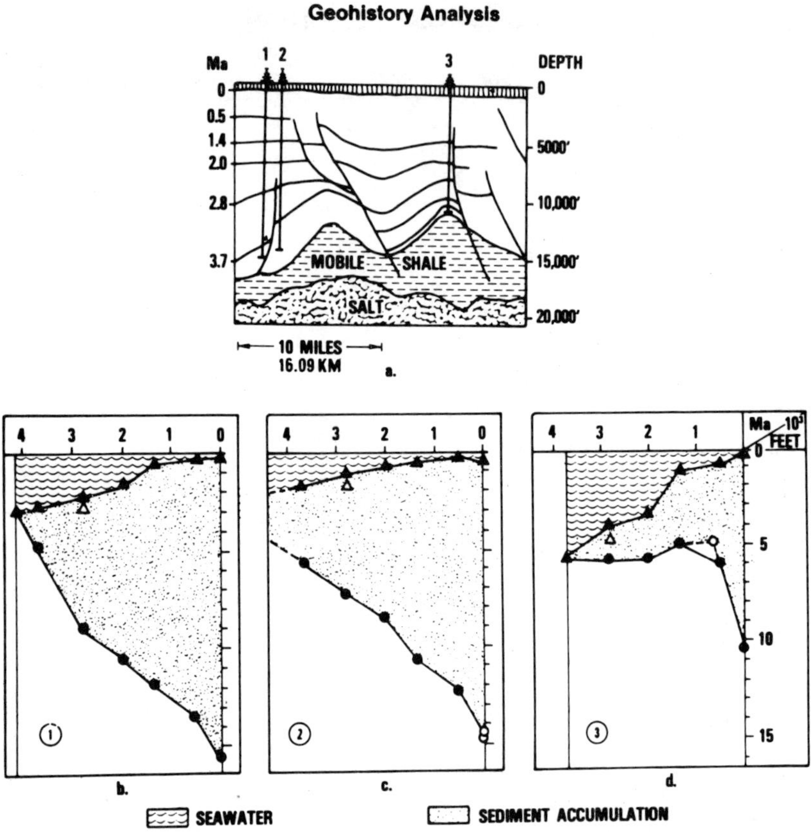

FIG. 6.3. (a) Cross section in Mississippi delta area of offshore Louisiana based on seismic data and three well data. Ma numbers are megayears (million years) before present. (b), (c) and (d) Burial history plots of wells showing that the dome is suspended in sediment column where its top remains at about the same level while the surrounding area subsides (from Van Hinte, 1978). (Reproduced by permission of the American Association of Petroleum Geologists.)

studies of three wells in this area that were underlain by mobile shale. Sites 1 and 2 subsided continuously at a rapid rate during the past 4 million years. Site 3, on the contrary, remained relatively high while the surrounding area was subsiding. During the past 0·5 million years, site 3 subsided at a rate more than 1·5 times greater than the regional rate at the other two sites. Van Hinte interpreted the cause of this excessive subsidence rate as either fluid release from the shales or shale withdrawal. Using the type of burial history plot shown in Fig. 6.3, one is able to interpret the time of structural development—information that is important for prospect evaluation.

Although information on paleobathymetry can significantly improve the burial history plot, such is not always available, particularly before drilling a well. However, structural timing is essential input for determining a well site.

Therefore, at the relatively early stages of exploration when seismic records are the only significant data, one would have to make a burial history plot without the depositional water depth. If there is no significant difference in the depositional water depth between an anticline and the nearby syncline, however, this simple burial history would indicate the structural growth, even though the water depth of the entire area may have changed with geologic time.

RATE OF LOADING (OR SEDIMENTATION) AND TRAP FORMATION

For a given geologic period, the rate of loading or sedimentation can be indicated for both the synclinal and anticlinal areas as follows,

$$R_S = \frac{\Delta D_S}{\Delta T} \tag{6.1}$$

$$R_A = \frac{\Delta D_A}{\Delta T} \tag{6.2}$$

where R_S and R_A are the rates of burial or sedimentation for the synclinal and anticlinal areas, and ΔD_S and ΔD_A the burial depth difference for both areas between the beginning and the end of a given geologic period of ΔT.

Rate of trap formation (R_T) may be given as the difference between the burial rates of both areas, or

$$R_T = R_S - R_A = \frac{\Delta D_S - \Delta D_A}{\Delta T} \tag{6.3}$$

From the burial history plots of Fig. 6.2, the rates of burial and trap formation were calculated and are shown in Table 6.1. Relatively large positive figures of R_T in Table 6.1 indicate periods of significant trap

TABLE 6.1
Rates of burial of anticlinal (R_A) and synclinal (R_S) areas, and of trap formation (R_T), for the case shown in Fig. 6.2

	Anticline R_A	*Syncline* R_S	*Difference* R_T
Stage 1	33·3	111·1	+77·8
Stage 2	66·6	66·6	0·0
Stage 3	46·2	53·8	+7·6
Stage 4	41·7	42·9	+1·2
Stage 5′ (before erosion)*	−200·0	42·9	+242·9
Stage 5 (after erosion)			
			unit: ft/million years

*Based on erosional period of 10 million years.

formation (see also Fig. 6.4 for the graphical presentation of the trap formation).

The above example shows a possible application for an anticlinal structure. Similar approaches may be used in interpreting the timings of domal and fault trap formations.

PETROLEUM GENERATION AND MATURATION

As described in Chapter 5, a burial history plot along with the geothermal gradient of the area can be used to estimate vitrinite reflectance—a measure of the level of petroleum generation and maturation. The data from the synclinal area is of particular importance, because most oil is considered to have derived from the syncline which usually covers area greater than the anticline.

If a burial history is relatively simple (or rate of burial was relatively uniform) for a syncline, the concept of oil genesis originally proposed by Connan (1974) may be applied for the estimation of petroleum generation and maturation. Hunt (1974) identified the oil and gas generation zones on the original Connan plot (see Fig. 6.5).

Hunt's interpreted lines of the oil and gas generation zones have been replotted on the burial history plot of Fig. 6.2 and are shown in Fig. 6.6.

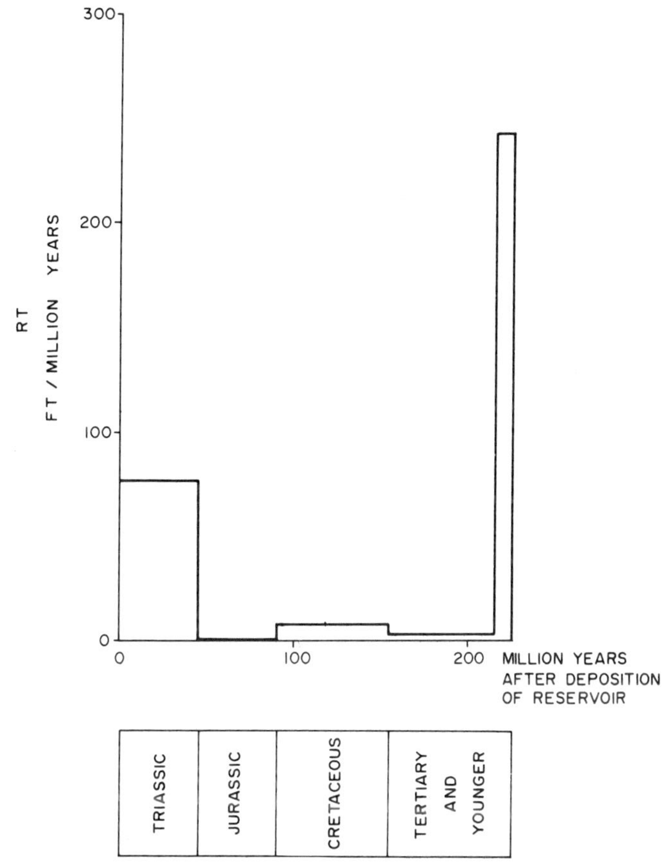

FIG. 6.4. Plot of rate of trap formation (R_T) against geologic time, based on data of Fig. 6.1.

The geothermal gradient in this case is assumed to be 1·5°F/100 ft. No compaction corrections were made. Using the plot of Fig. 6.6, one would be able to evaluate the time of trap development and petroleum generation, maturation and migration as follows:

1. Triassic Age (Stages 1 to 2)

The earlier trap was developed in the anticlinal area during this period, but the reservoir section did not reach a temperature high enough to generate a significant amount of oil.

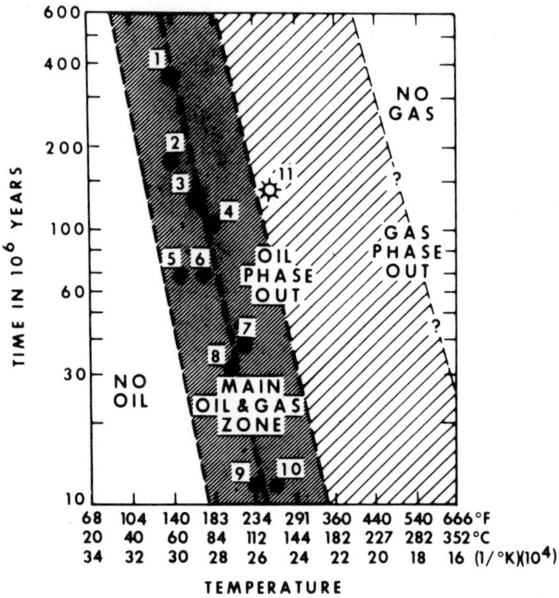

FIG. 6.5. Time–temperature relationship of petroleum generation (from Hunt, 1974). 1, Amazon Basin, Brazil; 2, Paris Basin, France; 3, Aquitaine Basin, France; 4, El Aaiun area, Rio de Oro, W. Africa; 5, Douala Basin, Cameroon; 6, Offshore Taranaki Basin, New Zealand; 7, Camargue Basin, France; 8, Offshore Taranaki Basin, New Zealand; 9, Los Angeles Basin, California; 10, Ventura Basin, California; 11, Aquitaine Basin, France. (Reproduced by permission of the Society of Petroleum Engineers.)

2. Jurassic Age (Stages 2 to 3)

The trap developed in the previous period was retained continuously. The reservoir section and the source rocks above and below reached the oil window in the synclinal area but not in the anticlinal. The most significant generation of oil took place in the syncline during the late Jurassic. The oil probably migrated from the syncline to the anticline updip, and rested there where the temperature was much lower (see the heavy line labelled 'oil' in Fig. 6.6).

3. Cretaceous Age (Stages 3 to 4)

During this period, even the reservoir section in the anticlinal area reached the stage of oil generation, and at the end of this period (stage 4) it reached the peak generation of oil. In the synclinal area, on the other hand, the source sections immediately above and below the reservoir

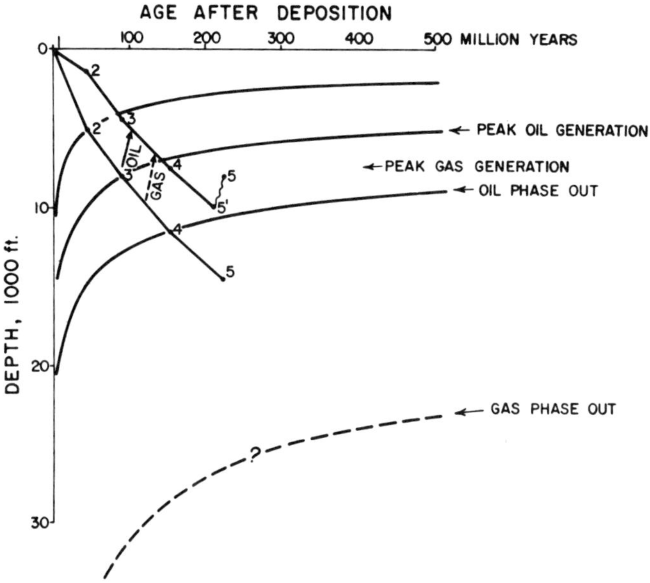

FIG. 6.6. Burial history plots of both synclinal and anticlinal locations (shown in Fig. 6.2) are superimposed on oil/gas generation zones derived from Hunt's (1974) chart (shown in Fig. 6.5).

reached the stage of significant thermal gas generation. The gas from the syncline probably moved to the reservoir in the anticline (see the dashed line labelled 'gas' in Fig. 6.6). In other words, by the end of stage 4 the reservoir section in the anticline probably contained both oil and gas.

4. Tertiary before Erosion (Stages 4 to 5′)

During this period, the oil trapped in the anticlinal area was gradually converted to gas, but a certain amount of oil still remained in the reservoir.

5. Tertiary during and after Erosion (Stages 5′ to 5)

The anticlinal area experienced significant uplift and erosion, which may have caused a better trapping configuration and a greater concentration of oil and gas in the anticlinal area (secondary migration). If no significant fractures developed in the anticline during this period, most of the gas could have been retained in the trap along with some oil.

The above estimates were made on the basis of Hunt's oil window as derived from Connan's (1974) time–temperature plot. The geothermal gradient is assumed to be 1·5°F/100 ft. If the area under study had a different oil window because the type of organic matter and thermal history were different, these estimates are not valid. In any case, applying the decompaction correction will alter the estimates of time of trap development and petroleum generation slightly. An example of the compaction-corrected burial history plot for the syncline is shown by the dashed line (1-2′-3′-4′-5) in Fig. 6.7.

In the above discussions, it has been assumed that there was no significant difference in depositional water depth between synclinal and anticlinal areas. If, however, there was some difference, we must take it into consideration; reliable paleobathymetric information is therefore necessary.

Although the paleo-water depth could influence the time of trap development, it may not cause a significant difference in evaluating the

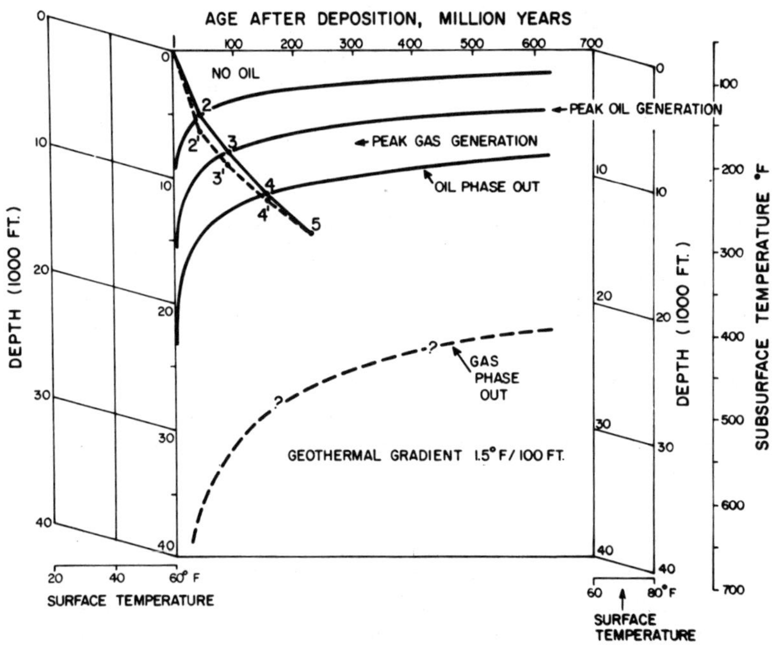

FIG. 6.7. Uncorrected (solid line) and compaction-corrected (dashed line) burial history plots of synclinal location of Fig. 6.1. Oil/gas generation zones from Hunt (1974) are also shown.

time of oil/gas generation, because the subsurface temperature commonly increases from the sea floor, but not from sea level. Therefore, the paleotemperature may be estimated reasonably well, on the basis of the temperature at the sea floor and the geothermal gradient below the sea floor at that time.

OTHER EXAMPLES

Figure 6.8 shows another example of a set of geologic columns. The burial histories for both syncline and anticline are shown in Fig. 6.9. The times of trap development and petroleum generation and migration can be interpreted using this schematic example as follows.

1. Triassic Age (Stages 1 to 2)
Both synclinal and anticlinal areas had an identical history, suggesting that there was no trap development. The reservoir section did not reach the maturation window in either area during this period.

2. Jurassic Age (Stages 2 to 3)
Burial continued at almost the same rate in both syncline and anticline. The source section immediately above and below the reservoir reached the oil-generation stage. However, as no trap had developed in this area, most of the liquid hydrocarbons generated must have either moved elsewhere, or stayed in the source rocks.

3. Cretaceous Age (Stages 3 to 4)
The source and reservoir rocks reached the thermal gas-generation stage, so that a certain amount of gas was probably generated from the liquid hydrocarbons remaining in the source rocks and/or directly from the organic matter these rocks contained. A trap gradually developed in the anticlinal area, so some gas may have accumulated there.

4. Tertiary Age and Younger (Stages 4 to 5)
The trap continued to grow during this period, with both syncline and anticline reaching the gas zone. A gas accumulation might be expected in the anticlinal area.

As mentioned earlier, the above estimates of the time of petroleum generation and migration have been made on the basis of the Connan–Hunt concept using the geothermal gradient of 1·5°F/100 ft. If these

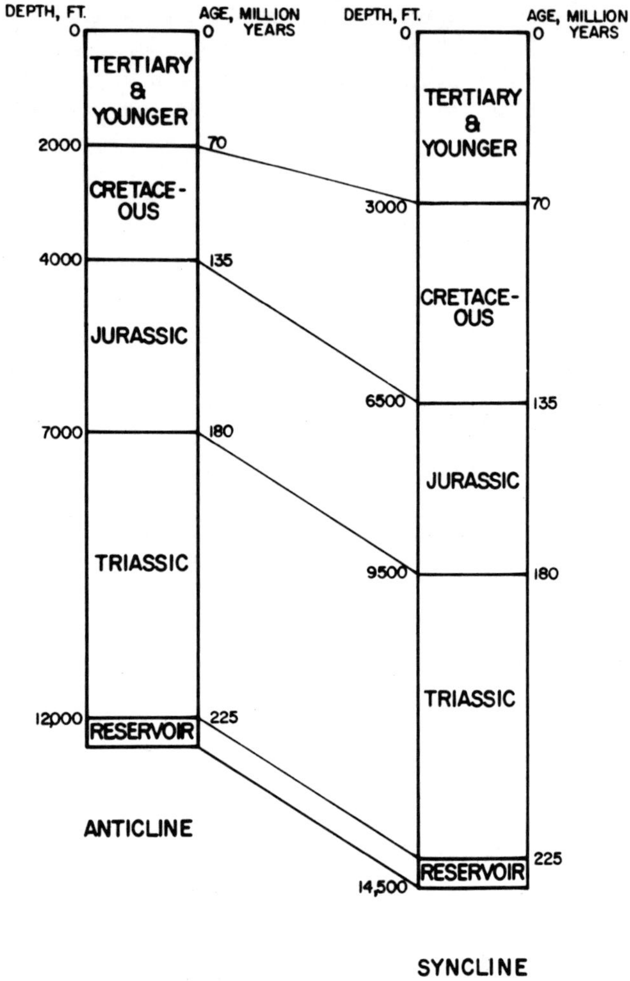

FIG. 6.8. Schematic geologic columns of both synclinal and anticlinal locations.

assumptions are not valid, the interpretation must be different. In case the burial history is more complicated particularly in the syncline, application is recommended of Waples' (1980) method described in Chapter 5 for estimating the level of petroleum generation.

A third example of the stratigraphic column and burial histories is

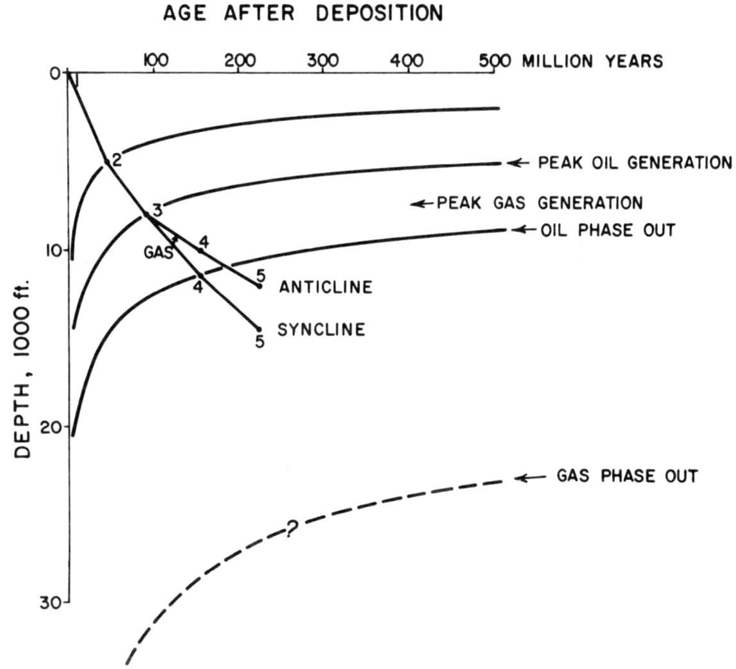

FIG. 6.9. Burial history plots of both synclinal and anticlinal locations (shown in Fig. 6.8) are superimposed on oil/gas generation zones from Hunt (1974).

shown in Figs. 6.10 and 6.11. In late Tertiary time, the anticlinal area experienced significant uplift and erosion, which removed about 2500 ft of the Tertiary section (Fig. 6.10). This erosional event is shown by a wavy line in Fig. 6.11. Refer to Chapter 7 for the methods of estimating thickness of erosion.

1. Triassic Age (Stages 1 to 2)

Both synclinal and anticlinal areas underwent uniform burial but did not reach the oil window. No trap was developed during this period.

2. Jurassic Age (Stages 2 to 3)

Burial of both syncline and anticline continued at almost the same rate. With no trap yet developed in the area, the hydrocarbons must have either moved elsewhere or stayed in the source rocks.

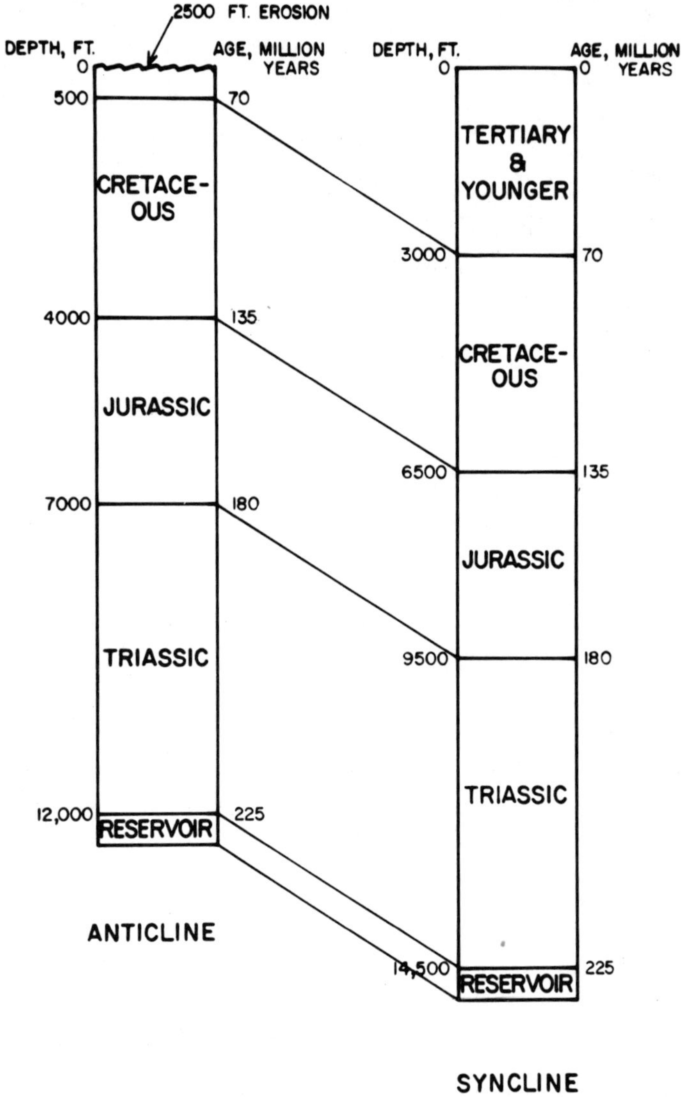

FIG. 6.10. Schematic geologic columns of both synclinal and anticlinal locations.

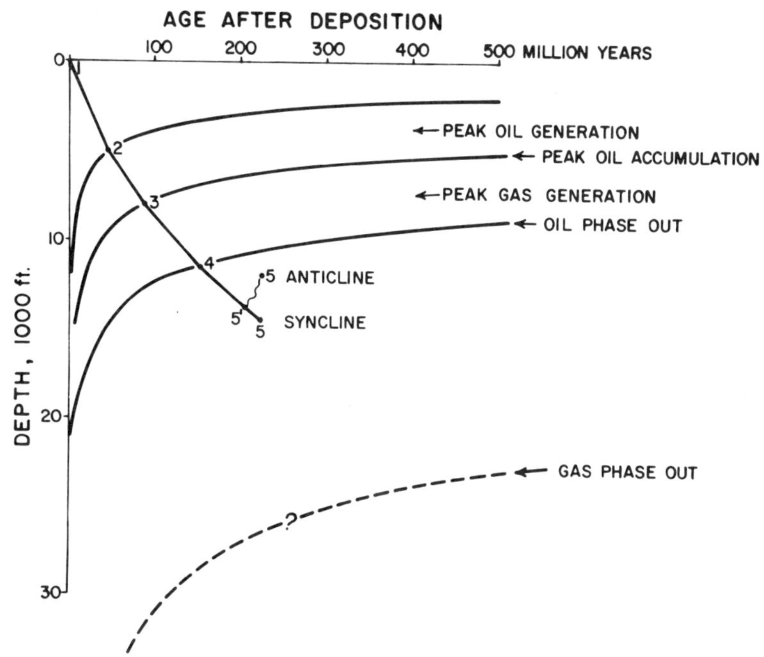

FIG. 6.11. Burial history plots of both synclinal and anticlinal locations (shown in Fig. 6.10) are superimposed on oil/gas generation zones from Hunt (1974).

3. Cretaceous Age (Stages 3 to 4)

The source and reservoir rocks in both synclinal and anticlinal areas reached the thermal gas-generation stage, but still no trap had developed. There would have been very little chance of gas accumulating.

4. Tertiary before Erosion (Stages 4 to 5′)

There is still no evidence of trap development. The source rocks reached the zone of complete gas phase, but the lack of a trap would make the accumulation of a significant amount of gas unlikely.

5. Tertiary after Erosion (Stages 5′ to 5)

This period shows the first evidence of trap development, associated with the erosion of a 2500-ft Tertiary section in the anticlinal area. Because the stages of peak oil and gas generation were long past, there would have been very little chance of oil or gas accumulating in this late anticline.

In applying the above-mentioned techniques to an undrilled location, one would require a seismic section through both anticline and syncline. Although information on the paleosurface temperature and paleogeothermal gradient is difficult to obtain, he may be able to guess these values from the paleontological information, which suggests the paleodepositional or surface temperature, and from the present geothermal gradient for the area close by. With the availability of seismic interval velocity data, which can be converted to porosity values, the estimates can be corrected for the decompaction effect.

SEAL FORMATION

Hydrocarbons that have migrated to the crest of a structure or the highest position of a stratigraphic trap must be sealed there by cap rocks. There are at least two possible sealing mechanisms: pressure seal and capillary seal. The pressure seal can be interpreted in terms of time of its formation and thus be incorporated with a burial history plot for an assessment of a petroleum prospect.

Figure 6.12A shows a plot of shale porosity–depth for a well in Nagaoka Plain, Japan (Magara, 1978). The calculated fluid-pressure profile of the shales is shown in Fig. 6.12B. There is a volcanic-tuff reservoir which contains natural gas and is overlain by the Tertiary shales (Miocene age). The shales between about 2250 and 2750 m are overpressured and overlie the gas reservoir below 2750 m. The excessive fluid-pressures in the shales are considered to be important in preserving the natural gas in the reservoir. These shales are called pressure seals.

Figure 6.13 shows a calculated fluid-pressure profile of the Miocene shale section in an area near the well in Fig. 6.12 (Magara, 1978). The Mitsuke Tuff, which is the main oil/gas reservoir in the Mitsuke field shown on the left-hand side of this profile, is overlain by shale of which the lower part is overpressured. These overpressured shales are considered to have acted as a seal for the hydrocarbons in the underlying reservoir section (pressure seal).

Figure 6.14 shows a few examples of pressure seals (Cretaceous and Tertiary ages) overlying reservoir sections in the Mackenzie Delta, Canada. In the left-hand plot, the pressure seals are unassociated with oil or gas, probably because some of the other important factors for petroleum accumulation, such as trap, source and hydrocarbon migration, are not suitable in this area.

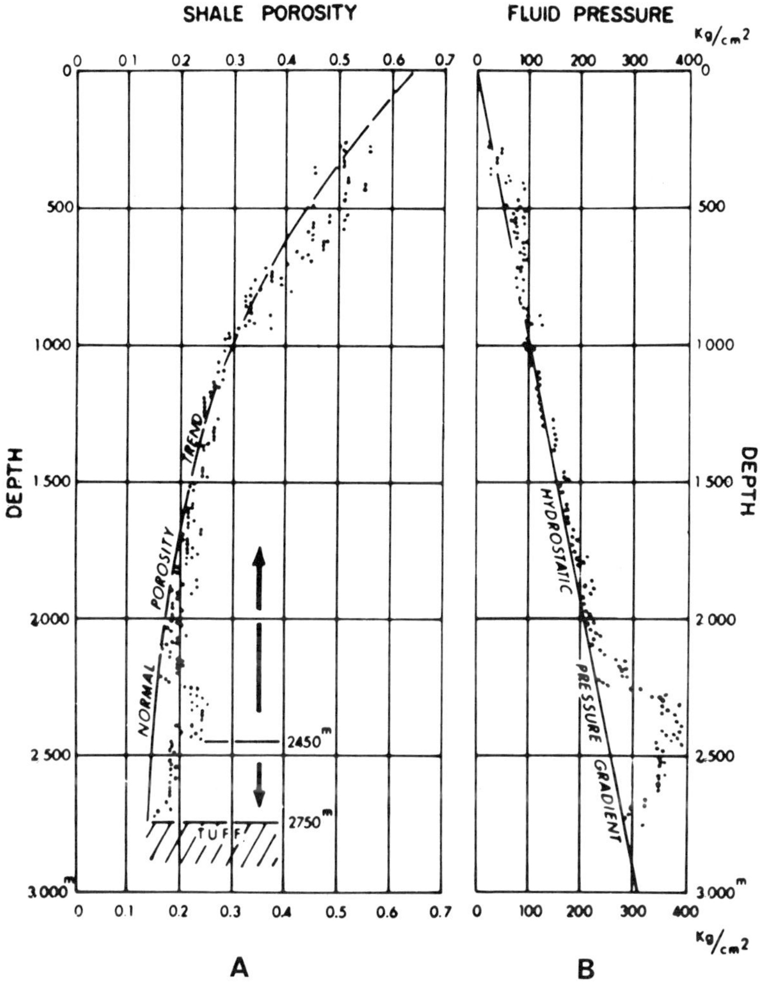

FIG. 6.12. Shale porosity–depth and fluid pressure–depth of a well in Shiunji Gas Field, Japan (from Magara, 1978). Slightly undercompacted and abnormally pressured shales between 2250 and 2750 m are called pressure seals.

PRESSURE-SEALING DEPTH AND TIME

It is believed that undercompacted shales generally underwent relatively normal compaction in the earlier stages of their history (Magara, 1978),

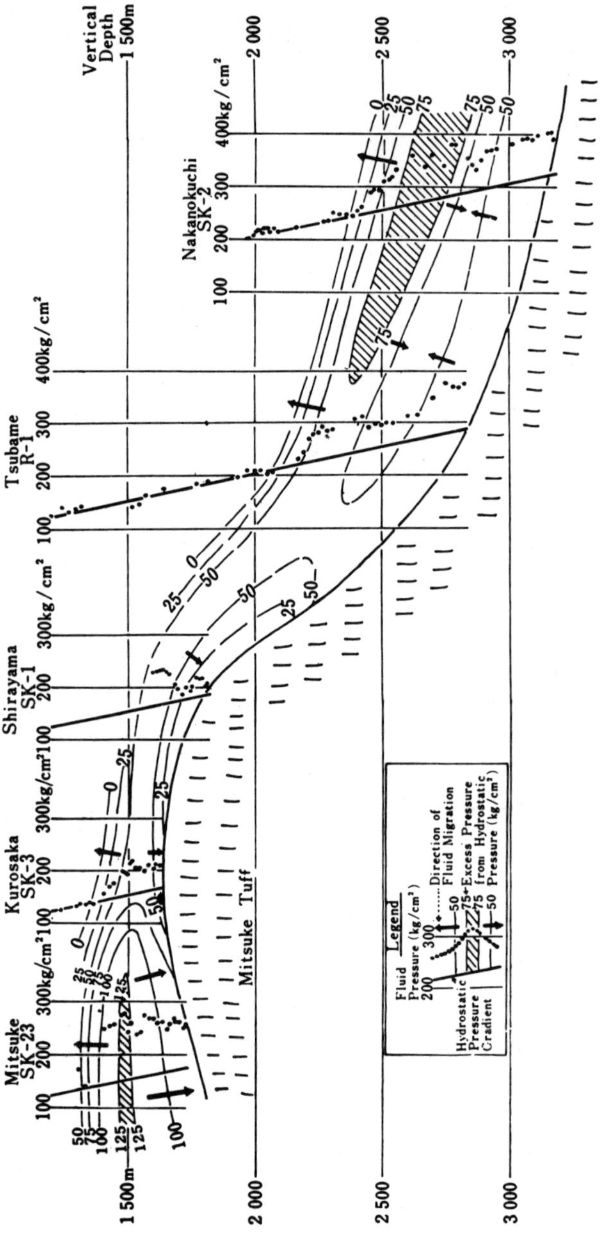

FIG. 6.13. Calculated fluid pressure profile in Nagaoka Plain, Japan (from Magara, 1978).

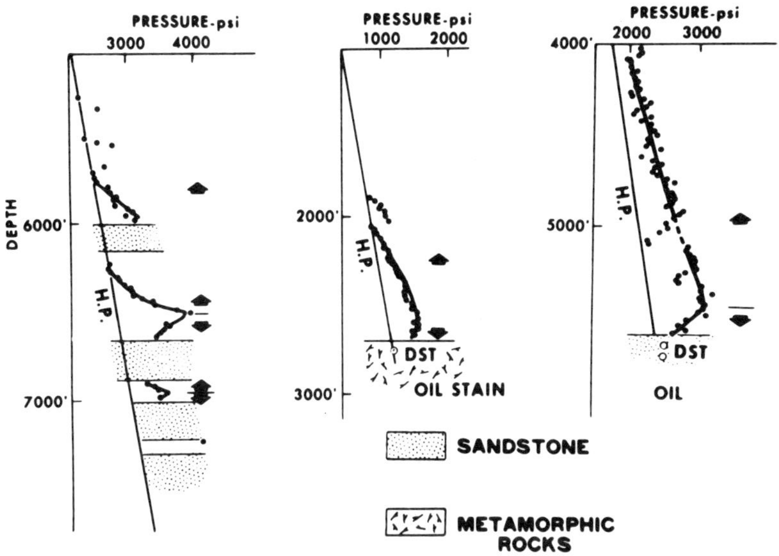

FIG. 6.14. Examples of pressure seals in Beaufort Basin, Canada.

until their permeability had been reduced to a level beyond which normal fluid expulsion and normal compaction became impossible. The shale compaction rate would then slow down, and undercompaction would result. Possible compaction histories for the undercompacted shales are depicted in Fig. 6.15.

The depth at which normal compaction and normal fluid expulsion for shale A, which is most undercompacted, ceased is shown as X in Fig. 6.15. This is the estimated depth at which shale A reached the stage of restricted fluid expulsion, or the sealing depth for shale A, which is the most important pressure-sealing zone for the underlying reservoir section (refer to Magara (1978) for detailed explanation of pressure-sealing depth).

If the termination of normal compaction were more gradual, one might interpret the sealing depth, X', for shale A as shown in Fig. 6.16 by drawing a more inclined line from A to X' (or X' to A). The resultant interpretation would be that the sealing depth for shale A was shallower than in the case illustrated by Fig. 6.15.

Once the sealing depth is estimated (either X or X' in the previous example), sealing time can be estimated quite easily by locating the

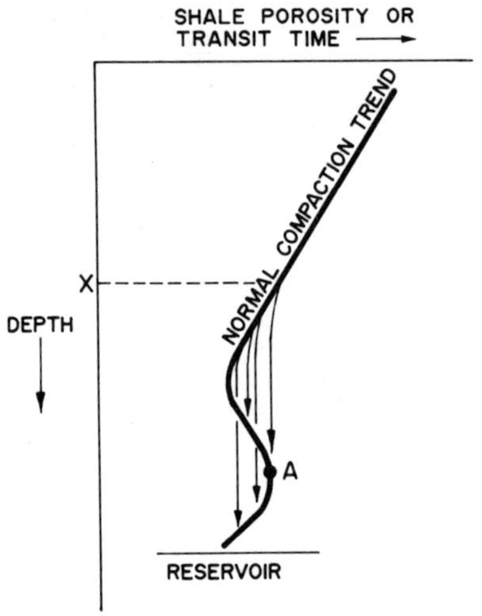

FIG. 6.15. Schematic diagram showing possible compaction history of undercompacted shale A and estimation of sealing depth *X*.

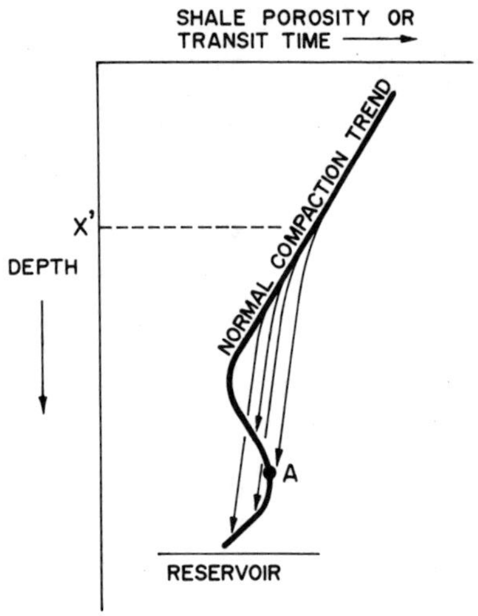

FIG. 6.16. Schematic diagram showing possible compaction history of undercompacted shale A and estimation of sealing depth *X'*.

sealing depth on a burial history plot and determining the corresponding geologic time (see Fig. 6.17). Note that both the sealing depth and time must be estimated at an anticline location rather than synclines because sealing capacity of the former will determine the trap's holding capacity of hydrocarbons. The sealing time corresponding to sealing depth *X'* would be earlier, geologically speaking, than that for *X*.

Because pressure seal is effective for any form of hydrocarbons, whether in aqueous solution or in a separate phase, it would be important for both oil and gas accumulations. Note that capillary seal is effective only for hydrocarbons in their separate phase, so that it would be more important for oil whose aqueous solubility is very low (oil tends to be in its separate phase; Magara, 1978).

PERCENTAGE ORGANIC CARBON

As described in Chapter 2, there are intimate relationships between the percentage organic carbon and sedimentation rate or burial rate in many

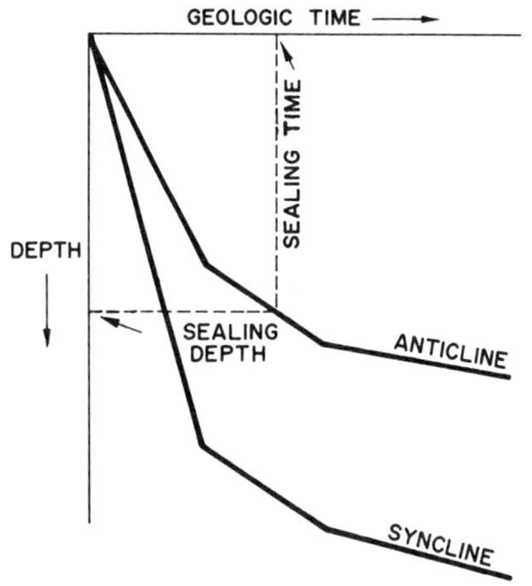

FIG. 6.17. Schematic burial history plots of both synclinal and anticlinal locations. Sealing time can be estimated from sealing depth.

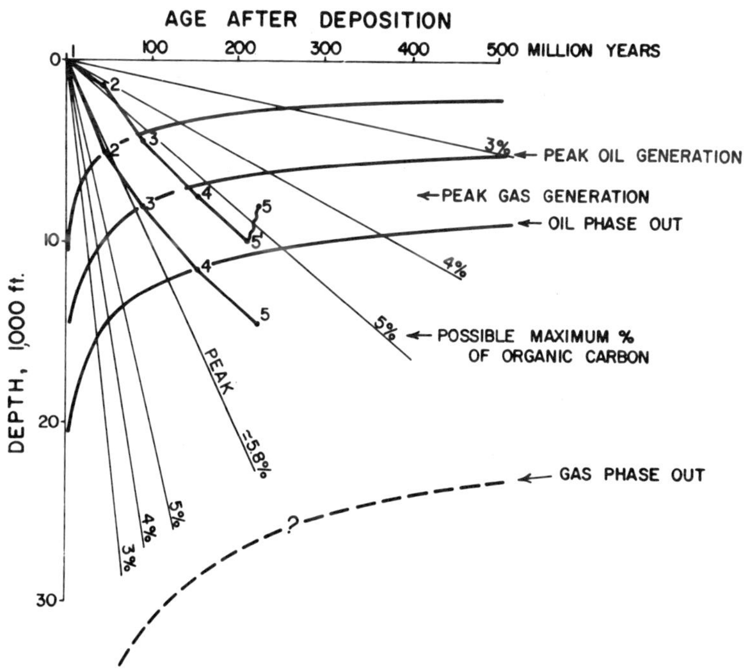

FIG. 6.18. Superposition of oil/gas generation zones and possible maximum percentage of organic carbon over schematic burial history plots of Fig. 6.2.

sedimentary basins. The burial rate of a given bed can be shown by the slope of a burial history line within a given geologic period, so that the organic concentration may also be predicted from the slope, provided that such a relationship has been established for the basin or area under study.

By using the Canadian Arctic data plotted in Fig. 2.3 (Chapter 2), one would be able to construct a burial history diagram with the possible maximum percentage of organic carbon (Fig. 6.18, schematic burial example shown in Fig. 6.6).

To estimate the possible maximum percentage of organic carbon, it is recommended that the slope of the earliest stages (between stages 1 and 2, in this case) be used, because it represents the sedimentation rate of the formations closest to the reservoir. It is also recommended that the slope used be from the synclinal location rather than the anticlinal, because the synclinal area usually covers a larger area. There is also a greater chance of advanced petroleum maturation in the syncline, because of the higher temperature there. In Fig. 6.18 the lower burial history plot between 1 and 2 (syncline) shows that there is a possibility of the highest organic carbon concentration being there.

Based on the empiricial relationships of percentage organic carbon and sedimentation rate from different parts of the world, it may be concluded that the optimum organic carbon percentage may be found between 25 and 300 ft/million years (sedimentation rate), and mid-values between 100 and 200 ft/million years. The four slopes representing these rates of burial or sedimentation are shown in Fig. 6.19, which is a plot of the reserves of the world's giant oil/gas fields by geologic age (Magara, 1978). The details of each giant field's burial history are not known, but the average rate may be deduced from these four lines. It is interesting to note that most giant-field reserves in the world are included within the range of the optimum rates of sedimentation determined for various basins described in Chapter 2. It may thus be useful to study the relationship between oil/gas reserve, organic carbon percentage, and average sedimentation on a global scale.

APPLICATION OF BURIAL HISTORY PLOT FOR PETROLEUM ASSESSMENT

The following hypothetical example summarizes the method of using a burial history plot for the assessment. Figure 6.20 shows a schematic

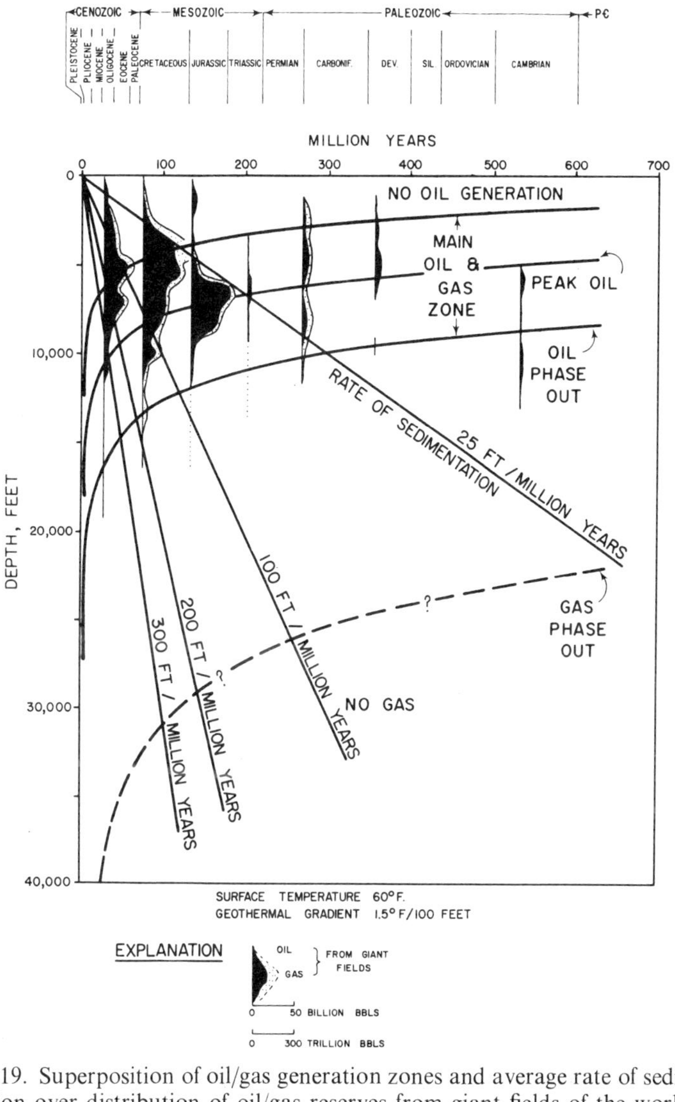

FIG. 6.19. Superposition of oil/gas generation zones and average rate of sedimentation over distribution of oil/gas reserves from giant fields of the world.

example of burial history plots of both anticline and syncline locations of a structure.

The slope of the initial portion of the burial history plot (A) of the syncline may be related to the possible maximum concentration of

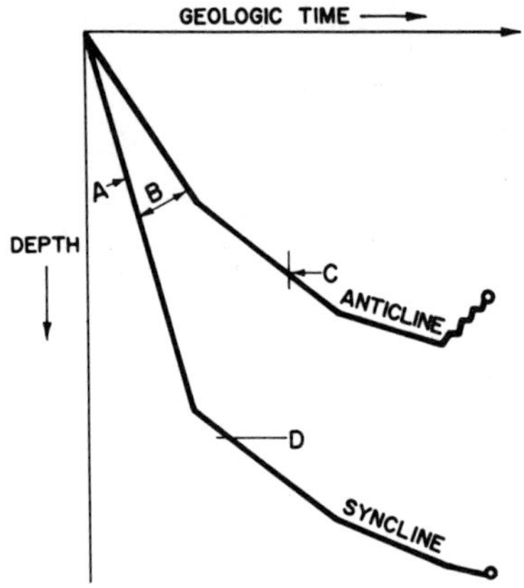

FIG. 6.20. Schematic burial history plots of both synclinal and anticlinal locations. Refer to text for explanation.

organic matter. Note that the plot from the syncline is more important because of the reasons mentioned previously. Hydrocarbons generated in the synclinal area would have moved toward the crest of the structure during the secondary migration stages.

The slope of the initial portion of the burial history plot may be related to percentage organic carbon of the section immediately above the reservoir, which can be quite important if a significant downward primary migration occurred (see for example Magara, 1968). If, on the other hand, the rocks below the reservoir are considered to be more important as source rocks, construction of a burial history plot for that section is recommended and relating it to the organic carbon concentration.

A period of trap development can be shown by the tendency of two burial history lines from both syncline and anticline to separate (see B in Fig. 6.20).

If pressure-sealing depth is estimated to be at depth *C*, insert this depth in the burial history plot for the anticline to read time of pressure-sealing.

Based on Waples' (1980) method described in Chapter 5, time–temperature index and vitrinite reflectance may be estimated in the syncline. The Connan–Hunt concept (1974) of oil genesis may also be applied if the burial history is relatively simple and uniform. The estimated peak stage of petroleum generation may be indicated in the burial history plot, shown as D in Fig. 6.20.

Comparison of relative timing of the above-mentioned geological events and the estimated organic carbon richness would provide valuable information with respect to the trap's potentiality of forming a hydrocarbon accumulation.

BURIAL MAPPING

If the concepts described above are shown in forms of a series of burial maps which indicate historical changes of trap formation, pressure-sealing, and petroleum generation and migration, they would have a great value for exploration. In constructing burial maps, burial history plots for many locations including synclines and anticlines must first be made and the geologic factors, such as trap, seal, petroleum generation and maturation, and its migration, must be evaluated for these locations.

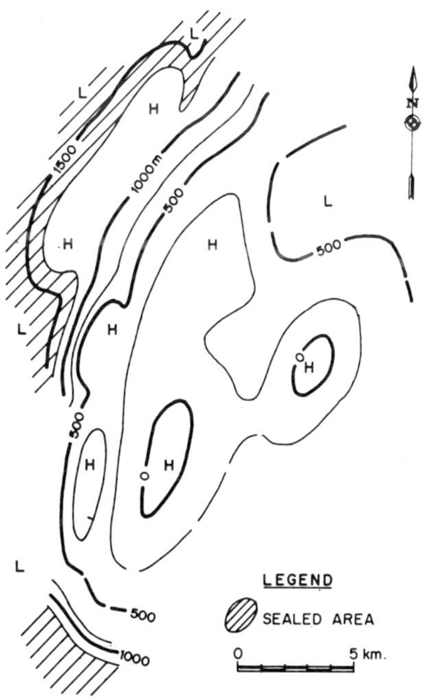

FIG. 6.21. Burial map of volcanic reservoir in Nagaoka Plain, Japan, 4 million years after deposition.

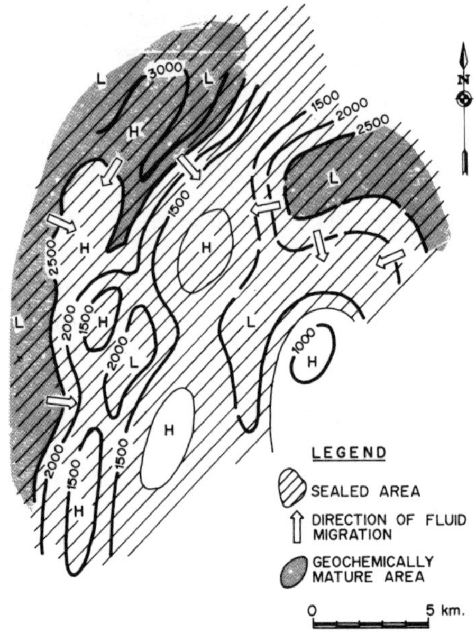

FIG. 6.22. Burial map of volcanic reservoir in Nagaoka Plain, Japan, 5 million years after deposition.

Figures 6.21–6.23 show burial maps of Nagaoka Plain, Japan. Nagaoka Plain is one of the most important oil and gas producing regions of Japan. Volcanic rocks (lavas, tuffs, and agglomerates) of the Pliocene and Miocene ages are the main reservoirs. Petroleum source rocks are considered to be mudstones overlying these volcanic reservoirs. Strong downward fluid migration from the mudstones into the reservoirs has been proposed as the most important mechanism in this area (Magara, 1968).

Figure 6.21 shows the burial map of the Pliocene reservoir (6 million years old) at about 4 million years after deposition of the reservoir. Early forms of most of the present anticlines are shown in this map. Only small areas in the northwestern and southwestern corners of the plain had developed pressure seals. Most compaction fluids including some earlier forms of hydrocarbons may have moved from structural lows to highs by differential compaction (Magara, 1968), but most of them may have been lost to the surface due to lack of effective seals at that time.

About 5 million years after the deposition of the reservoir, all the

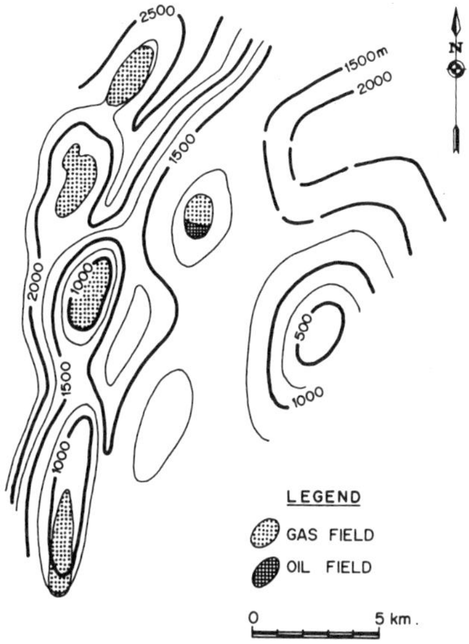

FIG. 6.23. Burial map of volcanic reservoir in Nagaoka Plain, Japan, at present (6 million years after deposition).

structures except for the two in the south-eastern portion of the plain were pressure-sealed (Fig. 6.22). Active petroleum generation occurred in most structural lows where temperature was relatively high. The type of organic matter here is mainly land-derived, which is primarily a source of gas.

Immediately after this stage, the entire area experienced a period of significant uplift and erosion, which resulted in the final structural forms as shown in Fig. 6.23 (6 million years after deposition of the reservoir). All the structures, which had been pressure-sealed at the earlier burial stage (5 million years after deposition of the reservoir, Fig. 6.22) now contain gas or small amounts of oil.

REFERENCES

Asakawa, T. and Fujita, Y. (1979) Organic metamorphism and hydrocarbon generation in sedimentary basins of Japan. In *Proceedings of the Seminar on Generation and Maturation of Hydrocarbons in Sedimentary Basins*, C.C.O.P., pp. 142–62.

Barker, C. (1977) Aqueous solubility of petroleum as applied to its origin and primary migration: Discussion, *A.A.P.G.*, **61**, 2146–9.

Connan, J. (1974) Time–temperature relations in oil genesis. *A.A.P.G.*, **58**, 2516–21.

Evans, C. R., McIvor, D. K. and Magara, K. (1975) Organic matter, compaction history, and hydrocarbon occurrence—Mackenzie Delta, Canada. *Proc. 9th World Petroleum Congress*, vol. 3, pp. 149–157 (Panel Discussion).

Hunt, J. M. (1974) How deep can we find economic oil and gas accumulations?, SPE 5177, 1974, Deep Drilling and Production Symp., Prepr., pp. 103–10.

Magara, K. (1968) Compaction and migration of fluids in Miocene mudstone, Nagaoka Plain, Japan, *A.A.P.G.*, **52**, 2466–501.

Magara, K. (1976) Thickness of removed sedimentary rocks, paleopore pressure, and paleotemperature, southwestern part of Western Canada Basin. *A.A.P.G.*, **60**, 554–65.

Magara, K. (1978) *Compaction and Fluid Migration—Practical Petroleum Geology*, Elsevier, Amsterdam, 319 p.

Powell, T. G. (1978) An assessment of the hydrocarbon source rock potential of the Canadian Arctic Islands, Geological Survey of Canada, paper 78-12, 82 p.

Ronov, A. B. (1958) Organic carbon in sedimentary rocks (in relation to the presence of petroleum). *Geochemistry*, **5**, 510–36.

Schmidt, G. W. (1973) Interstitial water composition and geochemistry of deep Gulf Coast shales and sandstones. *A.A.P.G.*, **57**, 321–37.

Van Hinte, J. E. (1978) Geohistory analysis—application of micro-paleontology in exploration geology. *A.A.P.G.*, **62**, 210–22.

White, I. C. (1885) The geology of natural gas. *Science*, **5**, 521–2.

Wilson, H. H. (1977) 'Frozen-in' hydrocarbon accumulation or diagenetic traps—exploration targets. *A.A.P.G.*, **61**, 483–91.

Chapter 7

Thickness of Erosion

INTRODUCTION

Previous sections have referred to the thickness of erosion. A realistic estimate of the thickness of the eroded interval is very important in evaluating the time of oil and gas generation as well as in understanding the timing of trap development. For estimating erosional thickness there are several possible methods, each using a particular type of data, as follows:

1. Standard stratigraphic column.
2. Extrapolated seismic reflection data.
3. Average denudation rate.
4. Moisture content of coal.
5. Colour of organic matter.
6. Sandstone porosity.
7. Extrapolated preceding rate of sedimentation.
8. Shale compaction
   (a) from well-log data;
   (b) from seismic interval velocity data.

STANDARD STRATIGRAPHIC COLUMN

In a region under consideration there would be a standard stratigraphic column, usually composed of surface and subsurface geologic information.

At a particular location within this region, assume that there is some evidence (e.g., from paleontological data) of lack of beds or formations.

In order to estimate a possible maximum thickness of erosion at this location, one simply subtracts the formation thickness there from that of the standard stratigraphic column. This approach is very simple, but the result always gives a possible maximum figure for erosional thickness—the true figure could be much less than the one predicted by this method.

EXTRAPOLATED SEISMIC REFLECTION DATA

If erosion is limited to a local area, a seismic section can be used to estimate the thickness of erosion, as in Fig. 7.1. This figure shows that most of formation C has been eroded at the crest of the structure. The commonest method of estimating the thickness of erosion is to draw line A parallel to B (base of formation C) and measure the distance between A and the erosional surface.

Although this method assumes that formation C had uniform thickness before erosion, the formation might well have thinned toward the crest of the structure. It may thus be reasonable to assume that in most cases the rate of sedimentation slowed down before significant uplift and erosion. A seismic section alone usually cannot indicate whether such thinning took place or not.

If erosion was not restricted to a local area but was fairly widespread,

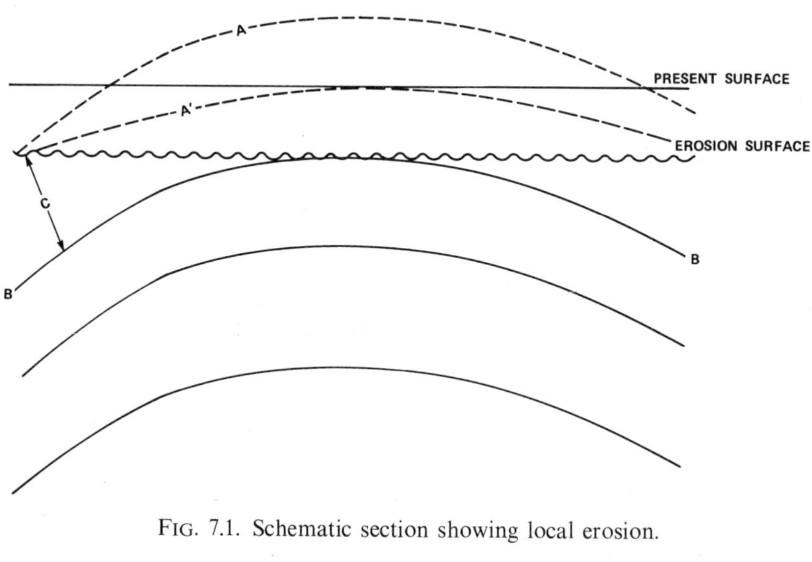

FIG. 7.1. Schematic section showing local erosion.

estimating erosional thickness by the seismic method would become more difficult.

AVERAGE DENUDATION RATE

This method applies the estimated present denudation (or erosion) rate to any erosional event that took place in the geological past. Currie and Nwachukwu (1974) used the same method in the western Canada foothills. It seems to involve at least two possible difficulties: (1) the validity of applying the present denudation rate to the geological past: the rate may have been completely different at that time; (2) knowing the exact geologic period of erosion in millions of years.

The method usually cannot indicate a significant variation of erosional thickness within an area, because a constant figure based on the present average denudation rate is usually applied for a large area. Nevertheless, it is reasonable to assume that the thickness of erosion within any area would vary considerably.

MOISTURE CONTENT OF COAL

The moisture content of coal decreases with increasing burial depth. By comparing the moisture content–depth relation of an uneroded area and that of a given area, one may be able to estimate the thickness of erosion in the latter. Figure 7.2 shows such a moisture content–depth relationship for the standard (uneroded) case. Different coal ranks are also shown in this figure (brown coal, lignite, etc.). This method was applied to near-surface coals by Hacquebard (1977) to evaluate the thickness of erosion in the Western Canada Sedimentary Basin. The estimated thickness of erosion is shown in Fig. 7.3. Figure 7.4 shows the regional changes of the moisture content of near-surface coal in Alberta, Canada. It also depicts the distribution of oil pools with reserves greater than 5 million bbl.

COLOUR OF ORGANIC MATTER

The colour of organic matter in sedimentary rock usually becomes darker with increasing burial depth because of thermal effect (Staplin,

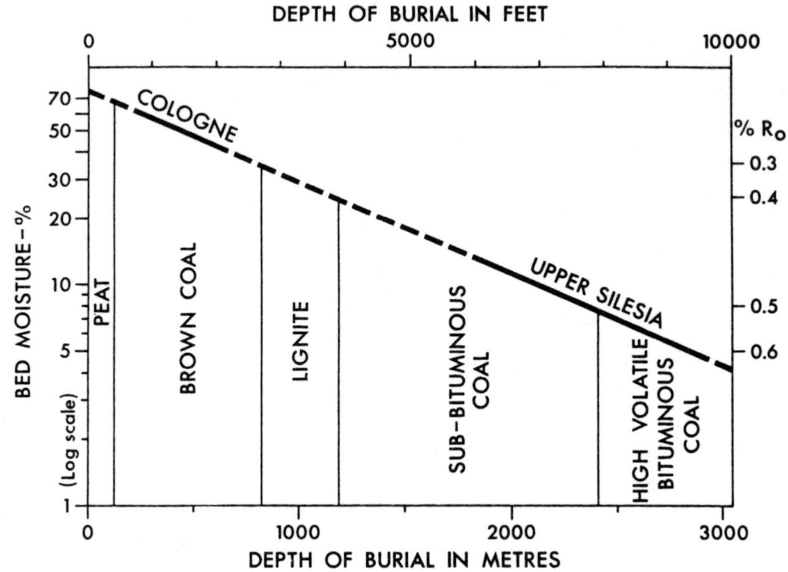

FIG. 7.2. Relationship of moisture content of coal and depth of burial (from Hacquebard, 1977). (Reproduced by permission of the Geological Survey of Canada.)

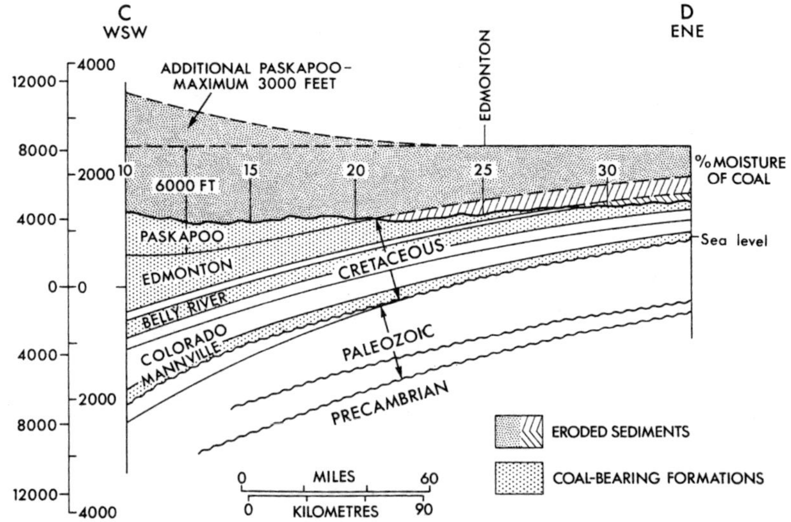

FIG. 7.3. Regional geological cross section through eastern flank of Alberta syncline, Canada (from Hacquebard, 1977). (Reproduced by permission of the Geological Survey of Canada.)

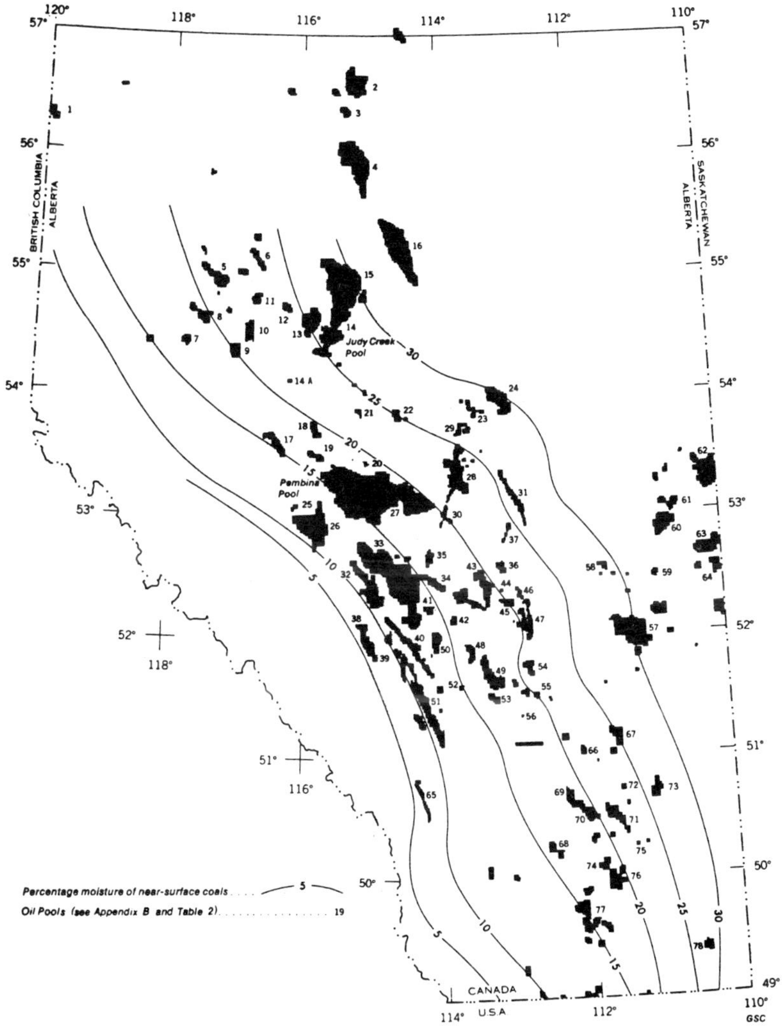

FIG. 7.4. Map showing isomoisture-content contours of near-surface coals and locations of oil pools in Alberta, Canada (from Hacquebard, 1977). (Reproduced by permission of the Geological Survey of Canada.)

1969). This concept based on vitrinite reflectance was used by Hacquebard (1977) for evaluating thickness of erosion in Alberta, Canada. Because the colour of organic matter depends not only on temperature but also on geologic time, estimating thickness of erosion without considering geologic time may be oversimplistic.

As discussed in Chapter 5, vitrinite reflectivity is one of the standard methods for indicating the level of thermal maturation of sedimentary rocks. Therefore, such information is relatively abundant in most exploratory wells, and may be used for estimating maximum burial depth and thickness of erosion.

Cooper (1977) showed a chart relating the vitrinite reflectivity, R_0, temperature and geologic age, which is compiled from data by Huck and Karweil (1955), Teichmuller (1971) and Bostick (1971). This chart is reproduced in Fig. 7.5. Figure 7.6 shows the location, depth and R_0 rank of the Mannville coal samples studied by Hacquebard. The R_0 rank increases westward. The figure also shows the isomoisture contours of near-surface coals of the Edmonton and Paskapoo Formations.

Hacquebard (1977) stated that 'the maximum depths of burial of the near-surface coals have been obtained from moisture-overburden correlations [see Fig. 7.2]. By adding this depth to the present subsurface position of the Mannville coals, the original overburden of these coals is

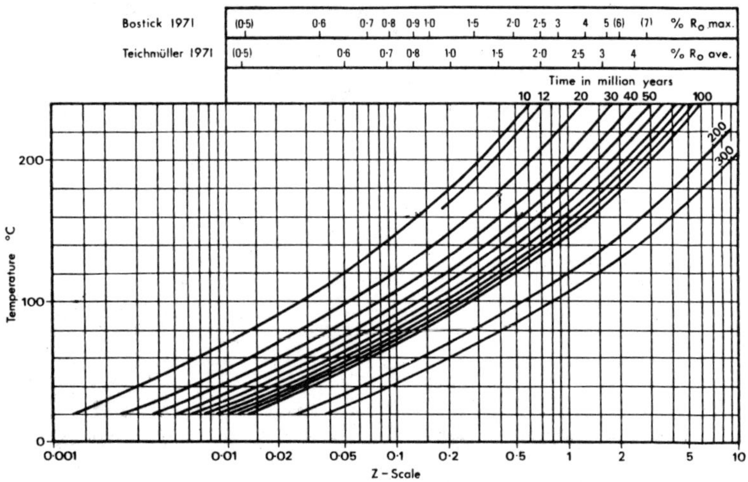

FIG. 7.5. Relationship between rank, temperature, and duration of coalification, based on Bostick's (1971) and Teichmuller's (1971) data (from Cooper, 1977).

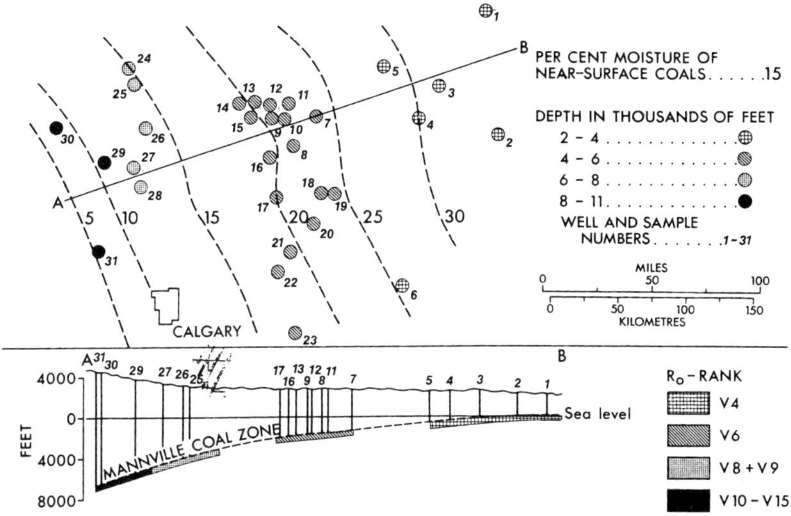

FIG. 7.6. Location, depth and rank of Mannville coal samples and isomoisture contours of near-surface coals of Edmonton and Paskapoo Formations, Alberta, Canada (from Hacquebard, 1977). (Reproduced by permission of the Geological Survey of Canada.)

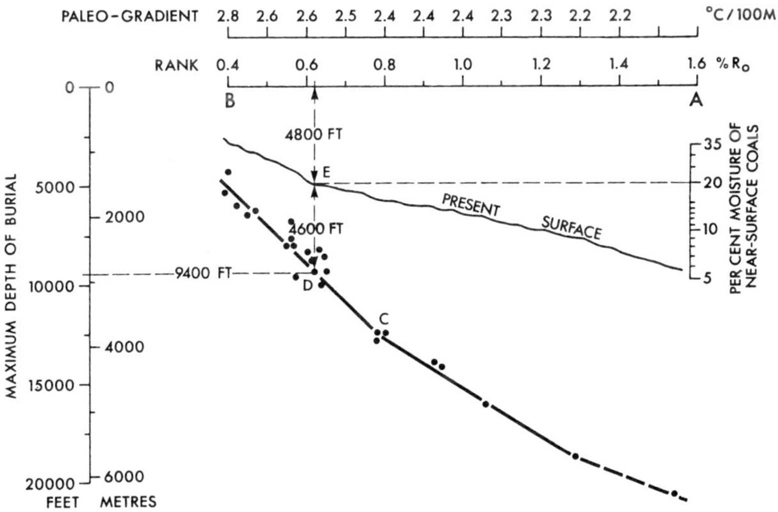

FIG. 7.7. Rank and maximum depth of burial of Mannville coals, Alberta, Canada (from Hacquebard, 1977). (Reproduced by permission of the Geological Survey of Canada.)

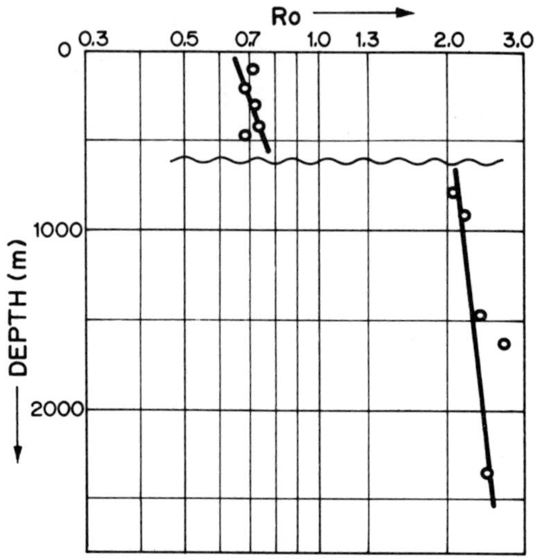

FIG. 7.8. Example of unconformity indicated by difference of vitrinite reflectance (R_0) values (from Hirai, 1979). (Reproduced by permission of the Japanese Association of Petroleum Technologists.)

obtained also. When this depth is plotted against the reflectance data of the coals, a smooth curve results'.

Figure 7.7 shows a cross-section of the R_0 rank and the maximum depth of burial of the Mannville coals and the moisture content of the near-surface coals.

Hirai (1979) reported an example of a large scale unconformity indicated by the difference in R_0 ranks (Fig. 7.8).

SANDSTONE POROSITY

Sandstone porosity usually decreases with burial, mainly from pressure solution and reprecipitation of minerals at and around the grain contacts. Therefore, if the relationship between sandstone porosity and burial depth is established where there was no significant erosion, the thickness of erosion for a particular area may be estimated by comparing its sandstone porosity with that of the uneroded area.

However, sandstone porosity is affected not only by burial depth but also by mean grain size, grain-size distribution, types of minerals that compose grains, geothermal gradient, and chemical composition of formation water, as well as pore-fluid pressure. In other words, the sandstone porosity is not a simple measure of the maximum burial depth, even if the pore-fluid pressure is hydrostatic and no secondary porosity due to dissolution and natural fracturing is involved. Therefore, sandstone porosity is not a reliable basis for estimating thickness of erosion.

On the other hand, the grain-to-grain contacts in shales are usually simpler than those in sandstones. The shale matrix or grain structure is usually weaker, so that shales are more sensitive to increasing overburden pressure, if pore-fluid pressure is hydrostatic.

EXTRAPOLATED PRECEDING RATE OF SEDIMENTATION

Given the geologic time interval that corresponds to the deposition of a formation removed by erosion in an area, and the average rate of sedimentation before that erosion, the thickness of erosion can be estimated. This method assumes that the same rate of sedimentation continued for some time until the area experienced significant uplift and erosion. However, the idea may be oversimplistic, because the rate of sedimentation may have slowed down before the erosional event. The concept was applied to actual cases by van Hinte (1978).

SHALE COMPACTION FROM WELL-LOG DATA

Magara (1976) proposed a method of estimating thickness of sedimentary rocks removed by erosion from shale-compaction data. Such an estimate may be possible because shale compaction is related to overburden load or burial depth, if pore pressure is hydrostatic. In most sedimentary basins hydrostatic pressure is commonly developed at relatively shallow depths. Therefore, if a shale bed at a given depth within an area's zone of hydrostatic pressure is compacted significantly more than a shale bed at the same depth in another area where there was no erosion, the former area is interpreted to have been buried more deeply in the geologic past and to have lost some of its uppermost section by erosion. If, however, the shales are undercompacted, this method cannot apply, because in an undercompacted-overpressured section shale compaction is not a simple matter of burial depth.

The level of shale compaction can be estimated from the study of sonic logs. Sonic transit time is a function of porosity, which in a zone of uniform lithology (e.g., shale) is a measure of level of shale compaction. In the relatively shallow intervals of many sedimentary basins, the relationship between the logarithm of the shale transit time (μs/ft) and depth may be approximated by a straight line (Fig. 7.9A). The pore pressure in these intervals is known to be near-hydrostatic. This normal compaction trend is extrapolated to the surface, giving a surface transit-time value of Δt_0 (Fig. 7.9A), for a situation without significant erosion.

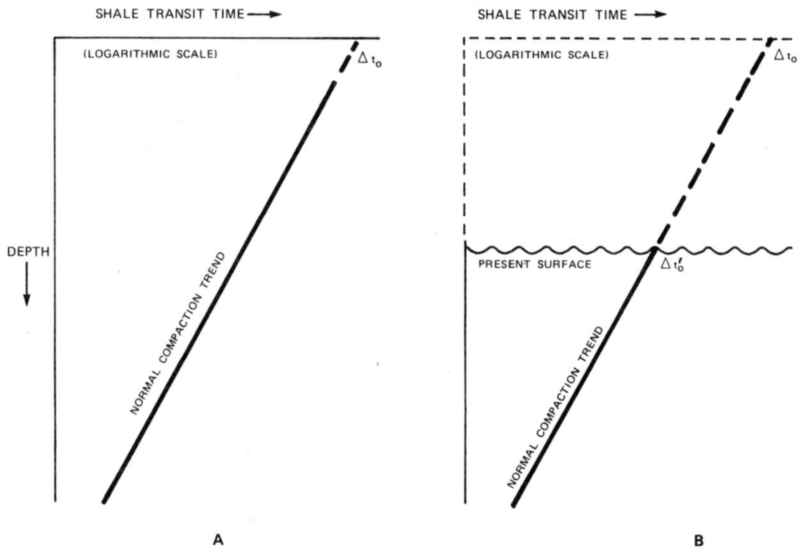

FIG. 7.9. Schematic diagrams showing normal compaction trends of shale transit time–depth plots without erosion (A) and with erosion (B).

Figure 7.9B, on the other hand, shows a schematic example of the transit time–depth plot where the uppermost part of the sedimentary sequence has been removed by erosion. If the normal compaction trend in the subsurface is extrapolated to the present surface, the surface transit-time value ($\Delta t'_0$) will be smaller than the value (Δt_0) for the case of no erosion. If the normal compaction trend is further extrapolated to Δt_0, the original surface of the sedimentary section can be determined (Fig. 7.9B). The distance between the present surface and the level at which the extrapolated value equals Δt_0 is the approximate thickness of sedimentary rocks that have been removed by erosion.

In this case, it is assumed that there was no significant rebounding or expansion of sediments during and after erosion that would invalidate such an erosion estimate.

There are two different methods of determining the surface transit time of Δt_0 for no erosion. The first method is to determine the normal compaction trends of as many wells as possible in the sedimentary basin in question, then extrapolate these trends to the present surface. The maximum transit-time value (the least compaction value) among these extrapolated values ($\Delta t'_0$) will be the value closest to Δt_0 for no erosion. On the basis of extensive studies of the normal compaction trends of the Cretaceous shales in the Western Canada Basin, Magara (1976) obtained a maximum surface transit time of about 200 μs/ft in the areas on the northeast; i.e., near the Canadian Shield where the amount of erosion is considered to have been relatively small.

The second method is based on the empirical relation between shale porosity and transit time. Magara (1976) derived the relationship shown in Fig. 7.10 from a study of 14 wells in the Western Canada Basin in which both sonic and formation-density logs were run. The shale porosity was calculated from the bulk density, on the basis of shale-grain density of 2·72 g/cm^3 (Youn, 1974) and water density of 1·02 g/cm^3.

The empirical porosity–transit time relationship shown by a thick line in Fig. 7.10 is terminated at the level of 200 μs/ft in the upper right-hand part of Fig. 7.10, because the transit-time values of a clay–water mixture should not exceed the value for water (approximately 200 μs/ft). The porosity value that corresponds to this termination point is about 62%. The relationship for porosity values from 62 to 100% is shown graphically as a thick horizontal dashed line.

The porosity–transit time relation shown in Fig. 7.10 may be explained in the following manner. The transit time for water, or 100% porosity, is about 200 μs/ft. A small amount of clay sediment added to the water causes no significant change in transit time, because the sound essentially will travel through the water, not the clay. The transit-time value will remain almost the same until the amount of clay in the water becomes 38% of the total bulk volume (or 62% porosity). After this critical stage the transit time decreases as the amount of clay increases (or porosity decreases).

Figure 7.11 shows schematic diagrams of shale porosity–depth and transit time–depth relationships in the subsurface. The porosity of clay

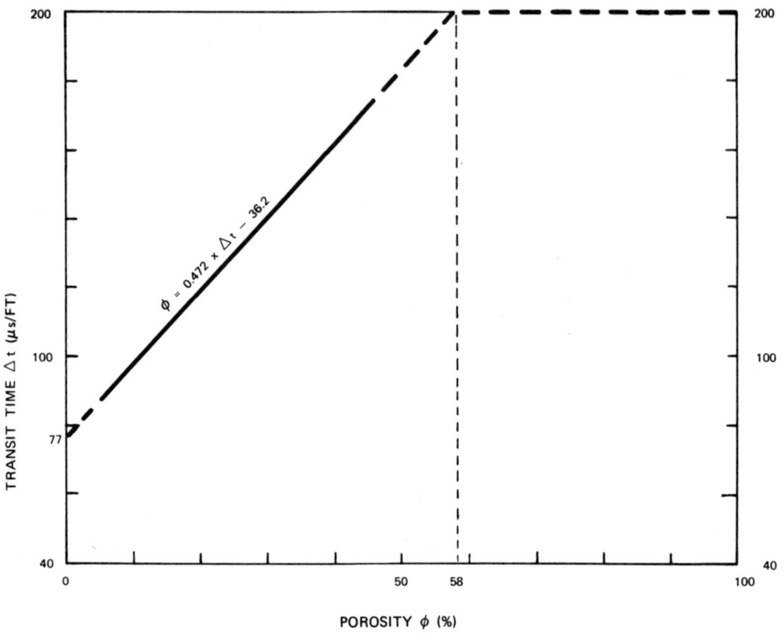

FIG. 7.10. Empirical relationship between porosity and transit time of shales of Cretaceous age in Western Canada.

on the sea floor is known to be 70–80%. Porosity decreases with burial. On the basis of Dickinson's (1953) shale porosity–depth curve of the Gulf Coast area, a porosity of 62% would be reached at about 100 ft (30 m) of burial. Above this depth, the transit time would be about 200 μs/ft as shown in Fig. 7.11B. This critical depth is not constant, but will change from one basin to another.

In the interval below this critical depth, the transit time decreases as the shale porosity decreases (Figs. 7.11A, B). When the normal compaction trend of transit time established below this critical depth is extrapolated to the surface, we will find a Δt_0 value slightly greater than 200 μs/ft. Shale compaction data from several different basins suggest that for most of them the value Δt_0 will not exceed 210 μs/ft.

On the basis of the above observation and interpretation, Magara concluded that the value of 200 μs/ft, as determined from the regional study, is a good approximation for the surface transit-time value when there has been no significant erosion. The calculated thickness of erosion

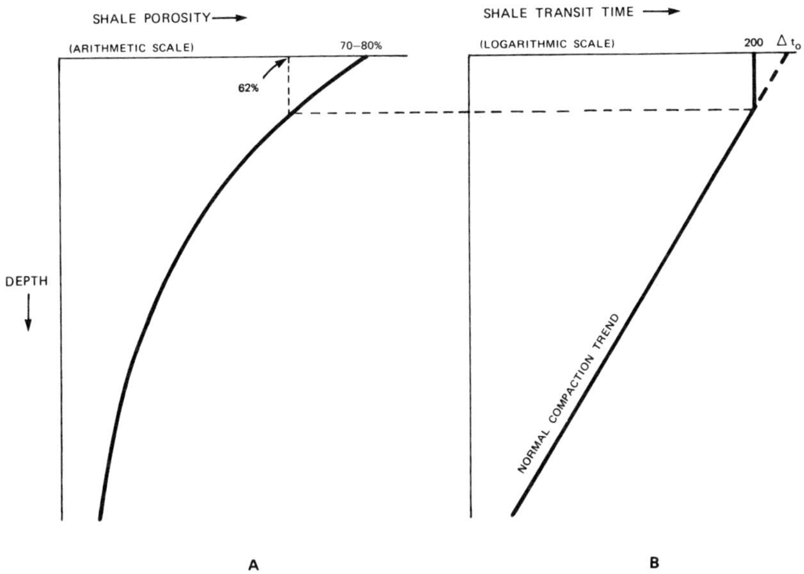

FIG. 7.11. Schematic diagrams showing shale porosity–depth relationship (A) and corresponding shale transit time–depth relationship (B) of normal compaction trend.

based on 200 μs/ft will, however, be a possible minimum estimate; the true figure can be greater.

Figure 7.12 shows a plot of the shale transit time–depth on semi-log paper for a well in the Canadian Foothills area. The shales above 3000 ft are compacted normally. The normal trend is extrapolated to the present surface at 116 μs/ft, indicating a significant amount of erosion which occurred in the geological past. The thickness of erosion is estimated to be about 4600 ft.

The shales below 3000 ft are undercompacted, suggesting that, during the continuous deposition and burial, the deeper section was undercompacted and overpressured.

Most of the overpressure may have since disappeared, because uplift and erosion can be expected to decrease pore pressure, mainly because of temperature drops during and after erosion, and slight expansion of rock framework.

In the previous discussions on estimating eroded thickness from sonic-log data, the normal compaction trend was assumed to be a straight line on semi-log paper. This assumption may be a matter for argument.

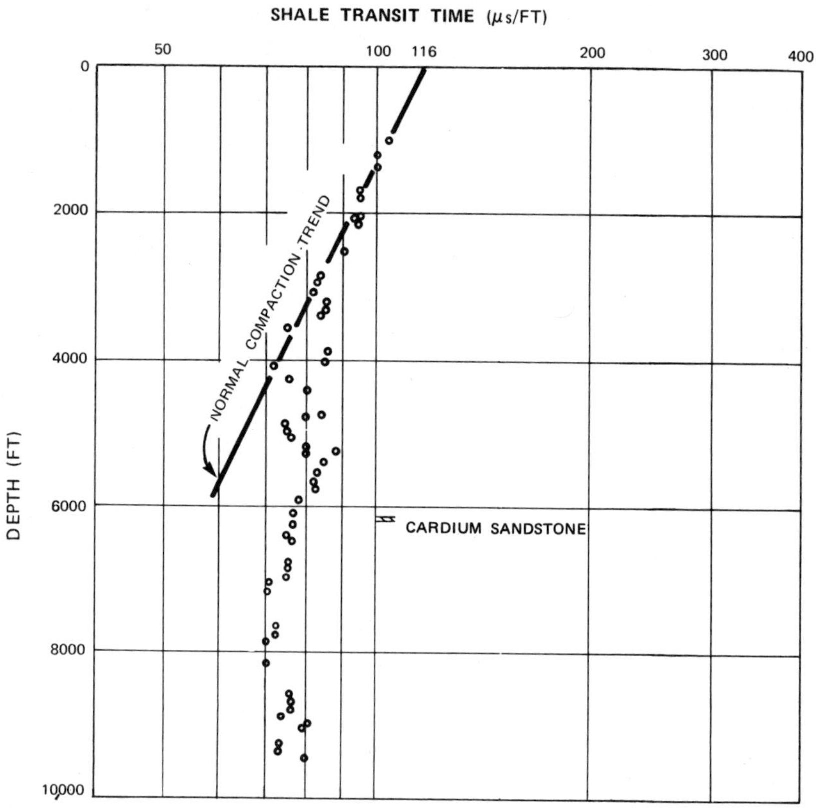

FIG. 7.12. Shale transit time–depth plot of Pacific Amoco Ricinus 16-29-34-8W5 well in Western Canada.

Since Rubey and Hubbert (1959) established an exponential function of shale porosity versus depth, it has been widely accepted by petroleum geologists, which means that the porosity–depth function can be log-linear. However, if the shale transit time is plotted versus depth on semi-log paper, it cannot be a straight line but tends to curve at depth (Fig. 7.13); that is, as the transit time approaches its matrix value, the trend will curve (see also Fig. 3.5 in Chapter 3).

Therefore, the log-linear function used in the current chapter is only an approximation which is usually acceptable in a shallow-to-intermediate depth range (from the original depositional surface to approximately 5000–8000 ft. I think that, in order to estimate thickness of erosion, this

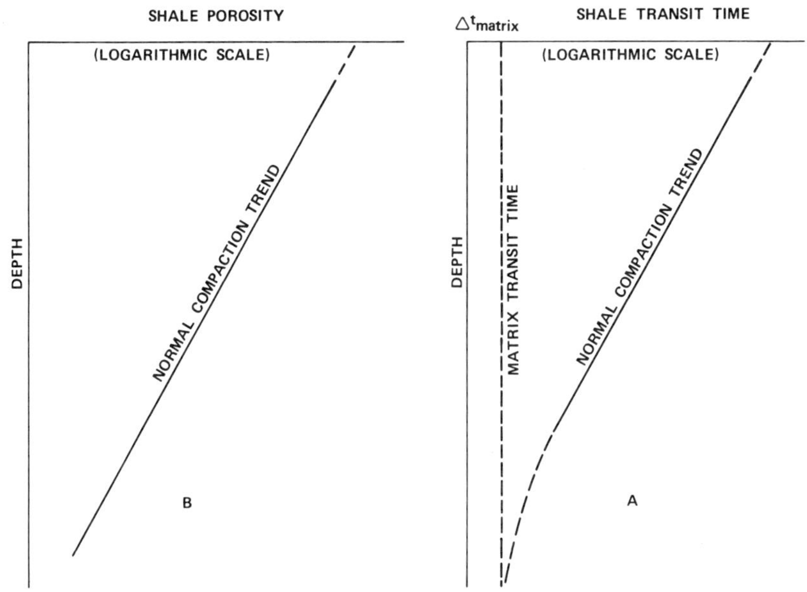

FIG. 7.13. Schematic diagrams showing transit time–depth (A) and porosity–depth (B) relationships of normal compaction trend.

shallow normal trend can be extrapolated to the original (depositional) surface.

I also believe that another reason for so-called normal compaction trends published in the past showing curved relations on semi-log paper had something to do with selection of data points. As shown in Fig. 7.14, in an intermediate depth range (3000–8000 ft) of many sedimentary basins, the mixed-compaction facies, a mixture of normally compacted and slightly undercompacted shales, is developed (Evans *et al.*, 1975). If one draws a least-square line without recognizing the undercompacted shales in the mixed-compaction facies, the trend will, of course, be bent. My normal-compaction trend is drawn through the most compacted shales, so that it is more nearly straight than the other curve. It must be noted that, although my trend is not absolutely straight because of the reason mentioned earlier, it is more nearly straight than the other curve which represents all the shales in the mixed-compaction facies.

It must also be remembered that, within the mixed-compaction facies, most sandstones are hydrostatically pressured and shales directly above

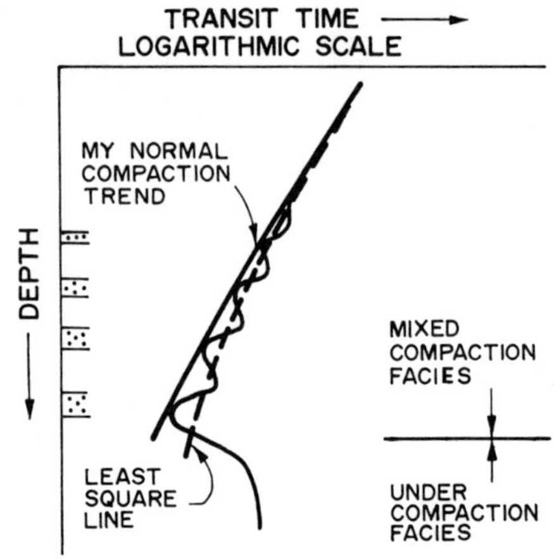

FIG. 7.14. Schematic shale porosity–depth plot showing normal, mixed, and undercompaction zones.

and below these sandstones are normally compacted. However, other shales in the centre of each shale section are slightly undercompacted. Therefore, the reservoir pressures in a mixed-compaction facies can be hydrostatic, but shales are not always normally pressured.

In summary, in order to make a realistic estimate of thickness of erosion, I recommend that a normal-compaction trend line be drawn through the relatively well compacted shales only within a shallow-to-intermediate depth range.

If the erosional surface is not at the present surface but in the subsurface, as shown in Fig. 7.15, the thickness of eroded sedimentary rocks must be evaluated differently. The value A in Fig. 7.15 is the difference between the present surface and the estimated original sedimentary surface. The erosional surface is at the level whose distance or depth from the present surface is B. This means that the new sediments, whose thickness is B, were deposited after erosion. The thickness of erosion in this case is the total of A and B. However, the maximum burial depth for the sandstone bed at the present depth C can be calculated by summing A and C.

If the erosional surface is much deeper than in the previous example, and if the thickness of sediments deposited after erosion is greater than that of the sediments eroded, as shown in Fig. 7.16, the two normal compaction trends in the new and old sedimentary sequences will be plotted on the same line and extrapolated to Δt_0 for no erosion. This

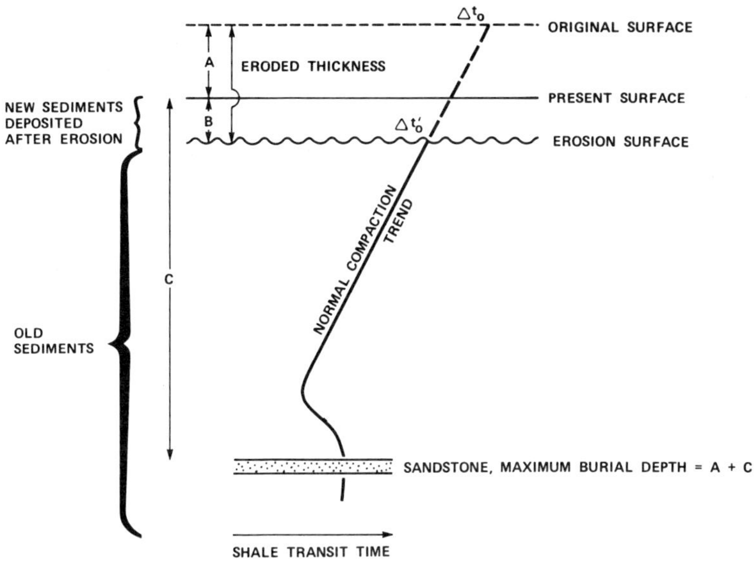

FIG. 7.15. Schematic transit time–depth plot where thickness of sediments deposited after erosion is less than thickness of erosion.

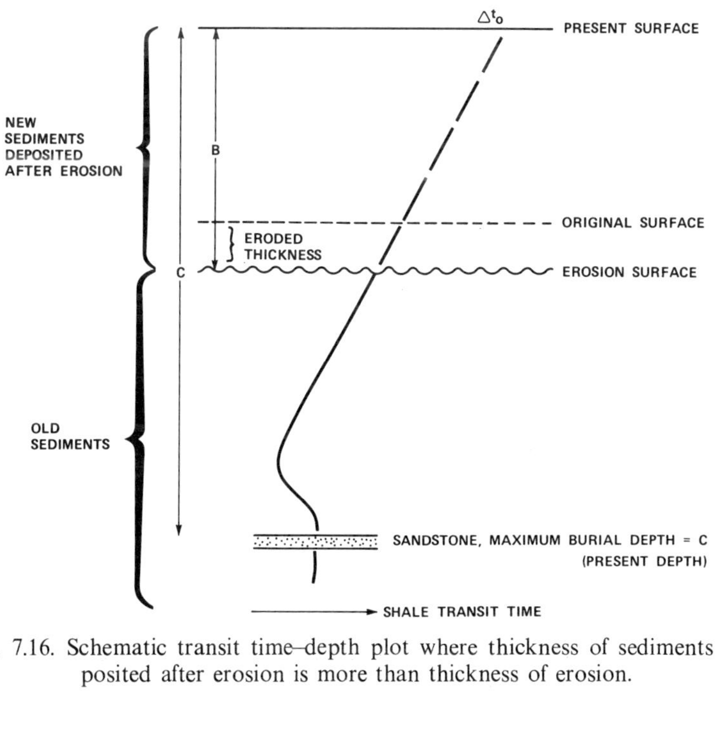

FIG. 7.16. Schematic transit time–depth plot where thickness of sediments deposited after erosion is more than thickness of erosion.

suggests that the record of shale compaction before erosion was completely removed by renewed sedimentation.

In summary, if the thickness of post-erosional sediments is greater than that of erosion, the shale compaction data cannot generally be used for estimating erosional thickness.

The following example (Fig. 7.17) is probably an exceptional one, for which an estimate of erosional thickness may be possible even though the thickness of post-erosional sediments is greater.

Figure 7.17 shows a transit time–depth plot of a well in the Mackenzie Delta, Canada. From the paleontological and seismic interpretations, an unconformity is known to exist at about 5600 ft. The section above

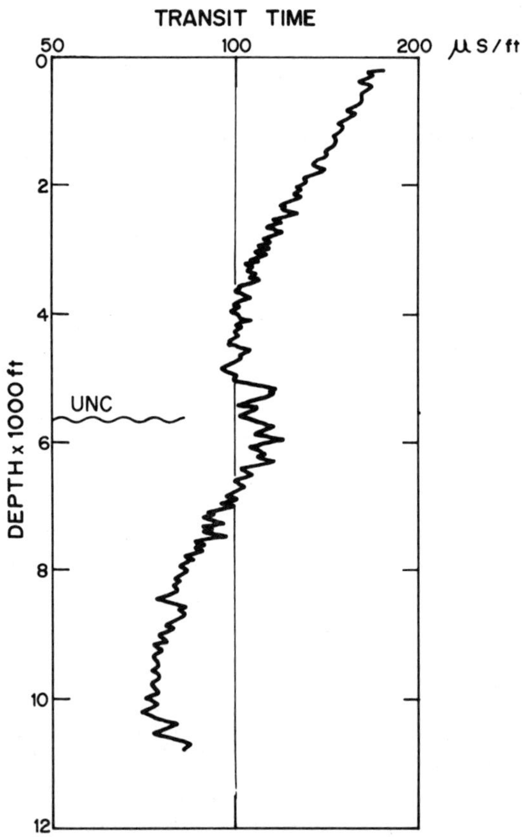

FIG. 7.17. Example of shale transit time–depth plot in northern Canada.

5600 ft is composed of a normal-compaction zone between 0 and 3400 ft, and an undercompacted zone between 3400 ft and 5600 ft. This deeper zone consists primarily of bituminous shales whose permeability seems extremely low.

The section below 5600 ft consists of a shallower zone (5600–8000 ft) with relatively straight compaction profile and a deeper zone (8000–11 000 ft). By using the transit-time plot of Fig. 7.17, one may construct a depositional and erosional history as follows.

1. Deposition

During this first depositional period, the sedimentary rocks between 5600 and 11 000 ft were laid down. At that time more sediments were present above 5600 ft depth points, but were eroded at a later stage. The relatively shallow zone (present depths between 5600 and 8000 ft) of this sequence was probably normally compacted, as indicated by a relatively straight compaction profile. The deepest part (present depths between 8000 and 11 000 ft) was undercompacted and abnormally pressured at that time.

2. Erosion

The erosional event removed a section above the 5600 ft level. The thickness of erosion may be estimated at about 2000 ft, by extrapolating the trend between 5600 and 8000 ft to 200 μs/ft. The fluid pressures of the entire section (5600–11 000 ft) would have become subnormal or near-normal hydrostatic, because of significant uplift and associated cooling and fluid contraction during erosion.

3. Deposition

The previous erosional event was followed by deposition of the low-permeability and bituminous shales (present depths between 3400 and 5600 ft) and of the normal shales above. It may be assumed that the permeability of the bituminous shales was so low that fluid expulsion and compaction of the deeper section (present depths between 5600 and 11 000 ft) were virtually terminated. Hence, the entire section below 3400 ft (present depth) would have been abnormally pressured.

4. Erosion

The second and final erosional event removed the shallowest part of the sedimentary column. The extrapolation of the shallow normal-compaction trend suggests that the thickness of erosion was about

1000 ft. During and after this erosional period, the entire geologic column would have become subnormally or normally pressured, from the cooling and fluid contraction caused by the erosion. No significant abnormal pressures are measured in this general area at present.

A point to be stressed is that the above estimate of 2000 ft erosion at the level of the unconformity (5600 ft) is valid only if the overlying bituminous shales covered the deeper sequences and restricted the expulsion of fluid from them almost entirely. If this assumption is not valid and if the deeper shales lost some fluid during the second depositional period, the estimate of erosional thickness reached by the above method will be greater than the true value.

The development of a completely closed environment in a deeper section is generally an unacceptable assumption, because even shales have some permeability, but such an assumption may be acceptable in this case only because the bituminous shales here seem to have extremely low permeability and cover a relatively large area. The same shales are considered to be important as source rocks also.

Well-log data seem to provide a valuable method of estimating the thickness of erosion. The method may provide a better understanding of burial history, petroleum generation and maturation, and trap development. Its only disadvantage is that it can be used only after drilling.

However, since in most cases it is preferable to have such estimates before drilling, interval velocity data from the seismic record can be valuable in this regard.

SHALE COMPACTION FROM SEISMIC INTERVAL VELOCITY DATA

If the reciprocal of the interval velocity, or transit time, is plotted versus depth on semi-log paper, the thickness of erosion and the maximum burial depth can be estimated by the method discussed above. Figure 7.18 shows an example of this type of plot. A weakness in the method is that change in rock composition is usually overlooked, because it is not easy to distinguish lithology from seismic data alone.

As none of the methods described above is absolutely reliable, it is recommended that different methods are tried to obtain the best estimate. The estimation of thickness of erosion and of maximum burial depth is quite important in evaluating the level of geochemical maturation and the timing of trap development. Figure 7.19 shows an

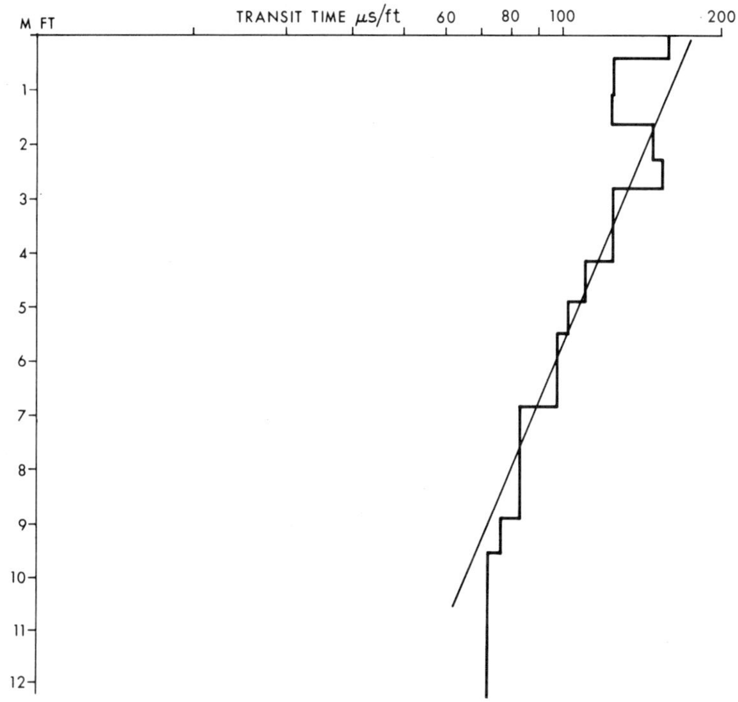

FIG. 7.18. Plot of transit time versus depth from seismic data.

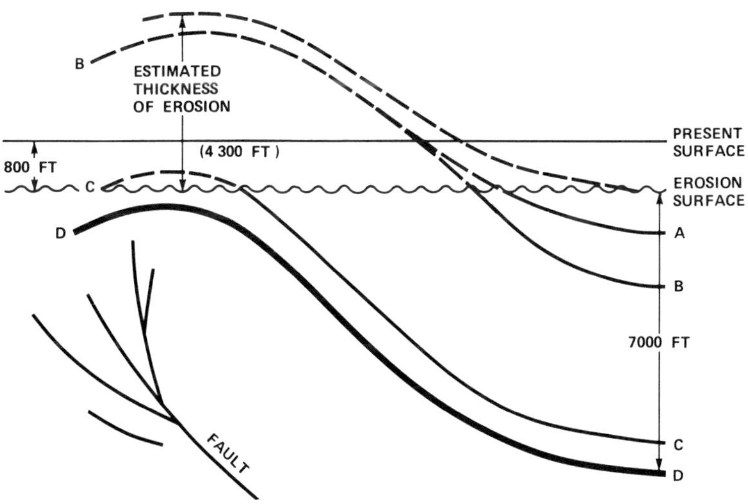

FIG. 7.19. Example of estimation of erosion by both seismic section and shale transit time plot of an anticline in Beaufort Basin, Canada.

example of a combined study on erosion, using the seismic cross section and sonic transit-time plot of an anticline in the Beaufort Basin, Canada. While the seismic section shows the thickness above D in the syncline as being about 7000 ft, the thickness of eroded section estimated by the transit-time plot at the anticline is only about 4300 ft. This means that a significant amount of formation thinning had taken place before erosion, suggesting that the time of trap development was earlier than that of erosion.

REFERENCES

Bostick, N. H. (1971) Thermal alteration of clastic organic particles as an indicator of contact and burial metamorphism in sedimentary rocks. *Geosci. Man.*, **3**, no. 83.

Cooper, B. S. (1977) Estimation of the maximum temperatures attained in sedimentary rocks. In *Developments in Petroleum Geology-1*, Hobson, G. D. (ed.), Applied Science Publishers, Chapter 4, pp. 127–46.

Currie, J. B. and Nwachukwu, S. O. (1974) Evidence of incipient fracture porosity in reservoir rocks at depth. *C.P.G.*, **22**, 42–58.

Dickinson, G. (1953) Geological aspects of abnormal reservoir pressures in Gulf Coast Louisiana. *A.A.P.G.*, **37**, 410–32.

Evans, C. R., McIvor, D. K. and Magara, K. (1975) Organic matter, compaction history and hydrocarbon occurrence—Mackenzie Delta, Canada. *Proc. 9th World Petroleum Congress*, vol. 3, pp. 149–157. (Panel Discussion.)

Hacquebard, P. A. (1977) Rank of coal as an index of organic metamorphism for oil and gas in Alberta. In *The Origin and Migration of Petroleum in the Western Canadian Sedimentary Basin, Alberta*, McCrossan, R. G., Powell, T. G., Tissot, B. and Deroo, G. (eds), Bulletin 262, Geological Survey of Canada, Chapter 3, p. 136.

Hirai, A. (1979) Vitrinite reflectivity. *Jap. Assoc. Petroleum Technologists Bull.*, **44**, 190–5 (in Japanese).

Huck, G. and Karweil, J. (1955) Physikalisch-chemische probleme der inkohlung. *Brennstoff, Chem.*, **36**, no. 1.

Magara, K. (1976) Thickness of removed sediments, paleopore pressure and paleotemperature, southwestern part of Western Canada Basin. *A.A.P.G.*, **60**, 554–65.

Magara, K. (1978) *Compaction and Fluid Migration–Practical Petroleum Geology*, Elsevier, Amsterdam, p. 319.

Rubey, W. W. and Hubbert, M. K. (1959) Role of fluid pressure in mechanics of overthrust faulting, II. *Geol. Soc. Amer. Bull.*, **70**, 167–206.

Staplin, F. L. (1969) Sedimentary organic matter, organic metamorphism and oil and gas occurrence. *C.P.G.*, **17**, 47–66.

Teichmuller, M. (1971) Anwendung Kohlenpetrographischer methoden bei der erdölund erdgasprospecktion. *Erdol Kohle/Brennstoff-Chemie*, **24**, 69–76, no. 69.

van Hinte, J. T. (1978) Geohistory analysis—application of micropaleontology in exploration geology. *A.A.P.G.*, **62**, 201–22.

Youn, S. H. (1974) Comparison of porosity and density values of shale from cores and well logs, M.A. thesis, University of Tulsa.

Chapter 8

Heat Flow and Subsurface Temperature

INTRODUCTION

Chapters 3–6 have described the importance of subsurface temperature in generating petroleum hydrocarbons from organic matter. An estimate of the maximum burial depth before erosion, discussed in Chapter 7, is another critical factor in evaluating the maximum temperature a sedimentary rock is likely to have attained. For estimating the present temperature at a given depth, it is important to understand the heat flow and geothermal gradient. The paleoheat flow and paleogeothermal gradient, however, do not have to be the same as the present ones, and must be estimated differently.

It is known that the movement of formation fluid influences heat-flow conditions in the subsurface. Temperature increase due to increasing burial, in turn, is considered to cause the movement of formation fluid because of aquathermal effect. There is a hypothesis that a relatively high rate of sedimentation tends to push cooler sediments downward very quickly, so that an equilibrium condition may not be attained. In such circumstances, the geothermal gradient may be subnormal.

In summary, then, there are at least six problems to be considered in connection with heat flow and subsurface temperature:

1. Determination of natural subsurface temperature at present.
2. Effect of lithological change and level of sediment compaction on heat flow and geothermal gradient.
3. Difference between present and paleoheat flow and geothermal gradient.
4. Influence of fluid motion on heat flow and geothermal gradient.
5. Effect of increasing temperature on fluid motion.

6. Influence of rate of sedimentation on subsurface temperature and geothermal gradient.

Although it is not intended to discuss all the above problems in this chapter, there are enough data to evaluate most of them.

DETERMINATION OF NATURAL TEMPERATURE IN THE SUBSURFACE

Subsurface temperature measurements are made during and after drilling operations and in the course of geophysical logging. It is well known, however, that these downhole temperature records do not indicate the true natural (or static) temperature, but may be significantly lower because of the cooling caused by the circulation of drilling fluid. A method of extrapolating the natural temperature from the measured temperatures has been used in the oil industry. To apply it, one must know the time interval (Δt) between the end of the circulation of drilling fluid and the actual measurement of the temperature, as well as the time occupied by circulation (t). The following value can thus be calculated (Fertl and Wichmann, 1977):

$$X = \frac{\Delta t}{t + \Delta t}$$

The measured temperature will then be plotted on semi-log paper against the value X. An example is shown as point A in Fig. 8.1 (the data are listed in Table 8.1).

Similar plots will be made for the temperatures measured at later stages (B and C in Fig. 8.1 and Table 8.1). A straight line will be drawn through points A, B and C, and the temperature value that corresponds to $X = 1$ will be read at approximately 235°F in Fig. 8.1. This is the static or natural temperature value determined by the extrapolation.

The reason the temperature is extrapolated to $X = 1$ (or $\Delta t/(t + \Delta t) = 1$) is that, as the elapsed time since the termination of drilling-fluid circulation (Δt) increases, X or $\Delta t/(t + \Delta t)$ approaches 1, because the circulation time t is given in each case. When the value Δt equals infinity, X becomes 1. Therefore, the extrapolated temperature that corresponds to $X = 1$ is considered to be the temperature closest to the natural or static temperature. Figure 8.2 shows other examples of such extrapolation by Fertl and Wichmann (1977).

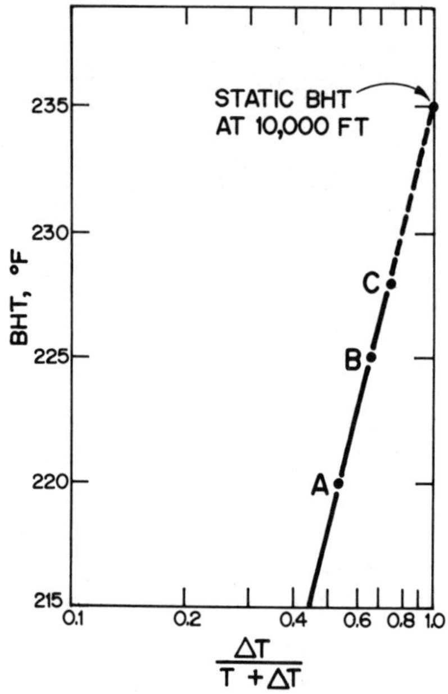

FIG. 8.1. Graph showing example of measured temperature-extrapolation to determine static or natural temperature (BHT). Circulation time of drilling mud is t, and time after circulation is Δt (from Fertl and Wichmann, 1977). (Reproduced by permission of the Gulf Publishing Company, Texas.)

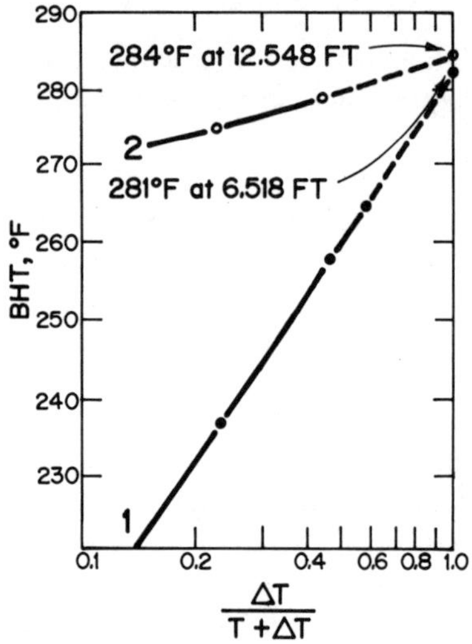

FIG. 8.2. Examples of extrapolation of measured temperatures to obtain static temperature (BHC) for wells 1 and 2 in the South China Sea region. Refer to Fig. 8.1 for other symbols (from Fertl and Wichmann, 1977). (Reproduced by permission of the Gulf Publishing Company, Texas.)

TABLE 8.1

Circulation time of drilling mud, t, and elapsed time after circulation, Δt, for three stages: A; B; and C shown in Fig. 8.1. Dimensionless value $\frac{\Delta t}{t+\Delta t}$ and measured temperature are also shown (Fertl and Wichmann, 1977)

Point	*Dimensionless time*	*Temperature* (°F)
A $\frac{\Delta t}{t+\Delta t}=\frac{7}{6+7}$	0·538	220
B $\frac{\Delta t}{t+\Delta t}=\frac{7+4{\cdot}5}{6+7+4{\cdot}5}$	0·671	225
C $\frac{\Delta t}{t+\Delta t}=\frac{7+4{\cdot}5+8}{6+7+4{\cdot}5+8}$	0·765	228

HEAT FLOW BY CONDUCTION

It is generally accepted that heat flows within the earth from its centre to the surface. The conduction of heat is shown by Lewis and Rose (1970) as

$$Q=(k_h/L)A\cdot\Delta T \tag{8.1}$$

where Q = heat flow
k_h = coefficient of thermal conductivity of the substance
L = length (thickness) of the substance
A = cross-sectional area of the substance
ΔT = temperature difference between both ends of the substance.

If we consider the heat flow through the substance of a unit cross-sectional area, eqn. (8.1) can be simplified as follows:

$$q=k_h\cdot\frac{\Delta T}{L} \tag{8.2}$$

where q = heat flow per unit cross-sectional area

$\frac{\Delta T}{L}$ = gradient of temperature change over distance.

Lewis and Rose (1970) showed the thermal conductivities of four substances as follows:

Water: 0·363 Btu/(h·ft·°F)
Slate: 1·138 Btu/(h·ft·°F)
Granite: 1·210 Btu/(h·ft·°F)
Sandstone: 1·330 Btu/(h·ft·°F)

There are relatively small differences among the thermal conductivities of slate, granite and sandstone, while the difference between water and the other (solid) substances is much greater. Water has only about one-third the thermal conductivity of the rock substances. This means that water has an important insulating capacity in sedimentary rocks.

Lewis and Rose (1970) mentioned that the coefficient of thermal conductivity of a water-saturated, porous sedimentary rock can be given by the following equation:

$$k_h=k_{hma}\left(\frac{k_{hw}}{k_{hma}}\right)^{\phi} \tag{8.3}$$

where k_h = thermal conductivity of the sedimentary rock
k_{hma} = thermal conductivity of the rock grains or matrix
k_{hw} = thermal conductivity of water
ϕ = porosity of the sedimentary rock.

To show the effect of porosity on thermal conductivity, they considered two different shales with porosities of 40 and 20% respectively, which consist of rock matrix similar to slate, and water (see eqn. (8.3)):

For 40% porosity $k_h = 1{\cdot}138\left(\dfrac{0{\cdot}363}{1{\cdot}138}\right)^{0{\cdot}4} = 0{\cdot}724$

For 20% porosity $k_h = 1{\cdot}138\left(\dfrac{0{\cdot}363}{1{\cdot}138}\right)^{0{\cdot}2} = 0{\cdot}906$

If there is a geologic sequence which is composed of two shaly formations of different porosities, 40 and 20% respectively, and if heat is flowing vertically upward, we will have the following heat-balance condition:

$$Q_{\phi=40\%} = Q_{\phi=20\%} \tag{8.4}$$

By combining eqns. (8.1) and (8.4), Lewis and Rose (1970) obtained

$$\left(\frac{k_h}{L} A \cdot \Delta T\right)_{\phi=40\%} = \left(\frac{k_h}{L} A \cdot \Delta T\right)_{\phi=20\%}$$

or

$$\left(k_h \cdot \frac{\Delta T}{L}\right)_{\phi=40\%} = \left(k_h \cdot \frac{\Delta T}{L}\right)_{\phi=20\%} \tag{8.5}$$

If the geothermal gradient through the shales of 20% porosity (thermal conductivity = 0·906) is 2°F/100 ft, then the gradient through the shales of 40% porosity (thermal conductivity = 0·724) can be given as

$$\left(\frac{\Delta T}{L}\right)_{\phi=40\%} = \frac{0{\cdot}906}{0{\cdot}724} \times 2°\text{F}/100\,\text{ft}$$
$$= 2{\cdot}5°\text{F}/100\,\text{ft}$$

The above sample calculation suggests that the geothermal gradient of an undercompacted section can be higher than that of a more normally compacted section, provided the lithology is constant. Even a slight difference in lithology between these two zones may have relatively little effect on the geothermal gradient, because of the comparatively small differences in thermal conductivity among the different rock materials already mentioned.

However, if there is no significant difference in the level of compaction, lithological variation can be the principal cause of a change in geothermal gradient.

Figure 8.3 shows an example of a temperature-vs-depth plot in the Gulf Coast. Below about 10 400 ft where both abnormal pressure and undercompacted shales exist, the geothermal gradient is 2·1°F/100 ft, while in the overlying normally pressured and compacted interval it is only about 1·3°F/100 ft.

In relatively young sedimentary basins such as the Gulf Coast and Mackenzie Delta, it is known that the shallower parts of the sedimentary column usually are normally compacted, while the deeper parts are sometimes undercompacted (see left-hand diagram of Fig. 8.4). Within

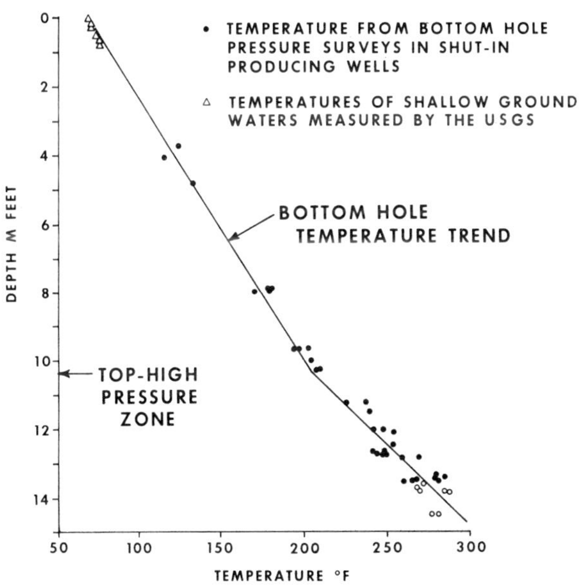

FIG. 8.3. Plot of temperature versus depth, which shows increased geothermal gradient at top of high-pressure zone (undercompacted zone) in Gulf Coast (from Schmidt, 1973). (Reproduced by permission of the American Association of Petroleum Geologists.)

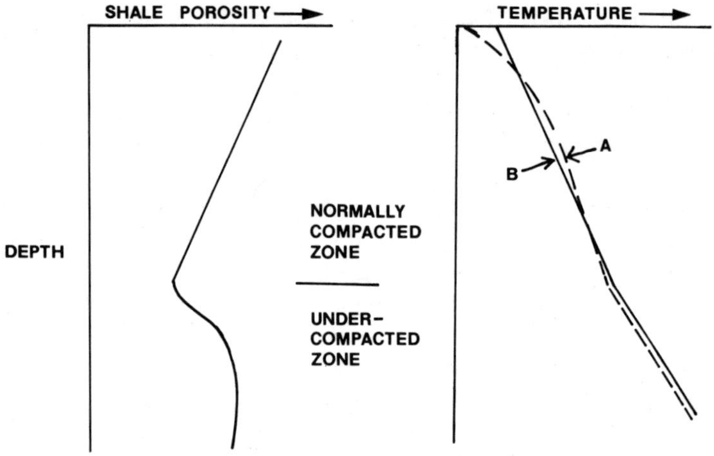

FIG. 8.4. Schematic diagrams showing shale porosity–depth relationship (left), and temperature–depth (right). Difference between dashed line A (theoretical conduction-heat-flow model) and solid line B (observed or real model) may indicate possible effect of fluid motion on geothermal gradient.

the normally compacted section, the shale porosity commonly decreases with depth. In the shallowest parts of the normally compacted zone, however, the porosity is probably higher than 25%, although these shales are called 'normally compacted'. Note that typical undercompacted shales at depth have porosities ranging from 20 to 25%. In other words, some 'normally compacted' shales at relatively shallow depths are less compacted than 'undercompacted' shales.

If, as Lewis and Rose (1970) point out, the geothermal gradient is higher in a high-porosity section, then the gradients in the shallowest parts of the normal compaction zone must be as high as, or even higher than, that of the deep undercompacted zone (see line A on right-hand side of Fig. 8.4).

However, we commonly observe a near-straight temperature–depth relationship such as shown in Fig. 8.3. This straight-line relationship is shown schematically by line **B** in Fig. 8.4.

What, then, makes the difference between lines A and B? I offer two possibilities, as follows:

1. Circulation of formation fluid, which tends to straighten the shallower temperature–depth curve.

2. Lithological variation. For example, if the percentage of sand increases upward, the average thermal conductivity of the rock matrix or grains increases upward (thermal conductivity of slate is 1·138 and that of sandstone is 1·330 Btu/h·ft·°F).

If the percentage of sand increases upward, then the chances of circulation of formation fluid would also increase upward, because of the higher permeability of those sands. In the subsurface sections of many sedimentary basins, the combined effects of these two factors probably caused the temperature–depth relationship in the shallower zones to be relatively straight.

REGIONAL HEAT FLOW

As discussed in the preceding section, the geothermal gradient in the undercompacted zone is higher than that in the overlying normally compacted zone. In this case heat is assumed to flow vertically upward at a constant rate.

If the distribution of undercompacted shales is not uniform, and if the

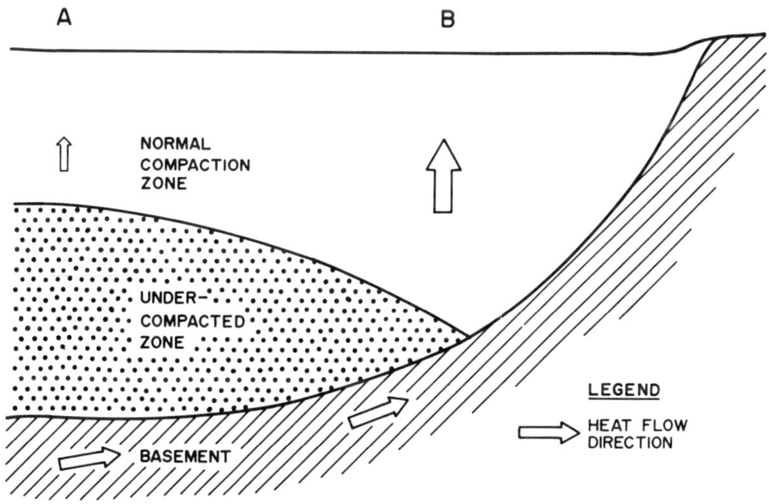

FIG. 8.5. Schematic model showing variable heat flow due to changing thickness of undercompacted zone (thermal insulator). Heat flow in area B is more than that in area A.

thickness changes in the lateral direction, as depicted in Fig. 8.5, we would not be able to expect a uniform vertical heat flow throughout the region. Heat flow through the areas of relatively thick undercompacted shales (see A in Fig. 8.5) would be greatly restricted, because of the insulating effect of these shales. Heat would preferably tend to move toward the areas of relatively thin undercompacted shales (see B in Fig. 8.5).

Now, let us make a schematic model of heat flow based on the diagram shown in Fig. 8.5. In area A, there is a thick undercompacted section, while in area B the shales are relatively thin. The thermal conductivities of these undercompacted shales are the same at these two locations, and are significantly lower than that for the normally compacted shales.

If the geothermal gradient of the normally compacted section in area A is compared with that of the normal compaction zone in area B, we would find a higher geothermal gradient in B (Fig. 8.6), because of the greater heat flow through that area. As shown by eqn. (8.2), the geothermal gradient ($\Delta T/L$) must be higher if the heat flow (q) is higher, so long as the thermal conductivity (k_h) is constant.

In the undercompacted zone of area A, the geothermal gradient is higher than that in the overlying normal compaction zone of the same

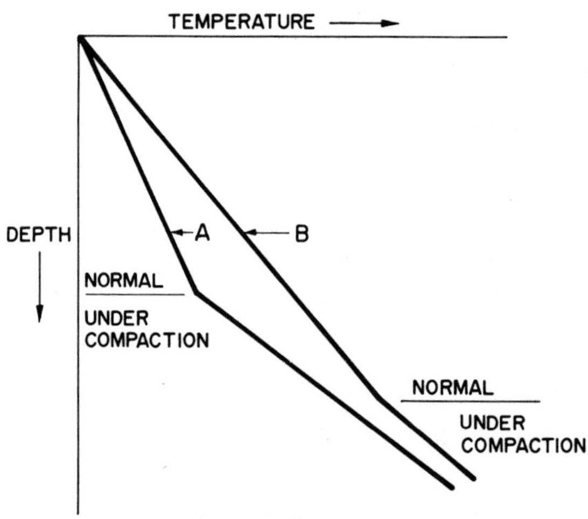

FIG. 8.6. Schematic plot of temperature versus depth for areas A and B, shown in Fig. 8.5.

area (Fig. 8.6), because of the lower thermal conductivity of the undercompacted shales. In the undercompacted interval of area B, the gradient may be even higher than that in the undercompacted zone of area A (Fig. 8.6), because of greater heat flow in area B than in area A.

If wells are drilled to intermediate depths in both areas, and if the bottom-hole temperatures are measured and extrapolated to the natural values, the values in area A would generally be lower than those in area B. If the average geothermal gradients are calculated by connecting the mean annual surface temperatures with the bottom-hole temperature (see Fig. 8.7), then the average gradients tend to be lower in area A, where the thick undercompacted shales are developed.

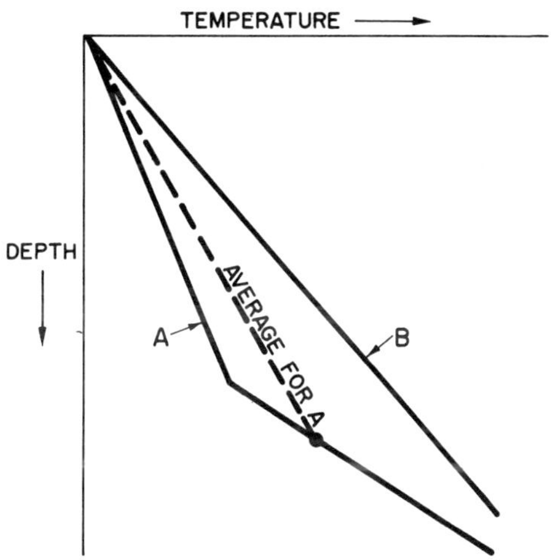

FIG. 8.7. Solid lines indicate geothermal gradient lines for areas, A and B, shown in Fig. 8.5. Dashed line indicates possible average geothermal gradient for area A.

The formation of undercompacted and abnormally pressured sections is considered to be the combined results of at least the following three factors (Magara, 1978):

1. A rapid rate of sedimentation.
2. Lithology, particularly a lack of permeable beds.
3. A thick accumulation of sedimentary column. (This means that, even if the above two factors are constant, the chances of undercompaction increase as sedimentation continues.)

In an area where a thick undercompacted shale section is developed, the rate of sedimentation has very probably been high. Therefore, we should be able to expect an apparent association of rapid rate of sedimentation and low geothermal gradient. However, this does not necessarily mean that such low gradient results solely from the rapid rate of sedimentation, in that cooler sediments have been buried too quickly to reach an equilibrium. Such an association may also result from a lower heat flow due to the presence of thicker undercompacted shales.

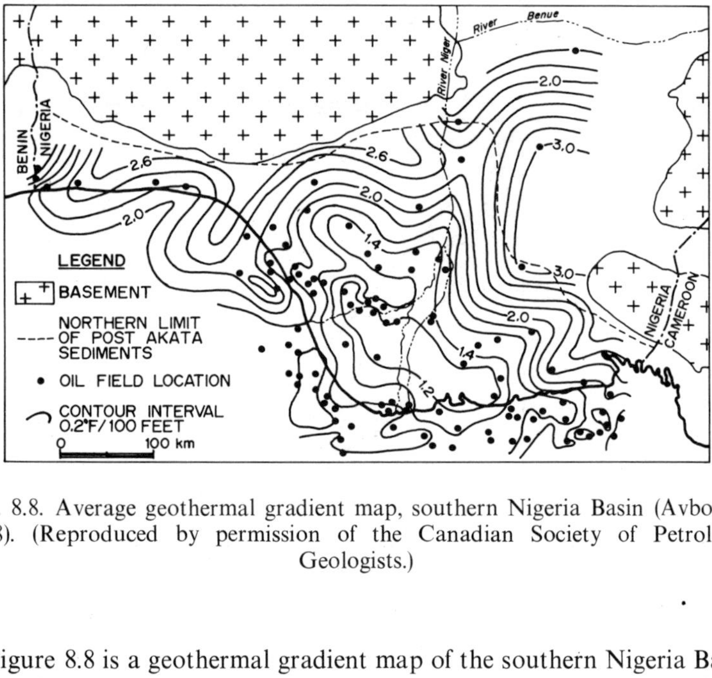

FIG. 8.8. Average geothermal gradient map, southern Nigeria Basin (Avbovbo, 1978). (Reproduced by permission of the Canadian Society of Petroleum Geologists.)

Figure 8.8 is a geothermal gradient map of the southern Nigeria Basin (Avbovbo, 1978). Avbovbo stated that, 'Low gradients of 1·20 to 1·40°F/100 ft occur in the Warri–Port Harcourt area in the Niger Delta'. He went on to say that this area roughly coincides with the nose of the Niger Delta, which shows a gravity low corresponding to a thick Tertiary section. Therefore, the areas of minimum geothermal gradient coincide with areas of maximum thickness of Tertiary section, and probably of maximum rate of sedimentation.

Figure 8.9 shows an example of a geothermal gradient map of the Gulf Coast area. The gradient generally decreases southward.

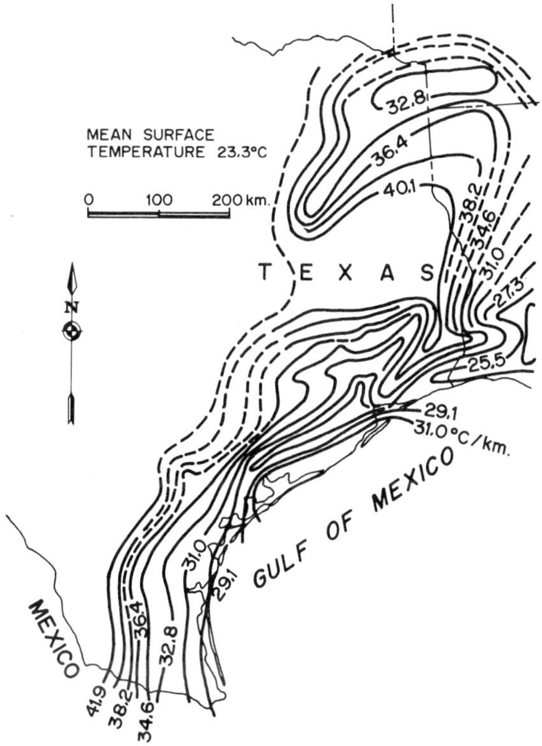

FIG. 8.9. Average geothermal gradient map in hydrostatic pressure zones of Gulf Coast (from Jones, 1975).

RELATIONSHIP BETWEEN THERMAL CONDUCTIVITY AND LEVEL OF SEDIMENT COMPACTION

Because of the fact that the difference in thermal conductivity is much greater between water and rock materials than it is among rock materials themselves, it seems reasonable to expect that an empirical relationship can be developed between level of compaction (or amount of pore-water) and thermal conductivity. Figure 8.10 shows such a relationship for the cores of Dunes and Mesa geothermal fields in California (Goss *et al.*, 1975). The lithology of these cores ranges from siltstones through graywackes to pebble sandstones.

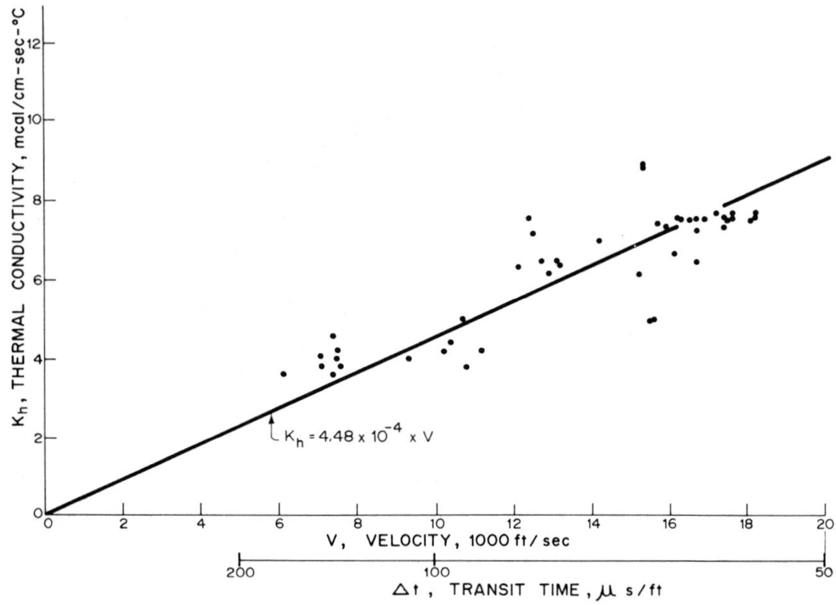

FIG. 8.10. Plot of thermal conductivity (k_h) versus interval velocity (V), or interval transit time (Δt). (Data are derived from Goss *et al.*, 1975.)

The widely scattered points in Fig. 8.10 are probably due to variations in rock composition and also to experimental errors. The relationship obtained by the least-squares method is as follows:

$$K = 4{\cdot}48 \times 10^{-4} \times V \tag{8.6}$$

where K = thermal conductivity in mcal/cm s °C
V = interval velocity in ft/s.

The thermal conductivity of a particular sedimentary rock is *not* a constant for the entire temperature range, but varies with temperature. Nevertheless, we usually assume that the conductivity of a given rock is constant regardless of temperature. I believe that this type of simplification is acceptable in most sedimentary sequences with petroleum accumulations.

TEMPERATURE MAPS

On the basis of regional temperature measurements, it is possible to construct a temperature (not geothermal gradient) map of a specified

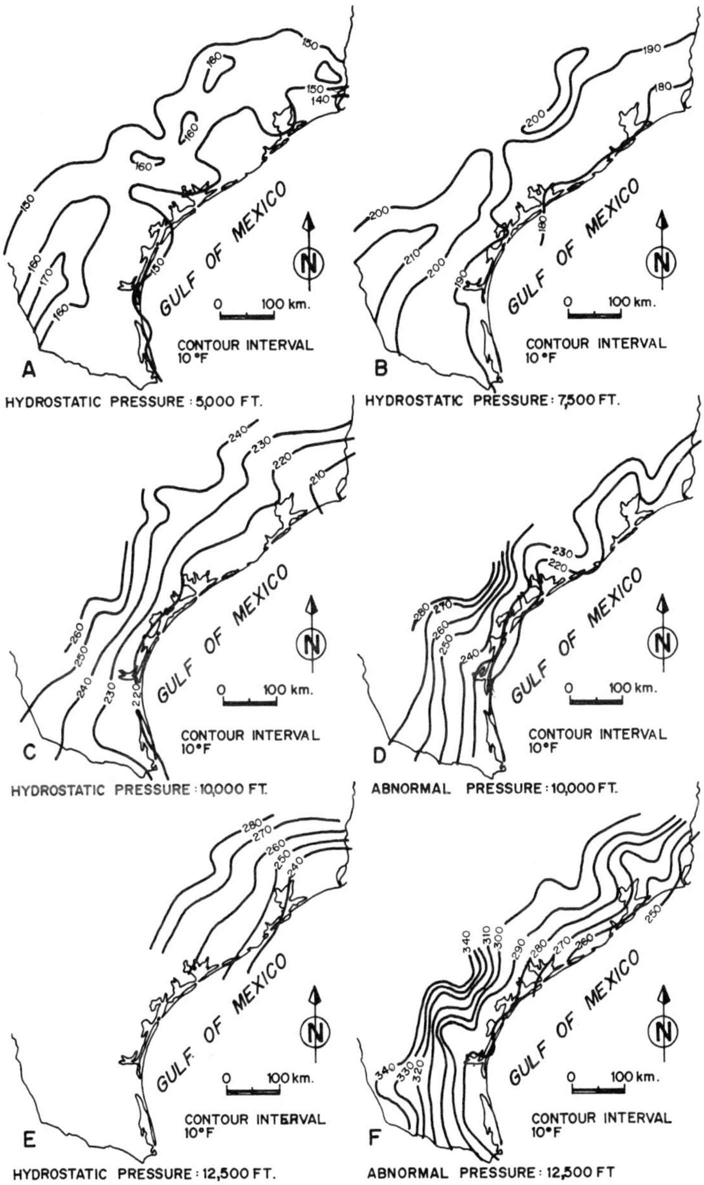

FIG. 8.11. Contour maps showing subsurface temperature (°F) at various depths shown (from Loucks *et al.*, 1979). (Reproduced by permission of the Bureau of Economic Geology, University of Texas at Austin.)

depth or horizon, which has a direct application for exploration, such as assessment of maturation of source rocks. Figures 8.11A–F show such examples in the Gulf Coast area, which depict the temperature maps at 5000 ft to 12 500 ft, with every 2500 ft increment (Loucks *et al.*, 1979).

REFERENCES

Avbovbo, A. A. (1978) Geothermal gradients in the southern Nigeria Basin. *Bull. Can. Pet. Geology*, **26**, 268–74.

Bebout, D. G., Weise, B. R., Gregory, A. R. and Edwards, M. B. (1979) Wilcox sandstone reservoirs in the deep subsurface along the Texas Gulf Coast, their potential for production of geopressured geothermal energy, Bureau of Economic Geology, University of Texas at Austin, 219 p.

Connan, J. (1974) Time–temperature relations in oil genesis. *A.A.P.G. Bull.*, **58**, 2516–21.

Fertl, W. H. and Wichmann, P. A. (1977) How to determine static BHT from well log data. *World Oil*, No. 1, 105–6.

Goss, R., Combs, H. and Timur, A. (1975) Prediction of thermal conductivity in rocks from other physical parameters and from standard geophysical well logs: SPWLA 16th Ann. Logging Symp., June 4–7.

Hunt, J. M. (1974) How deep can we find economic oil and gas accumulations?, S.P.E. 5177 in 1974 Deep Drilling and Production Symposium Preprints, pp. 103–110.

Jones, P. J. (1975) Geothermal and hydrocarbon regimes, northern Gulf of Mexico Basin. In Dorfman, M. H. and Delles, R. W. (eds.) First Geothermal Energy Conference, The University of Texas at Austin, Center for Energy Studies, Proceedings, pp. 15–97.

Lewis, C. R. and Rose, S. C. (1970) A theory relating high temperatures and overpressures. *J. Pet. Technol.*, **22**, 11–16.

Loucks, R. G., Dodge, M. M. and Galloway, W. E. (1979) Sandstone consolidation analysis to delineate areas of high-quality reservoirs suitable for production of geopressured geothermal energy along the Texas Gulf coast, Contract report, Bureau of Economic Geology, University of Texas at Austin, 97 p.

Magara, K. (1978) *Compaction and Fluid Migration—Practical Petroleum Geology*, Elsevier, Amsterdam, 319 p.

Schmidt, G. W. (1973) Interstitial water composition and geochemistry of deep Gulf Coast shales and sandstones. *Am. Assoc. Petroleum Geologists Bull.*, **57**, 321–37.

Chapter 9

Natural Fracturing and Faulting in a Geopressured Sedimentary Basin

INTRODUCTION

Differential loading, which is caused by variable rates of sedimentation in the geologic past, may result in significant differences in fluid potential or pressure in a sedimentary basin. Fluid migration due to differential loading was described by Magara (1977).

During the earliest stages of sediment compaction, most fluids tend to move vertically upward. At the following stages with reduced shale permeability, the principal direction of fluid migration may become stratigraphically controlled, particularly through the interbedded permeable beds (for example, sandstones and limestones).

The above-mentioned early fluid migration in many young sedimentary basins occurs in sections above about 5000 ft. Shale compaction at these stages is usually normal and the fluid pressure after loading reaches normal hydrostatic levels (Figs. 6.12–6.14).

At intermediate stages of compaction, slightly undercompacted and overpressured shales are developed. The fluid pressure in the interbedded sandstones stays nearly hydrostatic. The shales immediately above and below the sandstone beds compact normally, having near-hydrostatic pressure, but the shales away from the sandstones may become slightly undercompacted and overpressured. These undercompacted shales are called pressure seals (Evans *et al.*, 1975) and seem to retard the vertical escape of fluid. Refer to Figs. 6.12–6.14 in Chapter 6 for examples of pressure seals.

After the development of pressure seals, the directions of intergranular fluid flow would become stratigraphy-controlled, because pressure seals act as semi-barriers for vertical fluid flow. Even at this stage there may, however, be some local vertical fluid flow through fractures and intergranular pore spaces.

If a domal or closed-anticlinal structure in which a permeable reservoir is overlain by a pressure seal (shale) is formed in the subsurface, compaction fluids expelled from the shale into the reservoir move toward the crest of the structure, within the reservoir, from all directions (Fig. 9.1). This is because the fluids tend to move from points of greater loading (syncline) to a point of least loading (crest of the structure; see Magara (1977)). To continue this radial fluid flow and mass (fluid) balance, the fluids must eventually escape upward in the crestal area of the structure.

The most probable avenue for such fluid movement may be fractures, because it is generally quite difficult to reverse intergranular fluid flow patterns once they are developed in shales.

If the sediments are buried deeply with the development of some restricted fluid environments, the pressure in a relatively deep section may become abnormally high. If such high pressures are connected with shallow and low-pressured intervals, they may cause natural hydrofracturing in these intervals. Shale diapirs may also be formed.

At later stages, if the area experiences a period of significant uplift and erosion, the fluids in the sediment pore spaces will contract due to the cooling effect, causing significant reduction in pore-fluid pressure. The amount of pressure reduction can be quite significant; erosion of a 1-ft section, followed by a proper temperature adjustment based on the geothermal gradient of 1·4°F/100 ft (average value for Gulf Coast), could cause a pressure reduction of about 1·8 psi, or approximately 80% more pressure than that due to the reduction of overburden pressure (1 psi) due to this erosion.

Sediment grains also contract by cooling, causing some reduction in pore spaces. However, this effect is much less significant; the contraction of quartz grains is about 1/15 that of pure water. The value for shale grains is not known but may be assumed to be closer to quartz than to water.

Because thickness of erosion usually varies within a region, some local and lateral variations of fluid potential reduction may be developed. Lateral fluid flow may thus occur, and may further help to develop additional fracture systems.

Once significant subnormal fluid pressures are developed in the subsurface, the near-surface meteoric water, which usually has near-

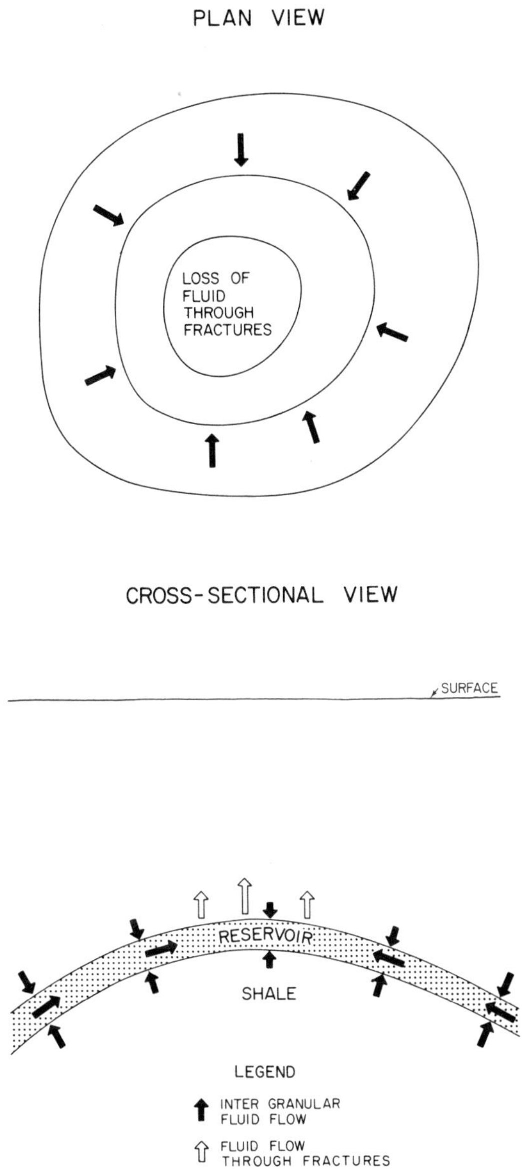

FIG. 9.1. Schematic diagrams showing cross-sectional and plan views of intergranular and fracture fluid flow.

hydrostatic pressure, may penetrate deeply and may cause some late-stage fracturing.

In summary, there are many possible causes of natural fractures. They are related to variable fluid pressures due to loading and unloading of a sedimentary column, to temperature, and also to mechanical deformation of rocks.

Using the data of the in-situ estimate of Poisson's ratio and abnormally high fluid pressure, the tectonic stress conditions necessary for faulting can be evaluated. In order for a fault to start, the stress must exceed the internal friction of the rocks involved. The minimum shear stress for slippage can also be estimated for different fluid pressure conditions.

GENERATION OF EXCESSIVE FLUID PRESSURE DURING LOADING

Magara (1977) stated that the excess fluid pressure, $(pl_0)a$, above hydrostatic pressure, generated by an instantaneous loading of a thin layer whose thickness is l_0, under water, is given as,

$$(pl_0)a = (\rho_{b0} - \rho_w) \cdot g \cdot l_0 \qquad (9.1)$$

where ρ_{b0} = water-saturated bulk density of the new sediment layer, l_0, and ρ_w density of water (see Fig. 9.2).

If sufficient geologic time is allowed and/or if the sediment permeability is high enough, the excess fluid pressure will be dissipated. The fluid pressure will reach a normal hydrostatic level. If, on the other hand, more sediment layers are added above this layer before the excess fluid pressure is completely dissipated, significant excess pressure will accumulate.

The increase of excess fluid pressure may also be influenced by the effect of thermal expansion of water, called the 'aquathermal effect'. Aquathermal pressuring becomes effective after sediments reach a stage of restricted fluid movement (isolation of pore-fluid) primarily due to the reduced permeability by increased compaction.

A theoretical consideration of the aquathermal pressuring effect revealed that the fluid pressure increases by 6000 psi for the burial increase of 3300 ft with perfect pore-fluid isolation (Fig. 9.3, and Magara (1975)). The rate of fluid pressure increase is about 1·8 psi/ft, which is significantly higher than that of overburden pressure. Geothermal gradient is

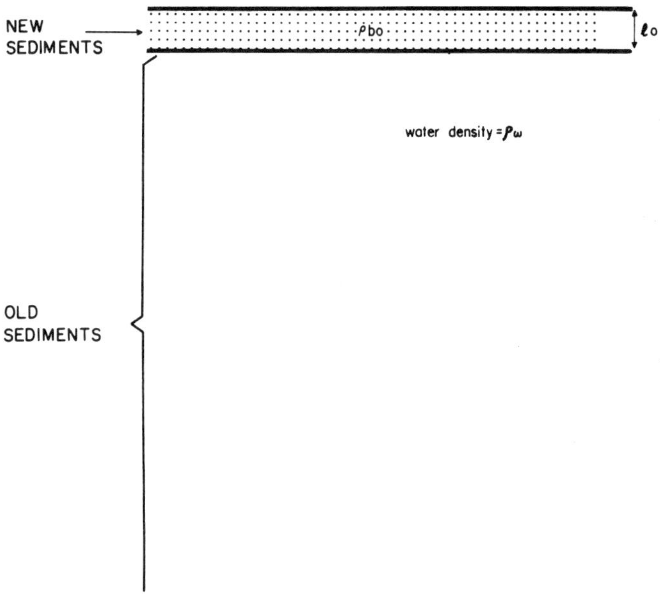

FIG. 9.2. Schematic diagram showing model in which new sediment layer is deposited on old sediment, ρ_{b0} and l_0 are bulk density and thickness of new sediment layer, and ρ_w is density of water saturating sediments. Model assumes deposition under water.

assumed to be in this case about 1·4°F/100 ft. Because shales usually have some permeability, it is not possible to have a perfect isolation as assumed above.

Magara (1975) estimated the actual rate of fluid pressure increase after the isolation to be approximately 1·4 psi/ft, on the basis of Hottman and Johnson's (1965) data on compaction and fluid pressures in the Gulf Coast area. This actual rate of pressure increase is about a mid-value between the theoretical aquathermal (1·8 psi/ft) and nonaquathermal (or overburden pressure gradient of about 1·0 psi/ft). In other words, about a half of the aquathermally generated pressure seems to have been dissipated in the geopressured intervals of the Gulf Coast because of the shales' permeability.

If both the restricted fluid environments and the aquathermal pressuring become effective at 8000 ft, which is the average depth for the Gulf Coast (Magara, 1975), and if the burial continues to a depth of 20 000 ft,

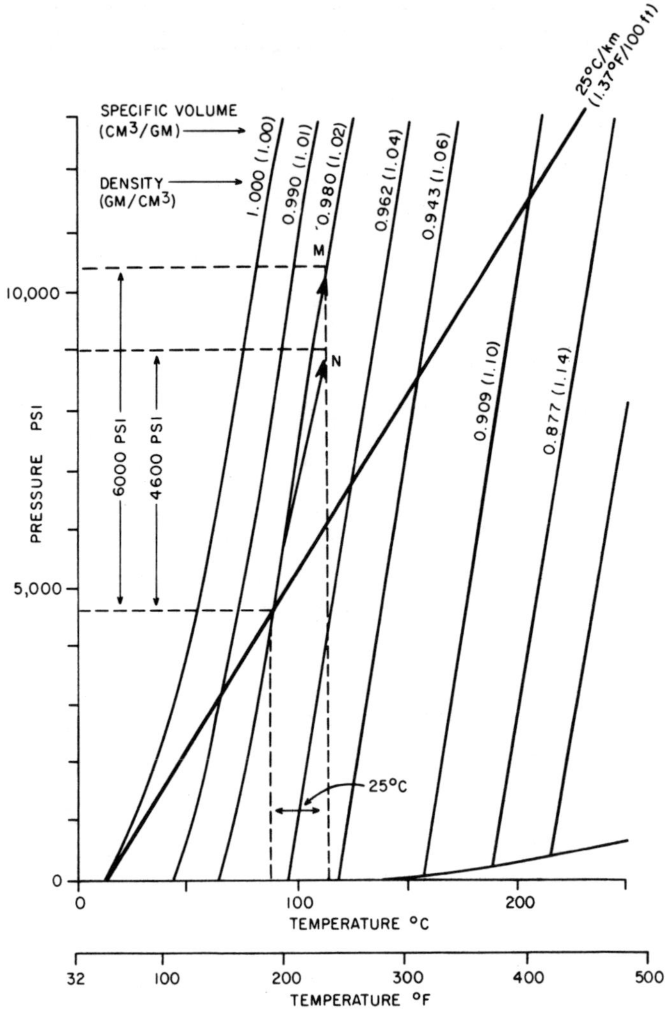

FIG. 9.3. Pressure–temperature–density (or specific volume) diagram for water. Geothermal line of 25°C/km for hydrostatically pressured fluids is superimposed on a basic diagram from Barker (1972, from Magara, 1975). (Reproduced by permission of the American Association of Petroleum Geologists.)

the fluid pressure could reach the level of overburden pressure (see Fig. 9.4B). The rocks there would thus become extremely unstable because of the virtually no internal friction of the rocks (Magara, 1978, Chapter 11). The overburden load is now totally supported by the fluid

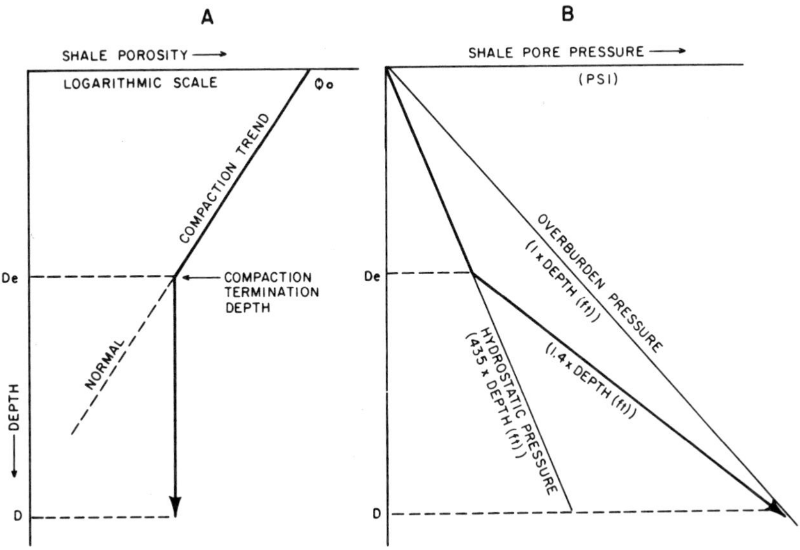

FIG. 9.4. Schematic shale porosity–depth and pore pressure–depth relationships at present in area where massive shales continued to be deposited (from Magara, 1978).

pressure in this case, causing conditions for possible fracturing, faulting, and/or diapirism.

REDUCTION OF GENERATED FLUID PRESSURE DURING UNLOADING

If the area experiences a period of major uplift and erosion, and if the subsurface temperature is reduced accordingly, a substantial contraction of pore-water would occur, causing semi-vacuum conditions and subnormal fluid pressure. A theoretical fluid pressure reduction due to erosion of a 1-ft section would be about 1·8 psi when the geothermal gradient is about 1·4°F/100 ft.

From the study of the Cardium sandstones in the Western Canada Basin using the results of both inclusion thermometry (of quartz crystals) and shale compaction data, Magara (1976) estimated that the pore-fluid pressure dropped from about 10 000 psi at the time of maximum burial to about 4000 psi at present, after erosion of about 4000 ft of section. This suggests that the actual rate of fluid pressure reduction by erosion was approximately 1·5 psi/ft. It is interesting to note that this figure (1·5 psi/ft) is quite close to the observed rate of fluid pressure increases (1·4 psi/ft) by the aquathermal effect in the Gulf Coast.

This observation suggests that, even in the case of sediment cooling by uplift and erosion (which may be termed the 'negative aquathermal or anti-aquathermal effect'), the shales cannot act as a perfect seal because of their permeability. The theoretical value of 1·8 psi/ft may be applicable only when the pore-water is completely isolated from outside.

The difference between these two figures (1·4 psi/ft of increasing rate in Gulf Coast and 1·5 psi/ft of decreasing rate in Western Canada) may be explained by the difference in the average geothermal gradients; 1·4°F/100 ft in the former and 1·8 °F/100 ft in the latter. There may also be some other geological factors causing such difference.

If subnormal pressure is generated in a relatively shallow section whose fluid system is relatively open, it may be normalized quickly due to the influence of near-surface ground water of meteoric origin. However, if such subnormal pressure occurs in a relatively deep and restricted fluid environment, it may be preserved for a long geologic period.

Because the thickness of erosion usually changes from location to location within a region, significant local variations of subnormal fluid pressure or potential could develop. This may cause some lateral fluid movement, possibly resulting in some natural hydrofracturing, mechanical aspects of which will be discussed in the following section. The principal direction of fluid flow would be from a less eroded area to a more eroded area.

Once significant subnormal pressures and potentials are developed at depth and the near-normal hydrostatic pressures and potentials are formed at shallow depths, the shallow water may penetrate downward, possibly causing some additional fracturing. Unloading itself may also cause some natural fracturing by reducing the stress around the sedimentary rocks. Mechanical aspects of this type of fracturing will be discussed later.

MECHANISM OF HYDROFRACTURING

The basic equations relating triaxial stress and strain are as follows (Price, 1959),

$$e_x = \frac{1}{E}[\sigma_x - \mu(\sigma_y + \sigma_z)] \tag{9.2}$$

$$e_y = \frac{1}{E}[\sigma_y - \mu(\sigma_x + \sigma_z)] \tag{9.3}$$

$$e_z = \frac{1}{E}[\sigma_z - \mu(\sigma_x + \sigma_y)] \tag{9.4}$$

where e_x, e_y, e_z are strains in mutually perpendicular directions, σ_x, σ_y, σ_z corresponding stresses, E is Young's modulus, and μ is Poisson's ratio.

Consider a small unit cube of rock in the subsurface, subject to a vertical stress σ_z. Lateral expansion is completely restricted by the pressure of the surrounding rock; i.e. $e_y = e_x = 0$. Therefore, the lateral stress is given by:

$$\sigma_y = \sigma_x = \frac{\mu}{1-\mu} \cdot \sigma_z \tag{9.5}$$

In this case, it is assumed that there is no additional lateral stress caused by tectonic forces in the subsurface.

On the basis of Terzaghi and Peck's (1948) basic concept of stress and fluid pressure, the value σ_z can be given as,

$$\sigma_z = S - p \tag{9.6}$$

where S is overburden pressure (or stress) and p fluid pressure. Fracture pressure (and its gradient) is given as a sum of (least) horizontal stress and fluid pressure (Hubbert and Willis, 1957). Thus we obtain,

$$\begin{aligned} \text{FPG} &= \frac{p_f}{D} = \frac{\sigma_h}{D} + \frac{p}{D} = \frac{\sigma_z}{D}\left(\frac{\mu}{1-\mu}\right) + \frac{p}{D} \\ &= \left(\frac{S}{D} - \frac{p}{D}\right)\left(\frac{\mu}{1-\mu}\right) + \frac{p}{D} \end{aligned} \tag{9.7}$$

where FPG is fracture pressure gradient, p_f is fracture pressure, D depth at which fracture occurs, and σ_h horizontal stress ($=\sigma_x = \sigma_y$ in this case) (Eaton, 1969).

The values of Poisson's ratio for various rocks in the laboratory conditions are available (Clark, 1966), but only a few under subsurface conditions have been reported. Phillips (1948) reported that Poisson's

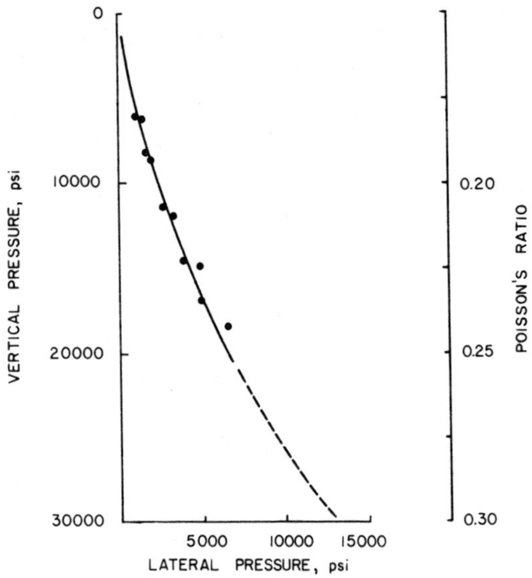

FIG. 9.5. Graph showing change of lateral pressure and Poisson's ratio with changing vertical pressure (from Price, 1959). (Reproduced by permission of the Geological Society of America.)

ratio increases with increase in vertical stress or pressure (Fig. 9.5, Price, 1959). Using Costley's (1967) fracture pressure data in the Gulf Coast area, Eaton (1969) estimated the in-situ values of Poisson's ratio under the subsurface, shown in Fig. 9.6. This curve also shows a tendency of the ratio to increase with vertical stress or overburden pressure.

As eqn. (9.5) shows, the lateral or horizontal stress increases as the vertical stress increases (the value $\mu/(1-\mu)$ is always positive). How the horizontal stress increases depends upon the value of Poisson's ratio.

As shown by eqn. (9.6), the vertical (effective) stress is abnormally low if the fluid pressure is abnormally high; the horizontal (effective) stress can also be abnormally low in such a section. Figure 9.7 shows changing horizontal stress with depth, using the estimated in-situ Poisson's ratio in the Gulf Coast (Fig. 9.6). The solid curved line shows the increasing trend of horizontal stress for a hydrostatically pressured section. A series of dashed curved lines in Fig. 9.7 indicates those when the aquathermally generated fluid pressure starts at 8000 ft, 10 000 ft, and so on.

Equation (9.7) means that the fracture pressure gradient is given as the total of fluid pressure and (least) horizontal stress. In an abnormal

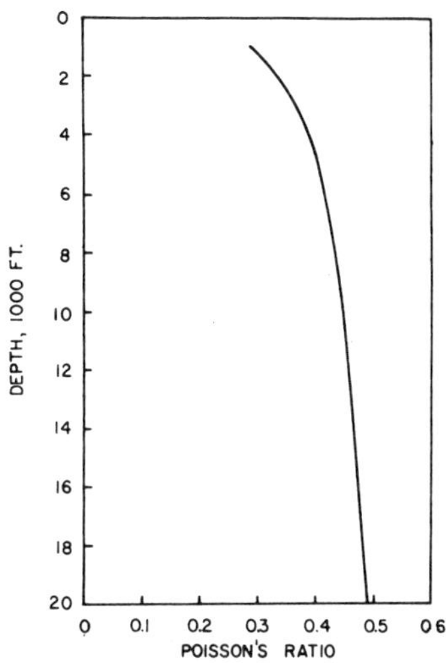

FIG. 9.6. Graph showing change of in-situ estimate of Poisson's ratio with depth in Gulf Coast area (from Eaton, 1969). (Reproduced by permission of the Society of Petroleum Engineers.)

pressure section, fluid pressure is higher than normal but both vertical (σ_z) and horizontal (σ_h) stresses are reduced (see eqns. (9.6) and (9.7)) for a given value of Poisson's ratio. Hence, the fracture pressure gradient may not increase drastically, even if the fluid pressure is abnormally high.

Figure 9.8 shows the estimated fracture pressure gradients in psi/ft when hydrostatic pressure exists (solid curved line) and when aquathermally generated fluid pressure exists below 8000 ft, 10 000 ft, and so on (dashed curved lines). These values are plotted in a relatively narrow range.

NATURAL HYDROFRACTURING DURING LOADING

Let us now assume geological models in which fluids of a deep and abnormally pressured reservoir charge a shallower section (Fig. 9.9). Charging may take place along a fault plane (Fig. 9.9A) or through the reservoir (Fig. 9.9B). If this type of charging occurs, the shallower section can be fractured quite easily, as demonstrated through simple calculations.

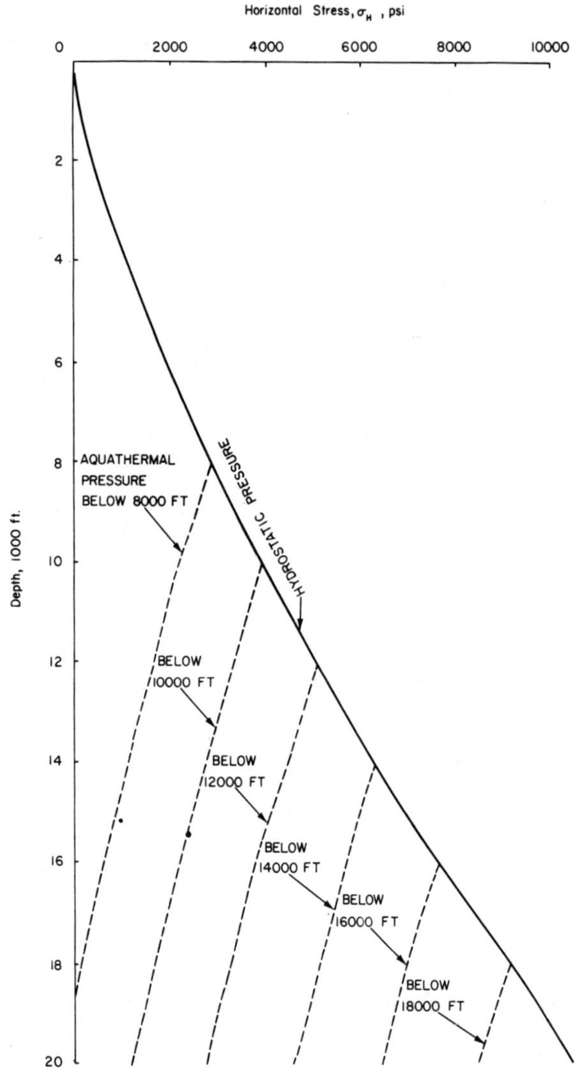

FIG. 9.7. Relation between calculated horizontal stress and depth for hydrostatically pressured interval (shown by solid line) and for abnormally pressured intervals (shown by dashed lines) in Gulf Coast. Dashed lines indicate aquathermally generated (abnormal) pressure at different depths (from Magara, 1981). (Reproduced by permission of the American Association of Petroleum Geologists.)

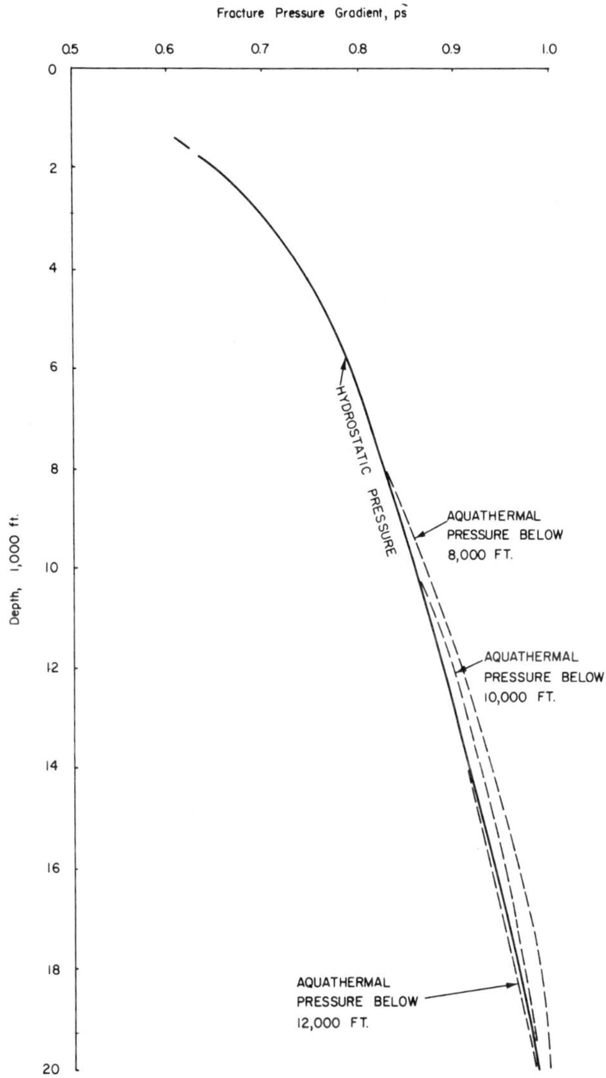

FIG. 9.8. Relation between calculated fracture pressure gradient and depth for hydrostatically pressured interval (solid line) and for abnormally pressured intervals (dashed lines) in Gulf Coast. Dashed lines indicate aquathermally generated (abnormal) pressure at different depths (from Magara, 1981). (Reproduced by permission of the American Association of Petroleum Geologists.)

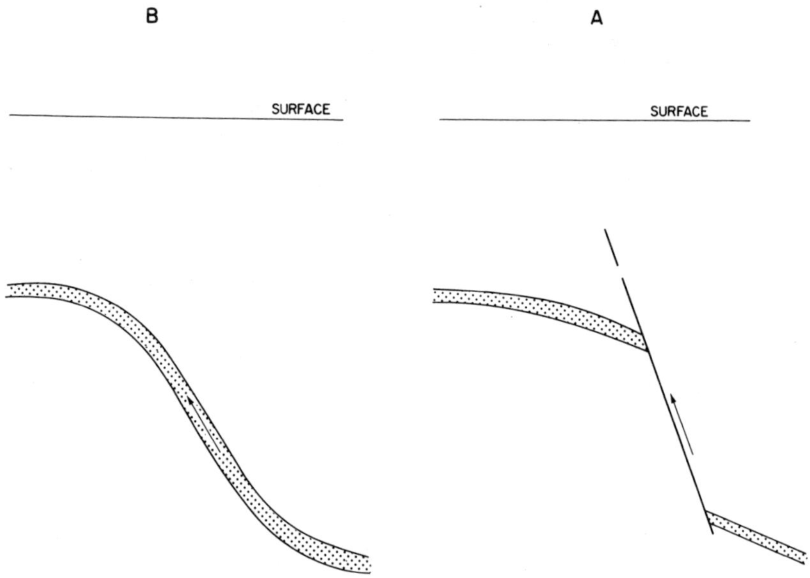

FIG. 9.9. Schematic diagrams showing fluid migration along fault plane (A) and through aquifer (B).

Let us assume that the fluid pressure at 10 000 ft is 7000 psi (or fluid pressure gradient is 0·7 psi/ft). The fracture pressure gradient at 10 000 ft is approximately 0·87 psi/ft (see Fig. 9.8), so that there would be no chance of fracturing at that depth. If the fluid charges a shallower section at 6000 ft, for example, there would be a chance of fracturing. As the deep fluid moves to shallow, its fluid pressure will decline on the basis of the weight of the water column; the pressure decline from 10 000 ft to 6000 ft would be 1860 psi based on 0·465 psi/ft gradient. Therefore, the fluid pressure at 6000 ft would become 5140 psi (= 7000 − 1860 psi), which corresponds to about 0·86 psi/ft gradient. The fracture pressure gradient at 6000 ft is about 0·79 psi/ft (see Fig. 9.8); it is most likely to cause some fractures there.

Figures 9.10–9.15 depict the calculated results of depths of fracturing by charging; Fig. 9.10 shows the cases when the charging originated at 10 000 ft, Fig. 9.11 at 12 000 ft, Fig. 9.12 at 14 000 ft, Fig. 9.13 at 16 000 ft, Fig. 9.14 at 18 000 ft, and Fig. 9.15 at 20 000 ft. The dashed curved lines in each figure depict the changing fluid pressure gradient from the

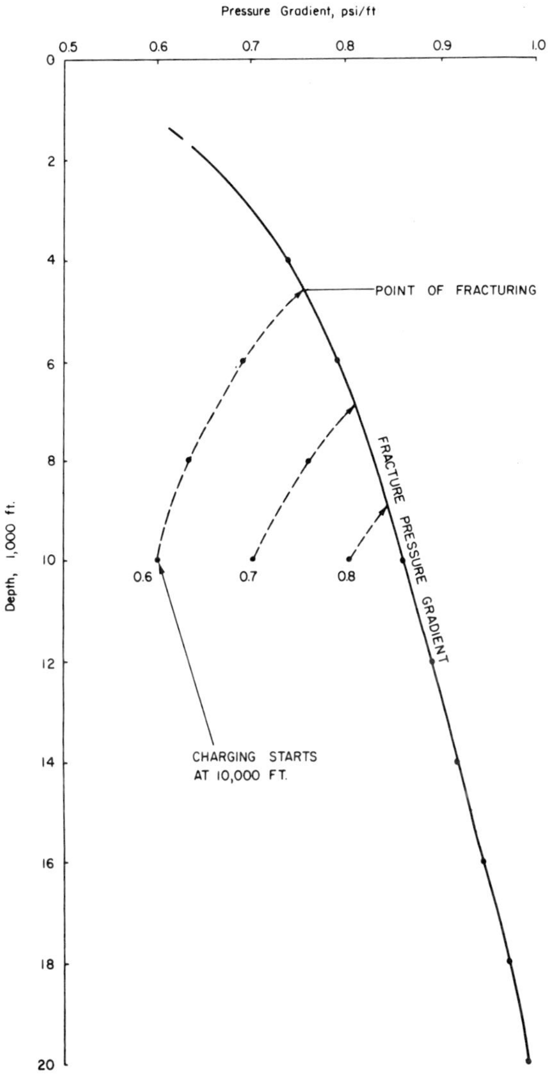

FIG. 9.10. Graph showing upward change of fluid pressure gradient when fluid charging starts at 10 000 ft, shown by dashed lines. Three curved, dashed lines show different pressure gradients at 10 000 ft (charging point). Solid curve shows calculated fracture pressure gradient for Gulf Coast. Arrows on curved lines indicate points of possible fracturing by charging (from Magara, 1981). (Reproduced by permission of the American Association of Petroleum Geologists.)

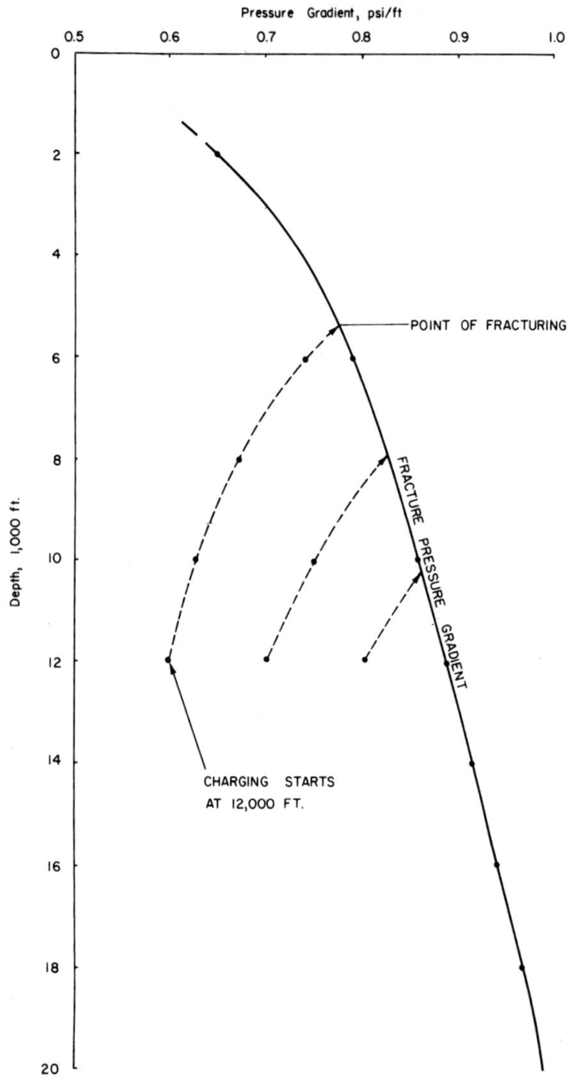

FIG. 9.11. Graph showing upward change of fluid pressure gradient when fluid charging starts at 12 000 ft, shown by dashed lines. Refer to Fig. 9.10 for explanation (from Magara, 1981). (Reproduced by permission of the American Association of Petroleum Geologists.)

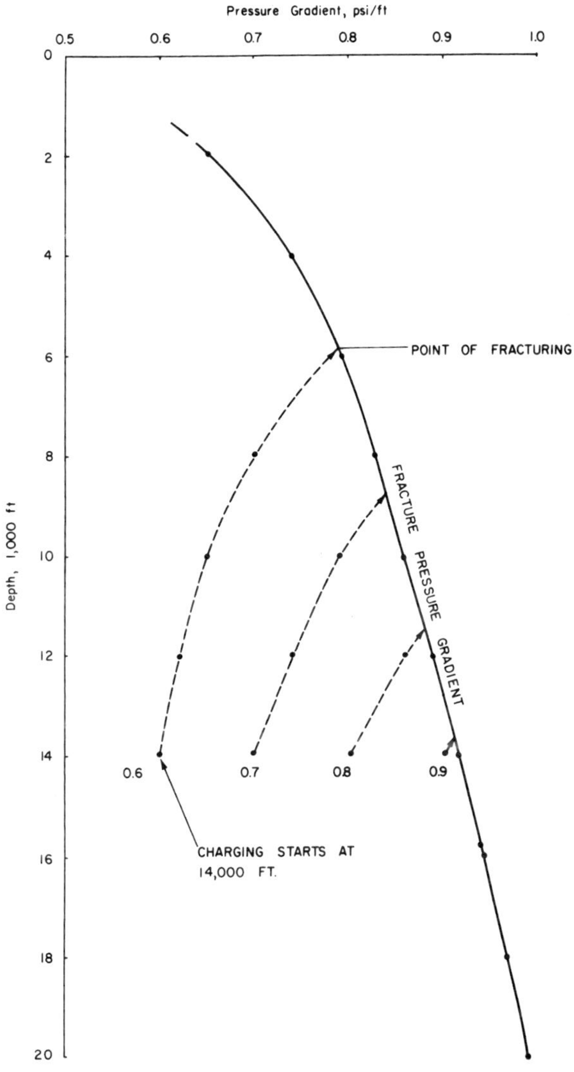

FIG. 9.12. Graph showing upward change of fluid pressure gradient when fluid charging starts at 14 000 ft, shown by dashed lines. Refer to Fig. 9.10 for explanation (from Magara, 1981). (Reproduced by permission of the American Association of Petroleum Geologists.)

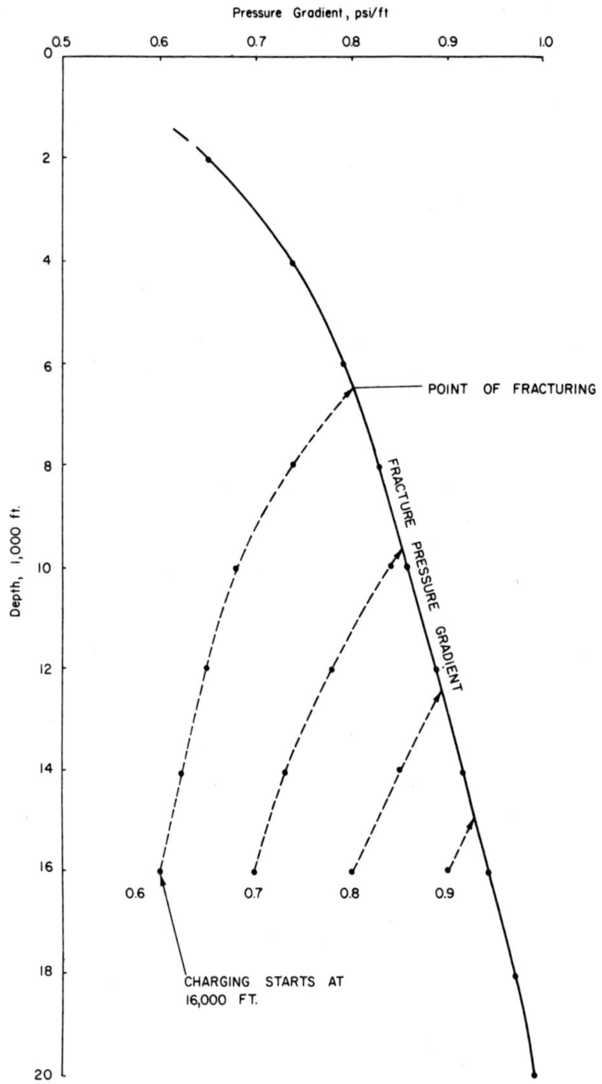

FIG. 9.13. Graph showing upward change of fluid pressure gradient when fluid charging starts at 16 000 ft, shown by dashed lines. Refer to Fig. 9.10 for explanation (from Magara, 1981). (Reproduced by permission of the American Association of Petroleum Geologists.)

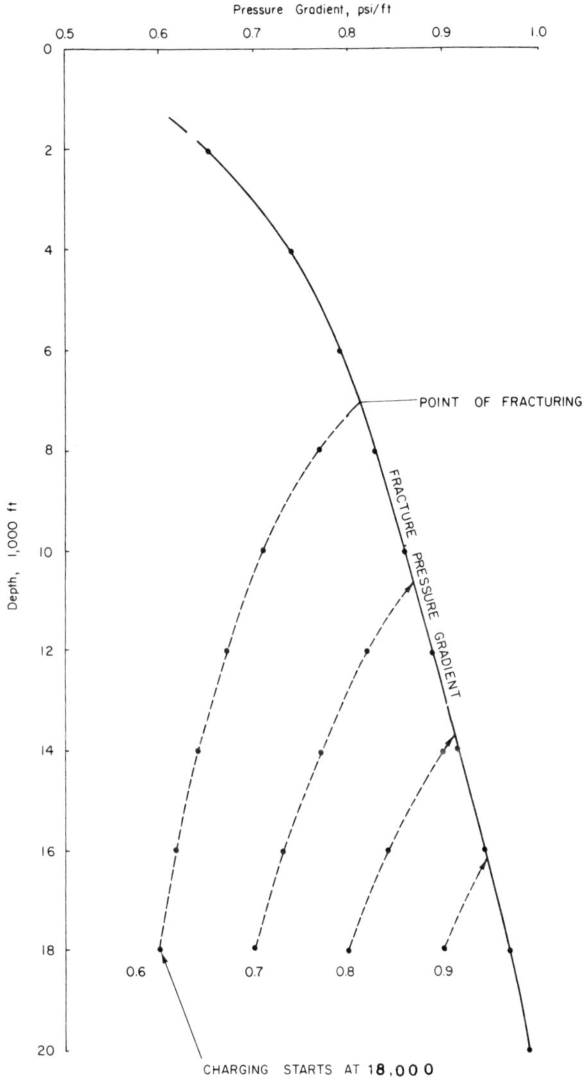

FIG. 9.14. Graph showing upward change of fluid pressure gradient when fluid charging starts at 18 000 ft, shown by dashed lines. Refer to Fig. 9.10 for explanation (from Magara, 1981). (Reproduced by permission of the American Association of Petroleum Geologists.)

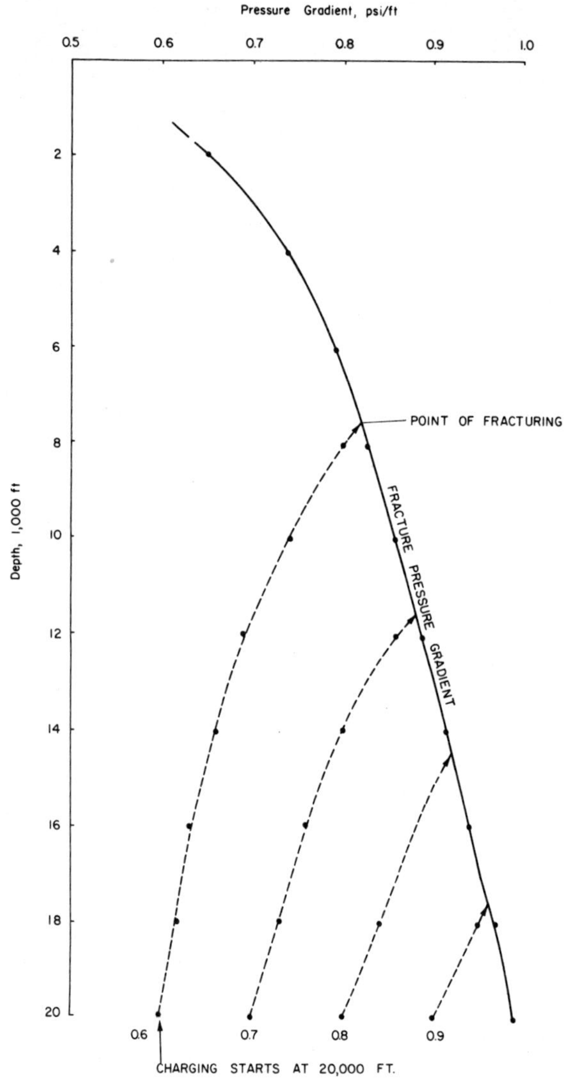

FIG. 9.15. Graph showing upward change of fluid pressure gradient when fluid charging starts at 20 000 ft, shown by dashed lines. Refer to Fig. 9.10 for explanation (from Magara, 1981). (Reproduced by permission of the American Association of Petroleum Geologists.)

original point in the upward direction. The solid curved line shows the calculated fracture pressure gradient. The upward arrow attached to each dashed line indicates a point of fracturing.

NATURAL FRACTURING DURING UNLOADING

An erosional event and the associated cooling could cause a significant subnormal fluid pressure by the negative aquathermal effect. Because the thickness of erosion usually varied locally, this event could have caused significant lateral variation in fluid potential after erosion, thus resulting in some lateral fluid migration. Erosion itself may also have caused some fracture or fault system, so that some fluid charging from an area of less erosion to an area of more erosion may have taken place.

The effect of erosion for a hydrostatic pressure section would be different from that for an abnormally pressured section. This is because, in the former section, the subnormal fluid pressure due to erosion would be normalized relatively quickly since the system is relatively open. In the latter section, on the contrary, the generated subnormal pressure may be preserved for a long geologic time because the fluid system there is relatively closed.

The hydrostatic section reduces the vertical and horizontal stresses as both the overburden and (hydrostatic) fluid pressures decrease after erosion. Reduction of the horizontal stress is shown by the two solid curved lines above 8000 ft (Fig. 9.16), representing the two cases of 2000-ft and 4000-ft erosion.

In the abnormally pressured section below 8000 ft, both vertical and horizontal (effective) stresses are abnormally small. If erosion occurs, it reduces the fluid pressure at a faster rate (1·4 psi/ft) than the overburden pressure gradient (1·0 psi/ft). The net result is an increase of the vertical (effective) stress, as shown by eqn. (9.6), thus causing increased horizontal stress. This situation is shown by the two dashed curved lines below 8000 ft in Fig. 9.16. In this figure, it is assumed that the aquathermally generated fluid pressures exist below 8000 ft of the original depth (before erosion).

Although there are significant changes in the horizontal stress, the fracture pressure gradient is relatively unchanged in most of the sections, because, where the horizontal stress increased, the fluid pressure decreased, and vice versa. As was pointed out previously, fracture pressure gradient is controlled by a combined effect of the stress and fluid pressure

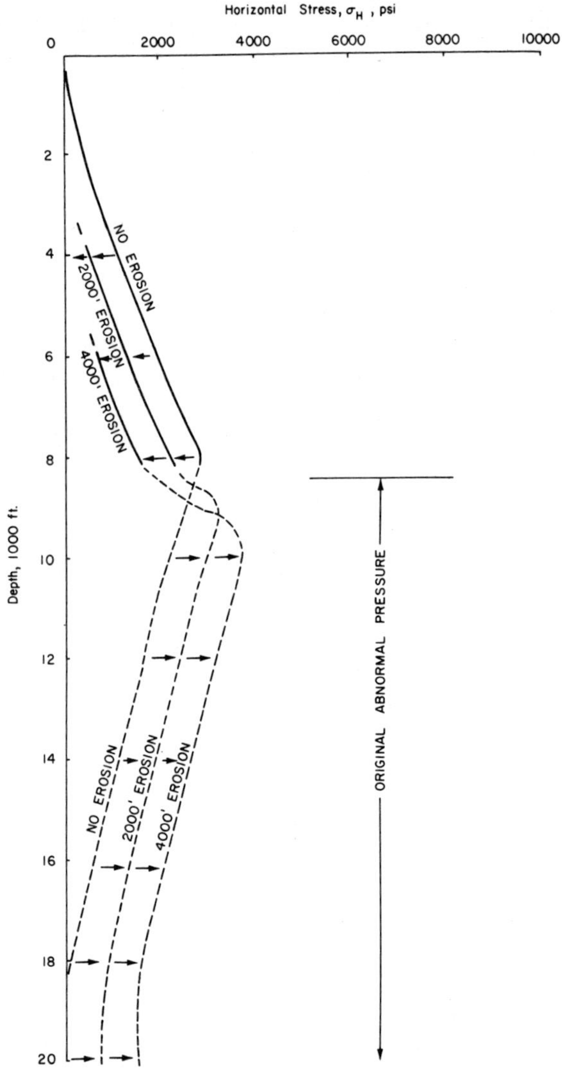

FIG. 9.16. Graph showing change of calculated horizontal stress by erosion in hydrostatically pressured (solid line) and abnormally pressured (dashed lines) intervals. Abnormal-pressure interval is assumed below 8000 ft of original depth before erosion. See text for further explanation (from Magara, 1981). (Reproduced by permission of the American Association of Petroleum Geologists.)

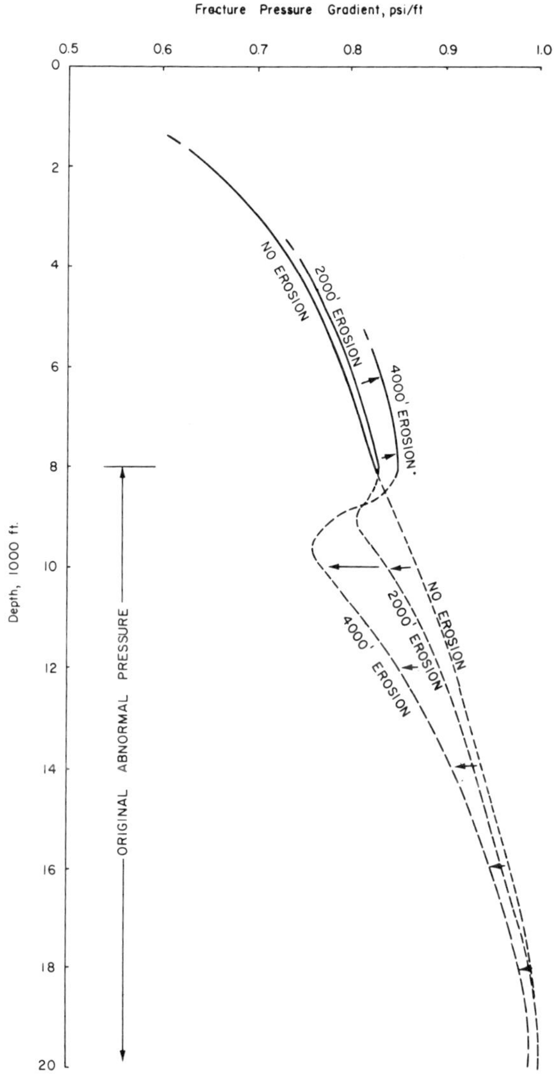

FIG. 9.17. Graph showing change of calculated fracture pressure gradient by erosion in hydrostatically pressured (solid line) and abnormally pressured (dashed lines) intervals. Abnormal pressure interval is assumed below 8000 ft of original depth before erosion. See text for further explanation (from Magara, 1981). (Reproduced by permission of the American Association of Petroleum Geologists.)

(see eqn. (9.7)), so that the net change in fracture pressure gradient may not be great.

Figure 9.17 shows that the gradient is relatively unchanged before and after erosion above 8000 ft and below 14 000 ft. However, the 10 000–12 000 ft section which is in the uppermost part of the original abnormal pressure zone shows reduction in fracture pressure gradient. Even in this section, the horizontal stress increased after erosion, but the fluid pressure drop was so large for this depth that the fracture pressure gradient dropped also. Fracture generation would result if this section were connected to deep high pressured sections through faults or continuous aquifers.

The values after erosion (Figs. 9.16 and 9.17) are shown at the original depth point for better comparison of values at different erosional stages. However, after the erosion, the depth of burial will, of course, be reduced by the thickness of erosion, so that the new values must be plotted at the shallow depth point after the erosion. In other words, a value of the horizontal stress at 10 000 ft of the original depth, for example, must move to 8000 ft after the erosion of a 2000-ft section. Figures 9.16 and 9.17 do not show this burial depth change.

POSSIBLE FRACTURING DUE TO RELAXATION OF STRESS AFTER EROSION

If an erosional event removes a very thick interval of sedimentary column, the horizontal stress will be reduced to a very low value. This is particularly true for a hydrostatically pressured section. However, in an abnormally pressured section, the horizontal stress would increase with erosion (Fig. 9.16).

The tendency of increase of horizontal stress with burial for the hydrostatic section is shown by the curved line (1) in Fig. 9.18. This trend is shown by a curved line because of continuous increase in Poisson's ratio with burial.

Once the area experiences significant erosional period, the horizontal stress of a hydrostatically pressured and deeper section would be reduced at a rate shown by line (2), which is almost tangential to that part of curve (1). In this projection by line (2), we assume that shale compaction and lithification is irreversible and Poisson's ratio is also irreversible. If this assumption is valid, the horizontal stress may become negative, when the thickness of erosion reaches approximately two-thirds of the

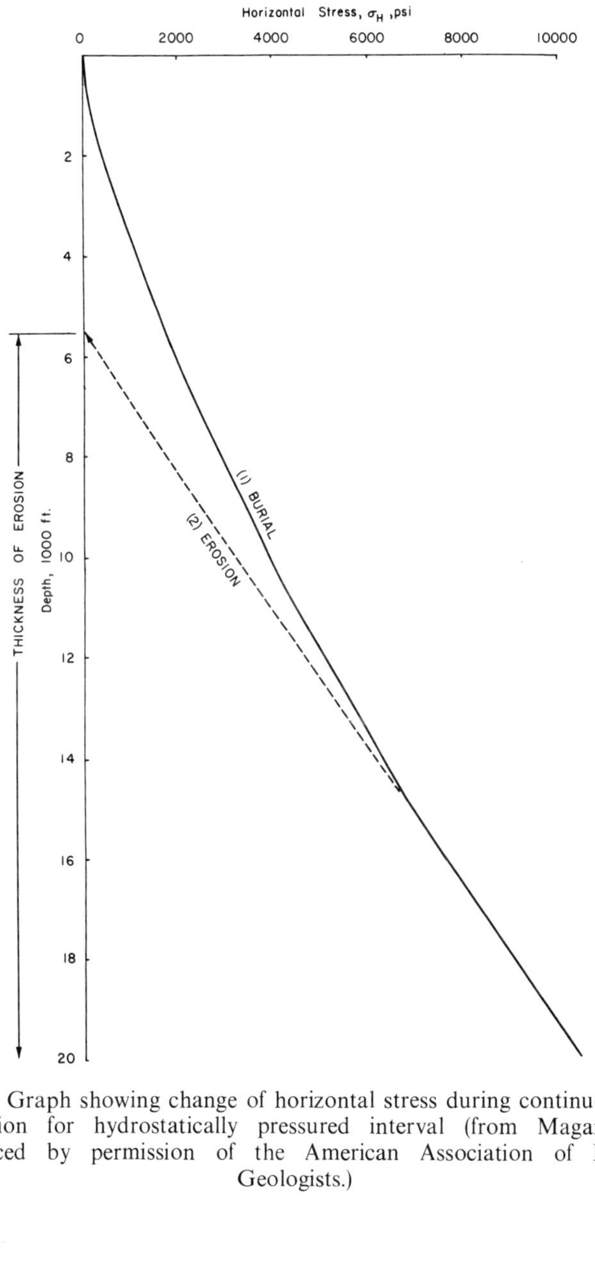

FIG. 9.18. Graph showing change of horizontal stress during continuous burial and erosion for hydrostatically pressured interval (from Magara, 1981). (Reproduced by permission of the American Association of Petroleum Geologists.)

total burial depth before erosion (see the arrow on line (2) in Fig. 9.18). As pointed out by Phillips (1948), if Poisson's ratio is decreased by the reduction of overburden pressure, the level of negative horizontal stress may be attained at a much later stage of erosion. In any case, if the negative horizontal stress is developed, sedimentary rocks will be greatly relaxed, possibly causing 'open' fractures.

In addition to the mechanisms of fracture generation already mentioned, the generation of hydrocarbons, particularly of natural gas, has been proposed as an agent for fracturing (Hedberg, 1974). Quantitative estimate of such effect in sedimentary rocks is not possible at the present time.

SIGNIFICANCE OF FRACTURING IN PETROLEUM MIGRATION AND ACCUMULATION

Natural fracturing seems to have been quite important in migration and accumulation of petroleum. Fractures could have created additional avenues for both primary and secondary migration of petroleum.

Fracturing in a reservoir section could have created porosity and permeability additional to intergranular porosity and permeability, thus improving the reservoir's productivity. This type of fracturing and the associated fluid migration may also cause some secondary porosity by leaching grains and cement (Schmidt and McDonald, 1979).

The only negative aspect of fracturing is potential loss of hydrocarbons through fractures above their accumulation. As pointed out before, in order to maintain material (fluid) balance during continuous burial and compaction, water at least must be able to move out of the crest of an anticline upward, possibly through fractures. However, it is important that hydrocarbons remain behind.

If fracture sizes are relatively small, such as those termed 'microfractures', then their capillary restrictions may be large enough to retain most hydrocarbons in a separate phase while most water can escape upward. If pressure seals are developed above a hydrocarbon accumulation, they may tend to shut some fractures almost instantaneously, geologically speaking, after some loss of water. This mechanism may be analogous to that of a puncture-proof tyre. Therefore, the presence of pressure seals may also help in preserving more hydrocarbons, while water moves upward intermittently.

SUBSURFACE STRESS CONDITIONS FOR NORMAL, WRENCH, AND THRUST FAULTING

Now let us discuss faulting in the subsurface. As described previously, the horizontal expansion in a confined sedimentary basin is restricted; that is $e_y = e_x = 0$. Hence, the horizontal stress, σ_h, is given by (see eqn. (9.5)):

$$\sigma_h = \sigma_y = \sigma_x = \frac{\mu}{1-\mu} \cdot \sigma_z \quad (9.8)$$

If an additional horizontal tectonic stress, C_y, is applied along the y axis of the established stress field, the new stresses, σ'_y and σ'_x, in the y and x directions can be given as follows (Price, 1959):

$$\sigma'_y = \frac{\mu}{1-\mu} \cdot \sigma_z + C_y \quad (9.9)$$

$$\sigma'_x = \frac{\mu}{1-\mu} \cdot \sigma_z + \mu \cdot C_y \quad (9.10)$$

Because Poisson's ratio of the rocks in the subsurface ranges from about 0·25 to 0·50, the value σ'_y is always greater than σ'_x.

The physical conditions of thrust, wrench, and normal faults may be expressed as follows:

Thrust: $\sigma'_x \geq \sigma_z$ (9.11)

Wrench: $\sigma'_y \geq \sigma_z \geq \sigma'_x$ (9.12)

Normal: $\sigma_z \geq \sigma'_y$ (9.13)

By combining eqns. (9.9) and (9.10) with eqns. (9.11), (9.12), and (9.13), we obtain:

Thrust: $C_y \geq \frac{1-2\mu}{\mu(1-\mu)} \cdot \sigma_z$ (9.14)

Wrench: $\frac{1-2\mu}{\mu(1-\mu)} \cdot \sigma_z \leq C_y \leq \frac{1-2\mu}{\mu(1-\mu)} \cdot \sigma_z$ (9.15)

Normal: $C_y \leq \frac{1-2\mu}{1-\mu} \cdot \sigma_z$ (9.16)

Although the stress conditions described above are requirements for faulting, the rocks may not slide if the internal friction of the rocks exceeds the stresses shown above. Hubbert and Rubey (1959) stated that the slippage along any internal plane in a rock should occur when the shear stress, τ, along the plane reaches a critical value. Using the Mohr–Coulomb law, they expressed this as

$$\tau = \tau_0 + \sigma \cdot \tan \phi \tag{9.17}$$

where σ = normal stress, τ_0 shear strength of the rock when σ is zero, and ϕ angle of internal friction, which has an average value of about 30° for a wide variety of rocks (Hubbert and Rubey, 1959).

Hubbert and Rubey further stated that 'once a fracture is started, τ_0 is eliminated', and therefore a simpler equation can be used:

$$\tau = \sigma \cdot \tan \phi = (S - p)\tan \phi \tag{9.18}$$

where S is overburden pressure and p fluid pressure.

If the subsurface stress exceeds the shear stress shown by eqn. (9.18) after generation of fractures and also satisfies one of the conditions expressed by eqns. (9.14), (9.15), and (9.16), a fault should occur.

APPLICATION OF CONCEPTS IN A SEDIMENTARY BASIN

As stated previously, on the basis of Costley's (1967) fracture pressure data in the Gulf Coast area, Eaton (1969) estimated the in-situ values of Poisson's ratio under the subsurface, shown in Fig. 9.6. This curve also indicates a tendency of the ratio to increase with vertical stress or depth.

The change of vertical (effective) stress, σ_z, can be estimated from the integrated bulk density or overburden pressure, S (Eaton, 1969), and fluid pressure, p, using the following equation,

$$\sigma_z = S - p \tag{9.6}$$

The vertical (effective) stress, when the fluid pressure gradient is hydrostatic (0·465 psi/ft), is shown by a solid curve in Fig. 9.19.

If the fluid pressure is abnormally high, the vertical (effective) stress is abnormally low under a given overburden pressure (eqn. (9.6)). A series of dashed curved lines in Fig. 9.19 shows the vertical stress when aquathermally generated fluid pressure starts at 8000 ft. 10 000 ft, and so on.

Using the data shown in Figs. 9.6 and 9.19 and eqn. (9.8), the horizontal stress, $\sigma_h (= \sigma_x = \sigma_y)$, when there is no superimposed tectonic stress, was also estimated (Fig. 9.7). The solid curved line in this figure indicates the horizontal stress when pressure is hydrostatic, and a series of dashed curved lines show the stress when the aquathermal fluid pressure exists below 8000 ft, 10 000 ft, and so on. The horizontal stress

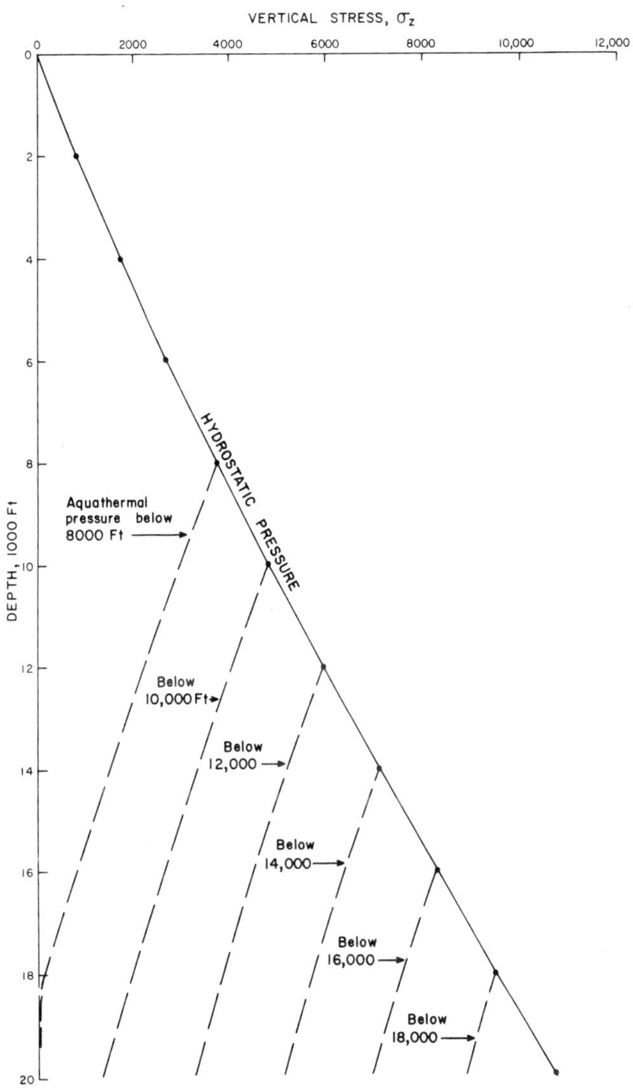

FIG. 9.19. Relationship between calculated vertical (effective) stress and depth for hydrostatically pressured interval (shown by solid line) and for abnormally pressured intervals (shown by dashed lines) in Gulf Coast. Dashed lines indicate cases where aquathermally generated (abnormal) pressure exists below 8000, 10 000,... and 18 000 ft, respectively (from Magara, 1983). (Reproduced by permission of the *Journal of Petroleum Geology*, Scientific Press, England.)

as well as the vertical stress is extremely small at depth, if the aquathermally generated pressure starts at a shallow depth.

Based on the data of vertical stress (Fig. 9.19) and of Poisson's ratio (Fig. 9.6), and on eqns. (9.14), (9.15), and (9.16), we are able to estimate the tectonic stress, C_y, which could cause thrust, wrench, or normal faulting in the subsurface. The estimate depends on the fluid pressure as well.

Three zones indicated in Fig. 9.20 provide the estimated tectonic stress requirements to cause thrust, wrench, or normal faulting in a sedimentary basin similar to the Gulf Coast basin. Unlike the Gulf Coast basin, it is assumed in this case that the fluid pressure is normal-hydrostatic from the surface to 20 000 ft.

The tectonic stress required to cause a thrust fault increases with depth in the interval from the surface to about 9000 ft, and declines signi-

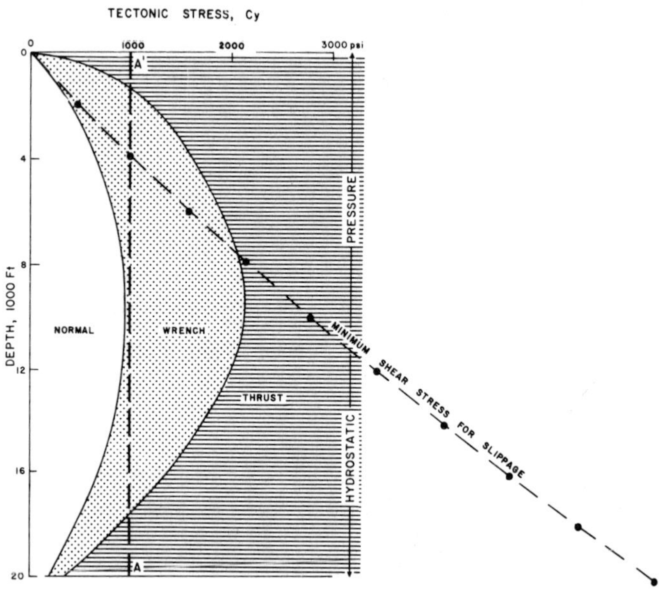

FIG. 9.20. Relationships between tectonic stress and depth for causing thrust, wrench, and normal faults, where entire sedimentary section has hydrostatic fluid pressure. Thin dashed line shows minimum shear stress for slippage. Thick vertical dashed line (A–A′) at 1000 psi shows an example of tectonic stress (from Magara, 1983). (Reproduced by permission of the *Journal of Petroleum Geology*, Scientific Press, England.)

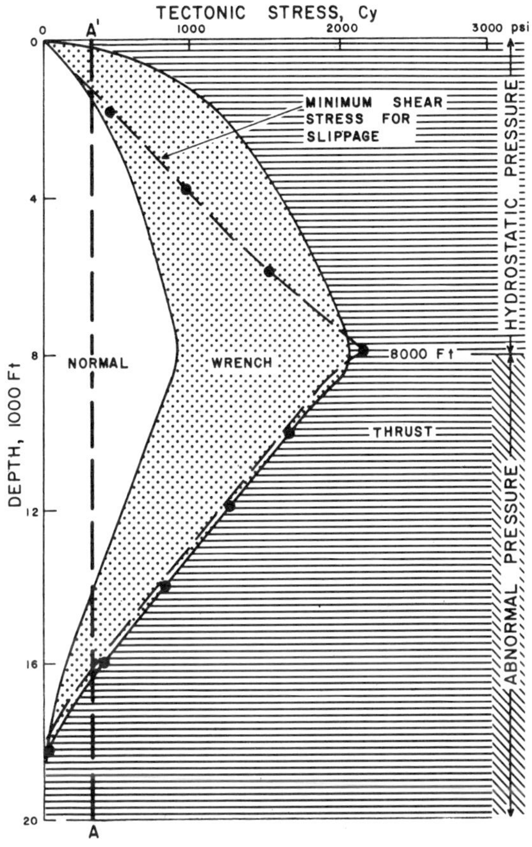

FIG. 9.21. Relationships between tectonic stress and depth for causing thrust, wrench, and normal faults, where hydrostatic pressure exists above 8000 ft and aquathermally generated (abnormal) fluid pressure below 8000 ft. Thin dashed line shows minimum shear stress for slippage. Thick vertical dashed line (A–A′) at 400 psi shows an example of tectonic stress (from Magara, 1983). (Reproduced by permission of the *Journal of Petroleum Geology*, Scientific Press, England.)

ficantly in the interval below 9000 ft. This is primarily because the value $(1-2\mu)/(\mu[1-\mu]).\sigma_z$ in eqn. (9.14) decreases in the deeper zone as the Poisson's ratio, μ, increases from about 0·25 to almost 0·5, even if the vertical stress, σ_z, continuously increases with depth in the hydrostatic pressure zone.

However, in this case, the critical shear stress to cause a slippage increases continuously with depth, as the effective (vertical) stress increases. The curved dashed line in Fig. 9.20 shows the estimated minimum shear stress for slippage, when the fluid pressure is hydrostatic from the surface to 20 000 ft. This figure suggests that a tremendous tectonic force is necessary to cause thrust faults in this hydrostatic case, because the internal friction of sedimentary rocks at depth is extremely high.

If a tectonic stress shown by line A–A′ (1000 psi) is applied, there seems to be an apparent chance of thrusting below 18 000 ft on the basis of the triaxial stress conditions. However, such thrusting is not possible based on the estimate of the minimum shear stress for slippage. The shear stress there is about six times as large as the tectonic stress (Fig. 9.20).

It may be concluded that thrusting is a very difficult process when the fluid pressure is hydrostatic and the sediment compaction is normal. Above 4000 ft, however, where the value shown by line A–A′ exceeds the minimum shear stress for slippage, there is a chance of thrusting and wrenching (Fig. 9.20).

If, on the other hand, the aquathermally generated pressure begins at a shallow depth (e.g. 8000 ft), the tectonic stress conditions for thrust, wrench, or normal faulting will be changed (Fig. 9.21). A more striking difference is that of the minimum shear stress for slippage (eqn. (9.18)) between the hydrostatic and abnormal pressure zones. The shear stress in the latter zone is abnormally low because the effective stress is abnormally low (eqn. (9.18) and Fig. 9.21). This suggests that a relatively small amount of horizontal tectonic stress would be sufficient to cause a thrust fault at depth (e.g. 20 000 ft), overcoming the internal friction there, if the aquathermally generated fluid pressure starts at a shallow depth (e.g. 8000 ft).

If, for example, a tectonic stress shown by line A–A′ (400 psi) is applied (Fig. 9.21), there would be a chance of thrusting below 18 000 ft, and also of thrusting and wrenching above 2000 ft.

Figure 9.22 summarizes the results of calculation where the aquathermal pressure starts at 8000 ft (A); 10 000 ft (B); 12 000 ft (C); 14 000 ft (D); 16 000 ft (E); 18 000 ft (F); and 20 000 ft (G), respectively. Each diagram in Fig. 9.22 shows the stress conditions for the three types of faulting, along with the estimated minimum shear stress for slippage. This figure suggests that chances of faulting, particularly of thrust faulting, increase at depth as the top of the aquathermally generated abnormal-pressure zone moves upward.

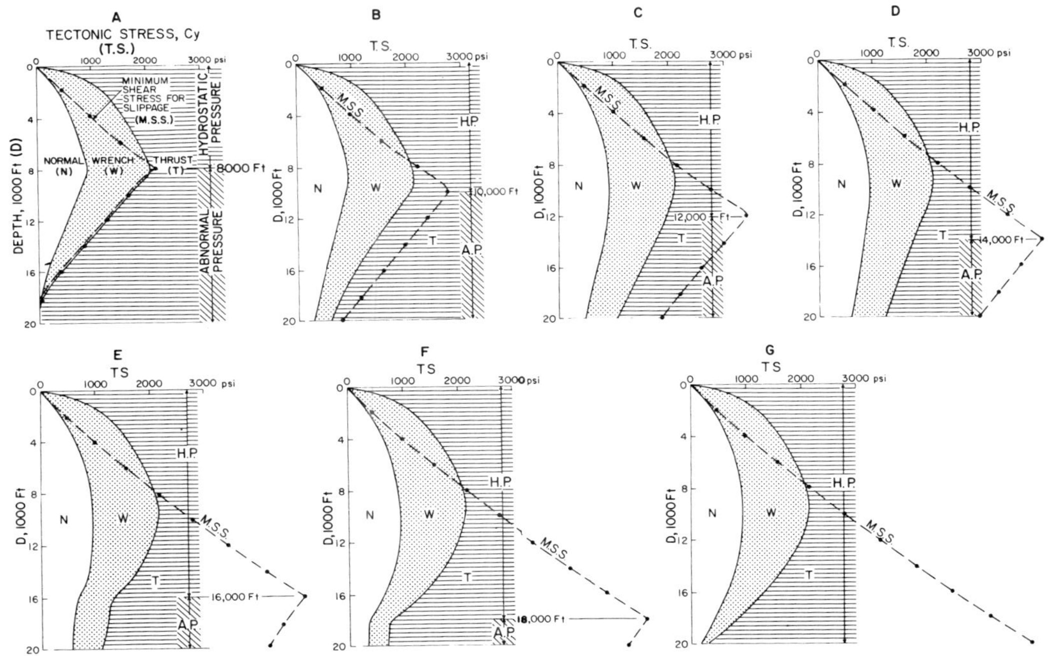

FIG. 9.22. Relationships between tectonic stress and depth for causing thrust, wrench, and normal faults with changing top depth of abnormal pressure zone; A at 8000 ft, B at 10 000 ft, C at 12 000 ft, D at 14 000 ft, E at 16 000 ft, F at 18 000 ft, and G at 20 000 ft (at G, entire zone is normally pressured) (from Magara, 1983). (Reproduced by permission of the *Journal of Petroleum Geology*, Scientific Press, England.)

In fact, in the case of Fig. 9.22A, where the abnormal pressure starts at 8000 ft, the minimum shear stress becomes zero at 20 000 ft. This suggests that the rocks there are quite unstable and mobile. The probability of any kind of faulting and sliding would be extremely high at this depth. Both Barker (1972) and Magara (1975) suggested the possibility of an extremely unstable condition of shales and of shale diapirism there.

REFERENCES

Barker, C. (1972) Aquathermal pressuring—role of temperature in development of abnormal-pressure zone. *A.A.P.G.*, **56**, 2068–71.

Clark, S. P., Jr. (1966) Handbook of physical constants (revised edition). Geological Society of America, Memoir 97, 587 p.

Costley, R. D. (1967) Hazards and costs cut by planned drilling programs. *World Oil*, October, 92.

Eaton, B. A. (1969) Fracture gradient prediction and its application in oil field operations. *J. Petrol. Technol.*, **21**, 1353–60.

Evans, C. R., McIvor, D. K. and Magara, K. (1975) Organic matter, compaction history, and hydrocarbon occurrence—Mackenzie Delta, Canada. *Proc. 9th World Petroleum Congress*, Vol. 3. p. 149–57.

Hedberg, H. D. (1974) Relation of methane generation to undercompacted shales, shale diapirs, and mud volcanoes. *Am. Assoc. Petroleum Geologists Bull.*, **58**, 661–73.

Hottman, C. E. and Johnson, R. K. (1965) Estimation of formation pressures from log-derived shale properties. *J. Petrol. Technol.*, **17**, 717–22.

Hubbert, M. K. and Rubey, W. W. (1959) Role of fluid pressure in mechanics of overthrust faulting, *Geol. Soc. Amer. Bull.*, **70**, 115–206.

Hubbert, M. K. and Willis, D. G. (eds.) (1957) Mechanics of hydraulic fracturing. *Transactions of AIME*, **210**, 153–66.

Magara, K. (1975) Importance of aquathermal pressuring effect in Gulf Coast. *Am. Assoc. Petroleum Geologists Bull.*, **59**, 2037–45.

Magara, K. (1976) Thickness of removed sedimentary rocks, paleo pore pressure, and paleotemperature, southwestern part of Western Canada Basin. *Am. Assoc. Petroleum Geologists Bull.*, **60**, 554–65.

Magara, K. (1977) A theory relating isopachs to paleo compaction-water-movement in a sedimentary basin. *Bull. Canadian Petroleum Geology*, **25**, 195–207.

Magara, K. (1978) *Compaction and Fluid Migration—Practical Petroleum Geology*, Elsevier, Amsterdam, 319 p.

Magara, K. (1981) Mechanisms of natural fracturing in a sedimentary basin. *Am. Assoc. Petroleum Geologists Bull.* **65**, 123–32.

Magara, K. (1983) Physical constraints of faulting in a geopressured sedimentary basin. *Petrol. Geology J.*, **5**, 417–26.

Phillips, D. W. (1948) Tectonics of mining. Sheffield University Mining Magazine.

Price, N. J. (1959) Mechanics of jointing in rocks. *Geology Magazine*, **16**, 149–67.

Schmidt, V. and McDonald, D. A. (1979) Role of secondary porosity in the course of sandstone diagenesis, Society of Economic Paleontologists and Mineralogists Special Publication No. 26, pp. 175–207.

Terzaghi, K. and Peck, R. B. (1948) *Soil Mechanics in Engineering Practice*, Wiley, New York, 566 p.

Chapter 10

Petroleum Generation and Migration in the Arabian Gulf Region

INTRODUCTION

As described in Chapter 1, there are two different types of water moving in a sedimentary basin; meteoric water and sediment-source (or compaction) water.

Out of the six items listed in Chapter 1 as the specific characteristics of each type, items 5 and 6 are of particular interest to many petroleum geologists.

(a) Meteoric water

5. Movement of this type of water is probably unimportant in primary migration, but it may affect the trapping condition of hydrocarbons in a pool. Meteoric water could alter the quality of hydrocarbons by biodegradation and water washing.
6. Movement of this type of water is a present event and may or may not have developed in the geological past.

(b) Sediment-source water

5. Movement of this type of water is probably important in the primary migration of hydrocarbons.
6. Most such movement occurred in the geological past.

In summary, the movement of the sediment-source water is mostly a past event and probably essential in understanding primary hydrocarbon migration, whereas that of the meteoric water persisted for a relatively short geological period near present, and may have affected late secondary migration and entrapment conditions of hydrocarbons.

For the Arabian Gulf region, Chiarelli (1973) reported the present

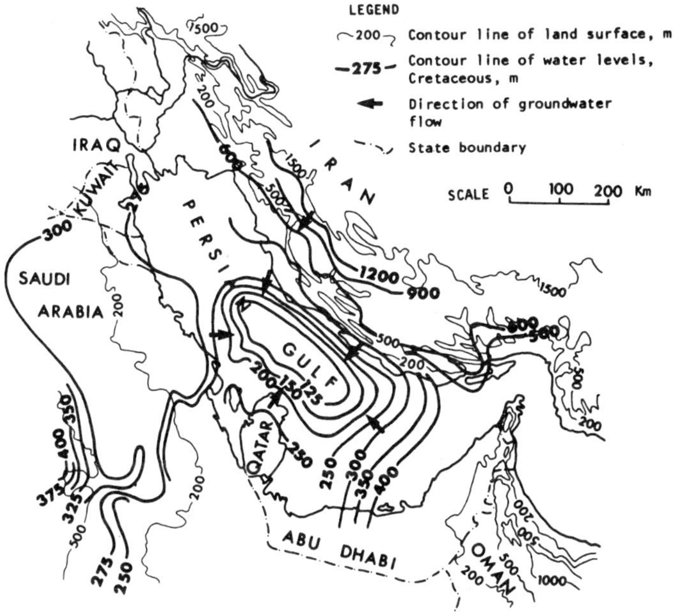

FIG. 10.1. Topography, potentiometric surface and lateral flow directions of ground water in Cretaceous formations, Arabian Gulf region (Chiarelli, 1973). (Reproduced by permission of the author.)

potentiometric surface and inferred directions of movement of the latter water in the Cretaceous formations (Fig. 10.1). No major work on the sediment-source (or compaction) water has so far been reported.

The purpose of this chapter is to present the result of the analysis of the sediment-source water movement in an attempt to understand the primary hydrocarbon migration of the region which took place in the geological past.

EXCESS FLUID PRESSURE DUE TO SEDIMENT LOADING

Let us suppose a sediment sequence in which sediment has reached a compaction equilibrium and within which the fluid pressure is hydrostatic (Stage A in Fig. 10.2). Such a compaction equilibrium for shales has been established in many sedimentary basins. According to Hubbert and Rubey (1959) who studied the Paleozoic shales of Oklahoma

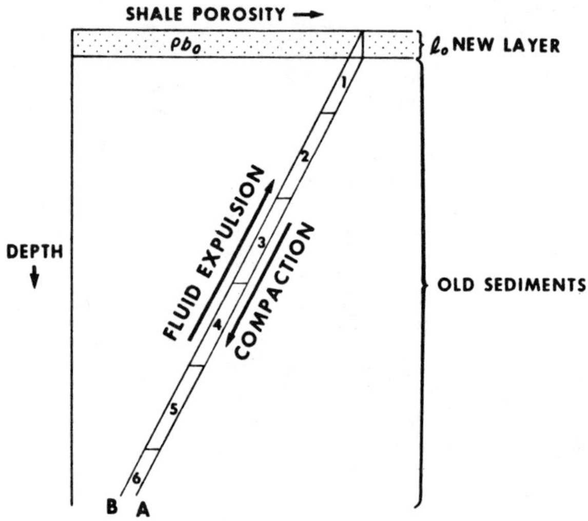

FIG. 10.2. Schematic diagram showing stepwise compaction of shales and fluid expulsion due to loading of new layer (Magara, 1978*a*).

analysed by Athy (1930), the compaction-equilibrium condition can be shown by an exponential function which is depicted by a straight line on semi-log paper (porosity-logarithmic, and depth-arithmetic, Stage A in Fig. 10.2).

Additional sediments, whose thickness is l_0, are deposited above the sequence under water. If the entire sequence reaches a new equilibrium condition of compaction within the time interval t, a porosity distribution such as shown by Stage B in Fig. 10.2 should be established.

Suppose that the outlet for fluid expulsion exists only at the surface, and the fluid is expelled vertically upward. Compaction of the sediment sequence from Stage A to Stage B in this case occurs from the shallower part to the deeper part in more or less a stepwise manner (Fig. 10.2).

The increase of fluid pressure (p) due to the instantaneous loading of a new sediment whose thickness is l_0 (original thickness before compaction) is given as,

$$p = \rho_{b_0} \cdot g \cdot l_0 \tag{10.1}$$

where ρ_{b_0} is water-saturated original bulk density (before compaction) of the new sediment layer l_0, and g the acceleration due to gravity.

When the fluid-pressure increase is resolved into two components (Hubbert and Rubey, 1959), we obtain

$$p = p_n + p_a \tag{10.2}$$

where p_n is normal or hydrostatic pressure increase, and p_a excess-pressure increase. The value p_n in this case is shown as,

$$p_n = \rho_w \cdot g \cdot l_0 \tag{10.3}$$

Rearranging eqn. (10.2) and introducing eqns. (10.1) and (10.3) into it, we obtain,

$$p_a = (\rho_{b_0} - \rho_w) \cdot g \cdot l_0 \tag{10.4}$$

If the thickness of the new sediment layer is not constant but changes at the rate of dl_0/dx horizontally (see Fig. 10.3), the rate of the horizontal excess pressure, which controls the horizontal fluid migration, is governed by the rate of horizontal thickness change of the new sediment layer. Most sediments in the subsurface are, however, not new but have been compacted, so that correction for the compaction effect must be applied if the horizontal fluid migration which took place in the geological past is being considered.

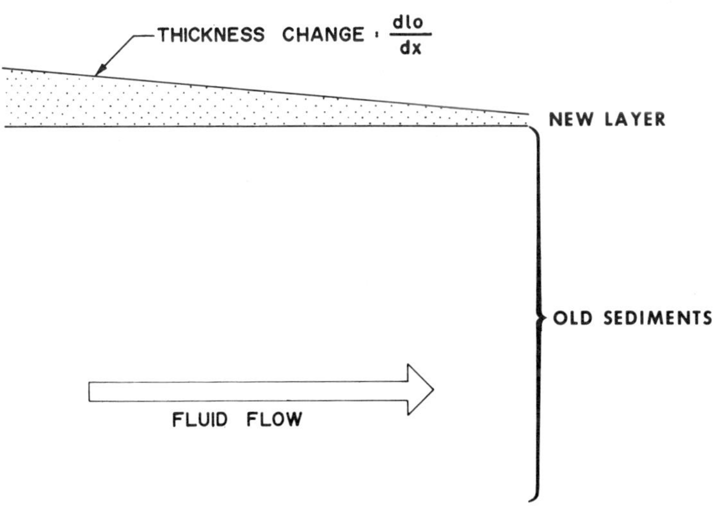

FIG. 10.3. Schematic diagram showing direction of fluid flow due to wedge-shaped sedimentary loading.

The relationship between the thickness, l_0, of a layer at the time of deposition (before compaction) and the thickness, l, at present (after compaction) is

$$l_0(1-\phi_0)=l\cdot(1-\phi) \tag{10.5}$$

where ϕ_0 = porosity at the time of deposition (before compaction), and ϕ = porosity at present (after compaction).

The porosity can be expressed in terms of density as,

$$\phi_0=\frac{\rho_m-\rho_{b_0}}{\rho_m-\rho_w} \tag{10.6}$$

and

$$\phi=\frac{\rho_m-\rho_b}{\rho_m-\rho_w} \tag{10.7}$$

where ρ_b is bulk density at present (after compaction), and ρ_m matrix (or grain) density of the sediment.

Introducing eqns. (10.6) and (10.7) into (10.5), we obtain

$$(\rho_{b_0}-\rho_w)\cdot l_0=(\rho_b-\rho_w)\cdot l \tag{10.8}$$

or

$$(\rho_{b_0}-\rho_w)\frac{dl_0}{dx}=(\rho_b-\rho_w)\cdot\frac{dl}{dx} \tag{10.9}$$

From eqns. (10.4) and (10.9), we obtain

$$\frac{dp_a}{dx}=(\rho_b-\rho_w)\cdot g\cdot\frac{dl}{dx} \tag{10.10}$$

An advantage of using eqn. (10.10) over eqn. (10.4) is that the horizontal fluid migration which occurred in the geological past can be estimated directly from the thickness change of compacted sediment layer (dl/dx).

The above discussion is based on the fluid pressure increase due to loading (hydraulic pressure increase). If the geothermal gradient is constant in a study area, the aquathermal pressure increase (Barker, 1972; Magara, 1974) would be almost proportional to the fluid pressure increase due to loading. In other words, aquathermal fluid migration can also be inferred from the study of sediment loading.

SIGNIFICANCE OF HORIZONTAL FLUID MIGRATION FOR HYDROCARBON ACCUMULATION

During the stages of secondary hydrocarbon migration in a reservoir, globules of oil or gas would be able to move from a structurally lower point to a higher, if the buoyancy is high enough to overcome the capillary restrictions of the reservoir rock. In other words, formation of an oil or gas accumulation is more or less a matter of structural development or trap formation during the secondary migration stages.

During the primary migration stages, however, buoyancy is usually considered to be insufficient to overcome relatively high capillary pressure of most fine-grained source rocks. Presence of structured water and both hydraulic and aquathermal forces may push oil or gas toward a trapping position (Dickey, 1975; Magara, 1978*a*, 1981). The structured water which is highly viscous or semi-solid is of particular importance for hydrocarbons to overcome relative permeability by pore water and capillary pressure in the source rock.

Under such a condition, the concentration of oil or gas may be possible if there are strong horizontal hydraulic and aquathermal forces related to sediment loading.

GENERATION OF EXCESS FLUID PRESSURE IN ARABIAN GULF REGION

Using the isopach maps of this region constructed by Kamen-Kaye (1970), excess fluid pressures due to sediment loading were estimated. The graphs shown in Figs. 10.4 and 10.5 depict changing average rate of excess fluid pressure increase (psi/MY (million years)) for two selected locations (A and B shown in Fig. 10.6).

At location A in the southeastern part of the region, for example, the rate gradually increased throughout the geologic period since Permian. At location B which falls in Iran, on the other hand, the average rate fluctuated in the geologic past and reached its highest value (about 60 psi/MY) during Tertiary.

Dashed curved lines in Figs. 10.4 and 10.5 show the estimated vitrinite reflectance values, R_0, at the bases of the Jurassic (J), Cretaceous (K), and Tertiary (T) formations (for this method, refer to Waples, 1980). According to the estimated reflectance values, the basal Jurassic and Cretaceous formations reached their early stages of petroleum matu-

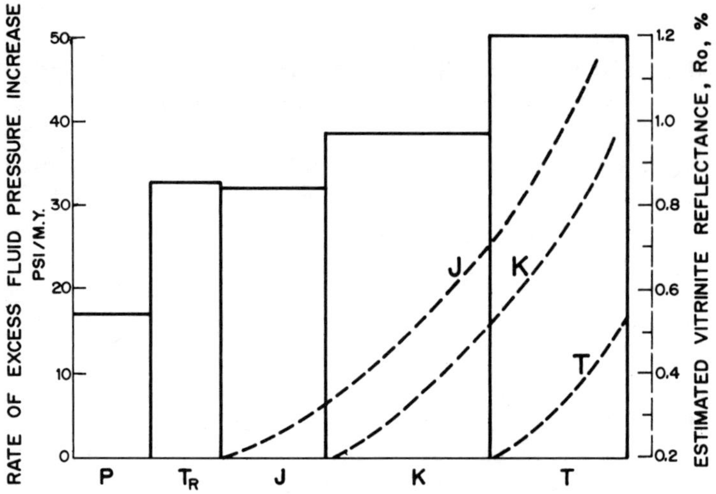

FIG. 10.4. Average rate of excess fluid pressure increase (psi/MY (million years)) for location A shown in Fig. 10.6. Dashed curved lines show estimated vitrinite reflectance, R_0, at bases of Jurassic (J), Cretaceous (K), and Tertiary (T) formations.

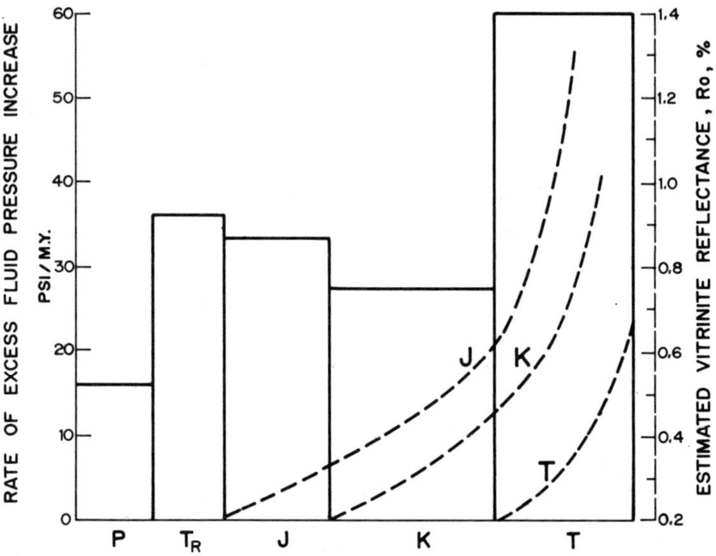

FIG. 10.5. Average rate of excess fluid pressure increase (psi/MY (million years)) for location B shown in Fig. 10.6. Dashed curved lines show estimated vitrinite reflectance, R_0, at bases of Jurassic (J), Cretaceous (K), and Tertiary (T) formations.

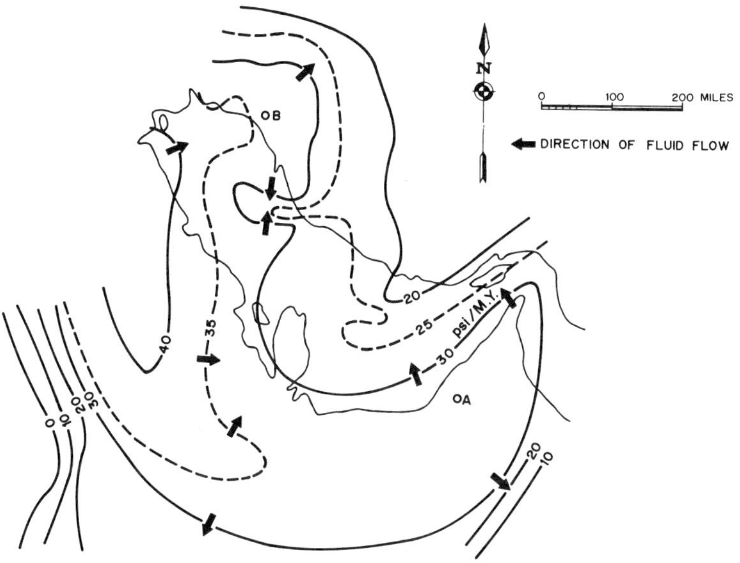

FIG. 10.6. Contour map showing calculated rate of excess fluid pressure increase (psi/MY (million years)) due to loading of Jurassic formations. Arrows show inferred directions of fluid flow.

ration ($R_0 = 0.5\%$) by the end of Cretaceous at these two locations (A and B). Primary hydrocarbon migration would have taken place at that time if the conditions described above existed in this region.

Figures 10.6, 10.7 and 10.8 show the calculated rates of fluid pressure increase (psi/MY) due to loading of the Jurassic, Cretaceous, and Tertiary formations, respectively. Arrows in these figures show the inferred directions of fluid movement.

As stated previously, for the formation of an effective oil/gas accumulation, the horizontal change of the excess fluid pressure increase (psi/MY/mile), which is the first derivative of the values shown in Figs. 10.6, 10.7 and 10.8, may be one of the controlling factors.

Figures 10.9, 10.10, and 10.11 depict the maps of the horizontal changes of the excess fluid pressure increase for the Jurassic, Cretaceous, and Tertiary formations, respectively. An interesting observation is that there are relatively low values north of the Qatar Peninsula during the Cretaceous time (see C in Fig. 10.10). The horizontal changes of the excess fluid pressure in the surrounding areas were relatively large,

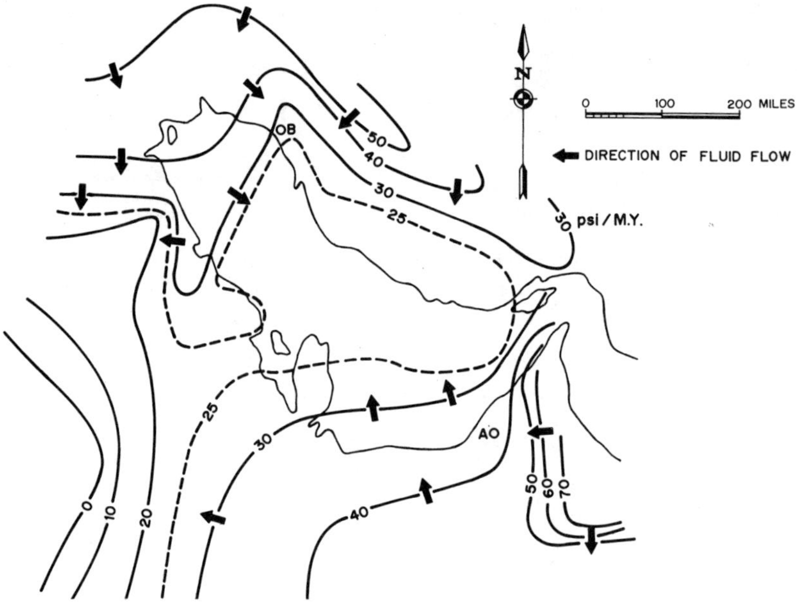

FIG. 10.7. Contour map showing calculated rate of excess fluid pressure increase (psi/MY (million years)) due to loading of Cretaceous formations. Arrows show inferred directions of fluid flow.

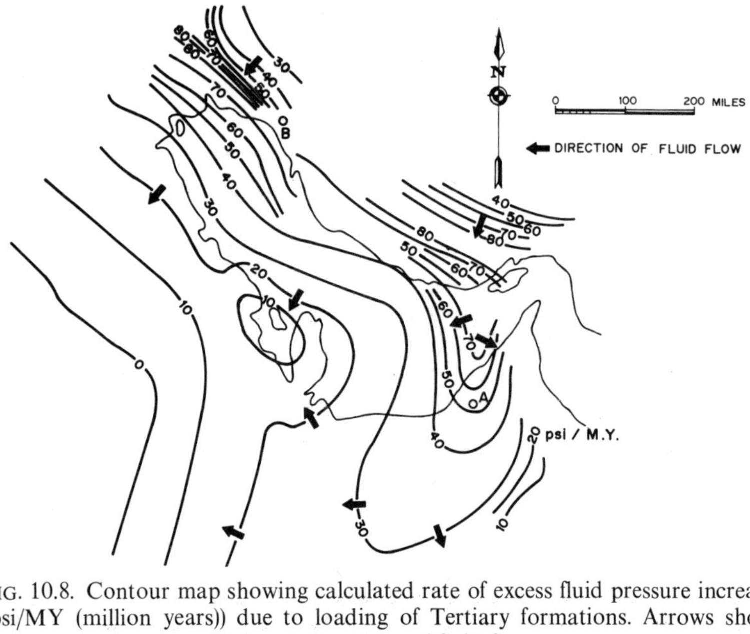

FIG. 10.8. Contour map showing calculated rate of excess fluid pressure increase (psi/MY (million years)) due to loading of Tertiary formations. Arrows show inferred directions of fluid flow.

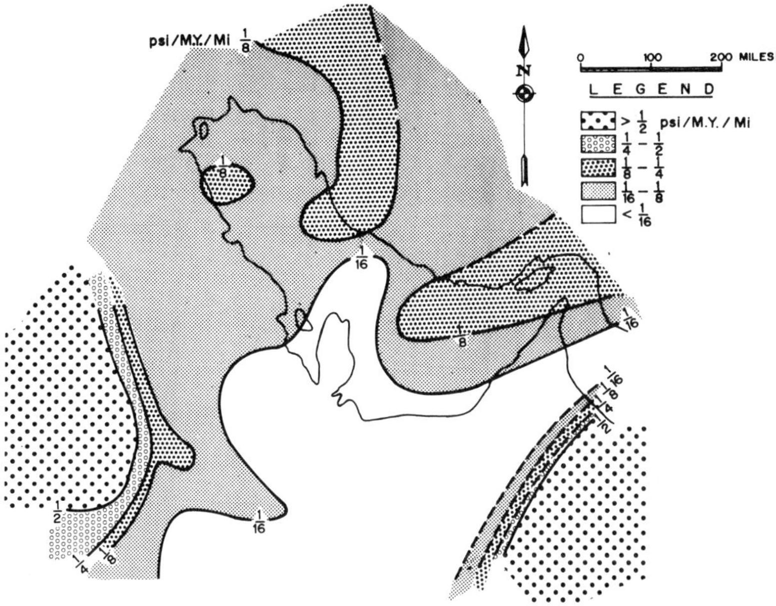

FIG. 10.9. Contour map showing horizontal change of excess fluid pressure increase (psi/MY/mile) for Jurassic formations.

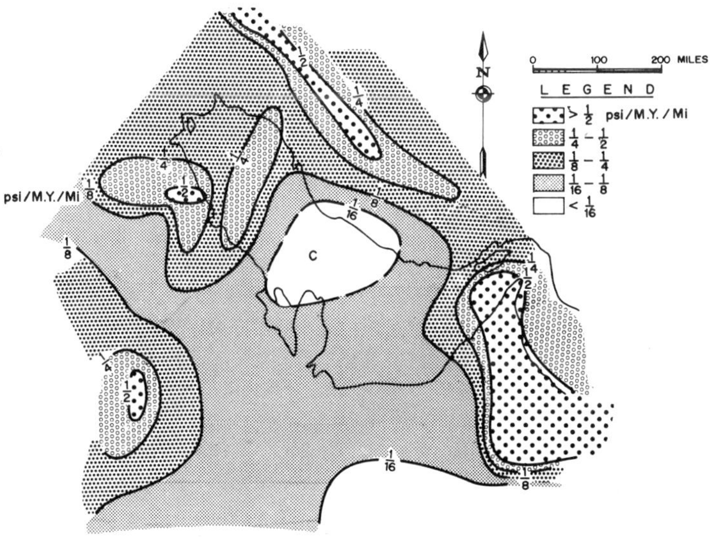

FIG. 10.10. Contour map showing horizontal change of excess fluid pressure increase (psi/MY/mile) for Cretaceous formations.

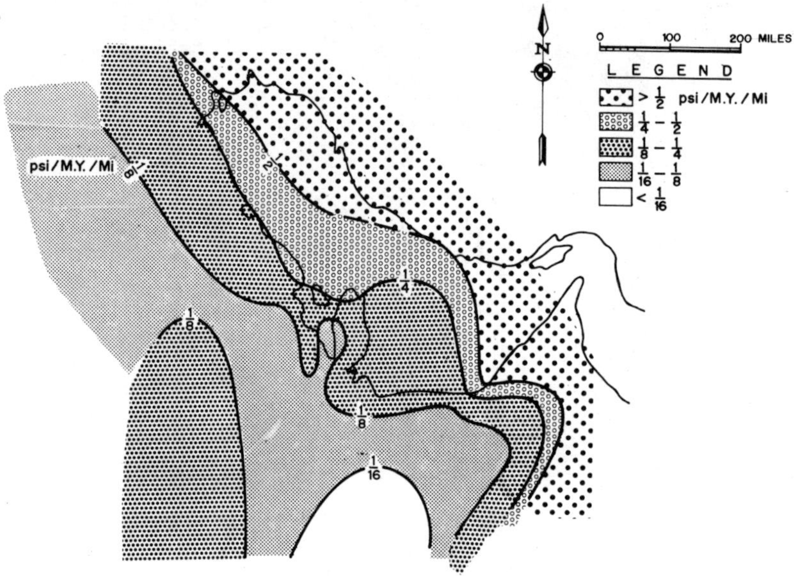

FIG. 10.11. Contour map showing horizontal change of excess fluid pressure increase (psi/MY/mile) for Tertiary formations.

suggesting effective concentrations of the generated oil or gas into traps during the primary migration stages.

Note that the early stages of petroleum maturation for both the basal Jurassic and Cretaceous formations were reached by the end of the Cretaceous period (Figs. 10.4 and 10.5). This is why the horizontal fluid migration during Cretaceous is important for further discussions in this chapter.

PETROLEUM GENERATION, MATURATION, AND MIGRATION IN THE ARABIAN GULF REGION

Based on Kamen-Kaye's (1970) regional isopach maps and Waples' (1980) method of estimating vitrinite reflectance, Sail and Magara (1985) constructed regional maps of hydrocarbon-generation zones. Figure 10.12 shows such a map for the basal Jurassic formations, superimposed with the oil-field distribution map of the Jurassic reservoirs. Figure 10.13

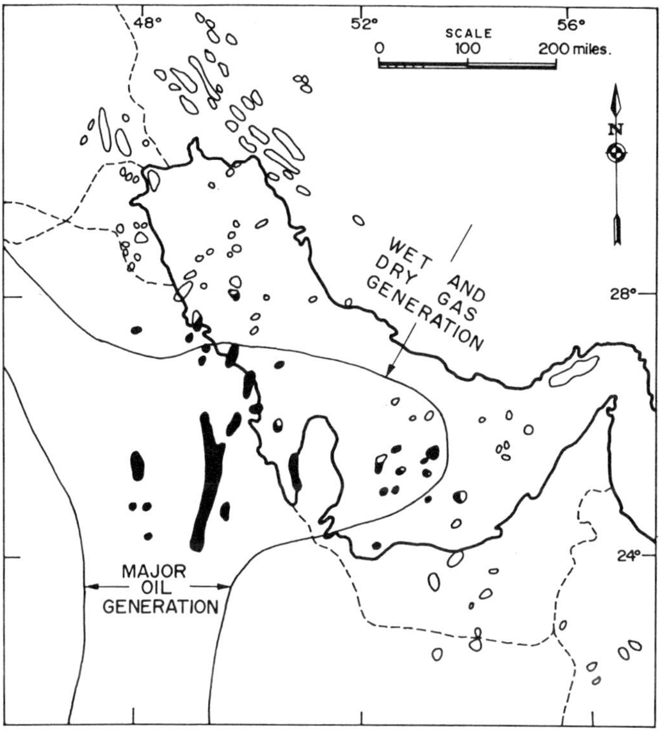

FIG. 10.12. Map showing Jurassic oil generation zones and Jurassic oil fields (solid areas). Open areas show oil fields of other geologic ages.

is a similar composite map of the Cretaceous formations. These maps show excellent matches between the estimated oil-generation zones and the oil-field distributions, except that virtually no oil field was discovered in the area north of the Qatar Peninsula (C in Figs. 10.10 and 10.13).

This area is a northern extension of the Qatar Arch where sedimentation rate was relatively slow. In fact some reservoir sections are missing due to both erosion and non-deposition. The lack of oil fields in this area may be at least partially attributed to this reason. However, there are also some porous sections of both the Jurassic and Cretaceous ages which are water-saturated (Schlumberger, 1981). Relatively weak horizontal fluid movement as shown in Fig. 10.10 may be a reason for the relative scarcity of oil fields in this area.

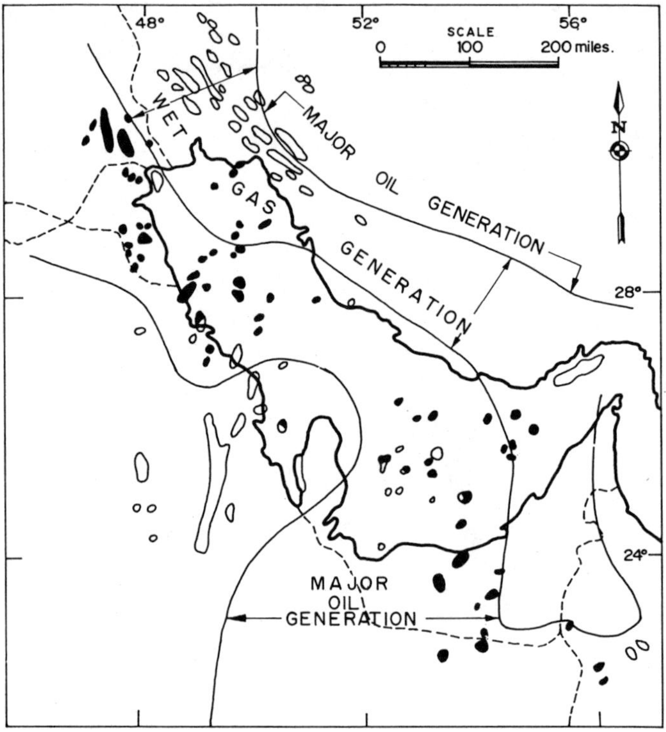

FIG. 10.13. Map showing Cretaceous oil generation zones and Cretaceous oil fields (solid areas). Open areas show oil fields of other geologic ages.

Figure 10.14 shows a similar composite map of the hydrocarbon generation and oil-field distribution for the Tertiary formations. In this case, there is a significant mismatch because most oils in the Tertiary reservoirs were derived from the older source rocks (Young *et al.*, 1977).

REFERENCES

Athy, L. F. (1930) Density, porosity and compaction of sedimentary rocks. *A.A.P.G.*, **14**, 1–24.

Barker, C. (1972) Aquathermal pressuring—role of temperature in development of abnormal-pressure zone. *A.A.P.G.*, **56**, 2068–71.

Chiarelli, A. (1973) Etude des nappes aquiferes profondes. Contribution de

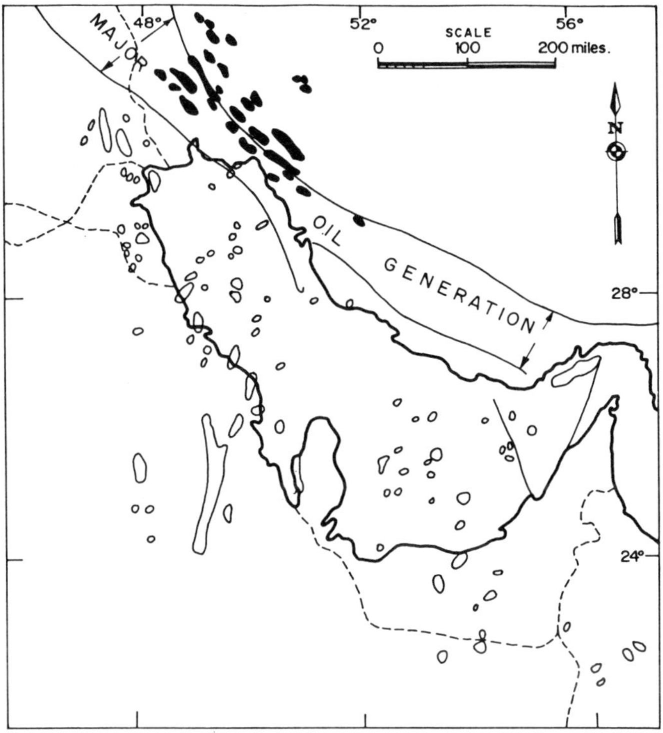

FIG. 10.14. Map showing Tertiary oil generation zones and Tertiary oil fields (solid areas). Open areas show oil fields of other geologic ages.

l'hydro-geologie a la connaissance d'un bassin sedimentaire et a l'exploration petroliere: D.Sc. thesis, no. 401, l'Université de Bordeaux I, Bordeaux, p. 187

Dickey, P. A. (1975) Possible primary migration of oil from source rock in oil phase. *A.A.P.G.*, **59**, 337–45.

Hubbert, M. K. and Rubey, W. W. (1959) Role of fluid pressure in mechanics of overthrust faulting. *Geol. Soc. Amer. Bull.*, **70**, 115–206.

Kamen-Kaye, M. (1970) Geology and productivity of Persian Gulf synclinorium. *A.A.P.G.*, **54**, 2371–94.

Magara, K. (1974) Aquathermal fluid migration. *A.A.P.G.*, **58**, 2513–16.

Magara, K. (1975) Importance of aquathermal pressuring effect in Gulf Coast. *A.A.P.G.*, **59**, 2037–45.

Magara, K. (1978*a*) *Compaction and Fluid Migration—Practical Petroleum Geology*, Elsevier, Amsterdam, 319p.

Magara, K. (1978*b*) The significance of expulsion of water in oil-phase primary migration. *Canadian Petroleum Geology Bull.*, **26**, 123–31.

Sail, I. A. N. and Magara, K. (1985) Petroleum generation and maturation of the Arabian Gulf region (in press), Faculty of Earth Sciences Bulletin, King Abdulaziz University, Jeddah, Saudi Arabia.

Schlumberger (1981) Well Evaluation Conference, United Arab Emirates/Qatar (1. Geology), Schlumberger Middle East S.A., 271 pp.

Waples, D. W. (1980) Time and temperature in petroleum formation: application of Lopatin's method to petroleum exploration. *A.A.P.G.*, **64**, 916–26.

Young, A., Monaghan, P. H. and Schweisberger, R. T. (1977) Calculation of ages of hydrocarbons in oils. *A.A.P.G.*, **61**, 573–600.

Chapter 11

Identification of Sandstone Body Types by the Computer Method

INTRODUCTION

Sandstone body types, such as those of prograding (coarsening upward), aggrading (coarsening downward) and blocky (relatively uniform grain size) sandstones, inferred by the shapes of SP, gamma or resistivity logs, are useful information for interpreting the depositional environments of these sandstones (Saitta and Visher, 1968; Fisher *et al.*, 1969; Visher *et al.*, 1971; Ruoff, 1976). Figure 11.1 illustrates representative examples of these three types: (1) prograding—stream-mouth bar, (2) aggrading—point bar and (3) blocky—beach sandstone shown by gamma log.

The gamma-ray log measures the natural radioactivity of sedimentary rocks. In clastic sequences it usually reflects the clay content, because clays contain relatively high concentrations of radioactive elements. Pure sand or sandstones, and carbonate rocks, usually have very low levels of radioactivity.

Therefore, it is commonly believed that a lower gamma-ray reading means a sandier (or more carbonate) rock. However, such a low gamma-ray reading has theoretically nothing to do with the grain size.

Geologically speaking, it has been known that clastic sediments having low gamma-ray readings are usually rich in coarser grains such as sands and silts.

The SP log may also be used for interpreting sandstone body types.

Interpretation of sandstone body types in the sense of coarsening upward or downward (or cleaning upward or downward) is important from the standpoint of petroleum migration, reservoir quality, and sealing efficiency. I judge that in this regard a prograding sandstone has a better chance of accumulating petroleum than an aggrading sandstone, provided other factors controlling petroleum accumulations are the same. My reasons are as follows.

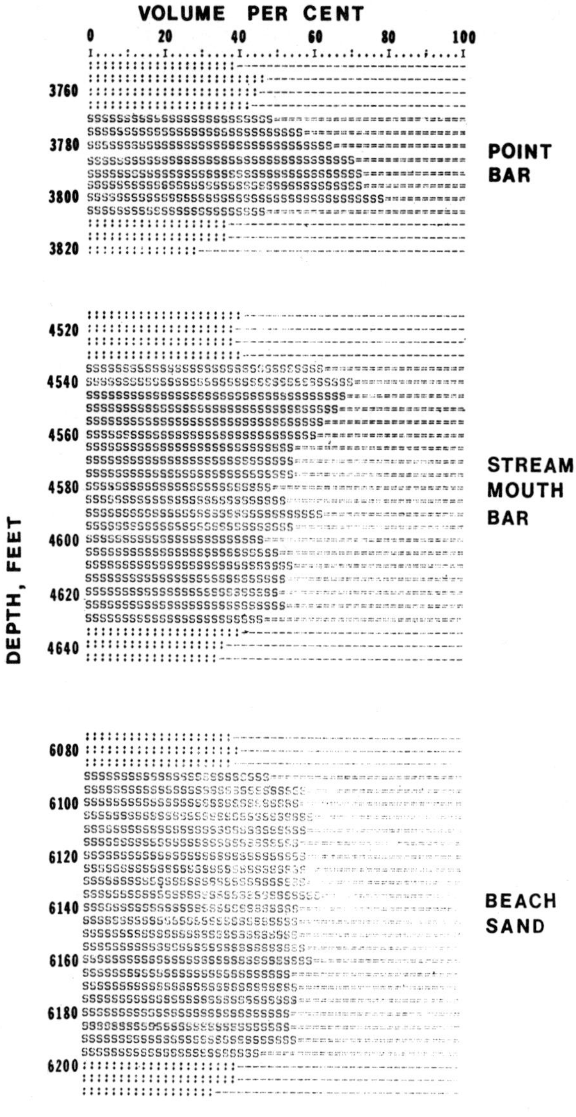

FIG. 11.1. Example of compositional plot of a Beaufort well, Canada. (Reproduced from Magara (1979) by permission of the Plenum Publishing Corp.)

1. Petroleum Migration

A prograding sandstone has a sharp contact with the overlying shale and a gradational contact with the underlying shale. Fluid migration from shale to sandstone would be easier with the gradational contact.

Petroleum maturation is known to be a function of geological time and temperature (Connan, 1974). This means that the chances of petroleum generation usually increase with increasing depth within an intermediate depth range. In other words, the underlying shale usually has a better chance of maturation than the overlying shale.

Because of the better conditions for fluid migration from the underlying shale to the prograding sandstone, the generated petroleum will move to the sandstone. This upward movement may also be supported by a buoyant force acting on the petroleum.

2. Reservoir Quality

The reservoir quality of a prograding sandstone improves upward. Most petroleum tends to move upward as a result of buoyancy until it is sealed by the cap rock (overlying shale in this case). There is thus a high probability of a petroleum accumulation in the best part of the prograding sandstone.

3. Sealing Efficiency

Capillary sealing capacity can be shown by the following equation (Berg, 1975),

$$P_c = 2\gamma\left(\frac{1}{r_t} - \frac{1}{r_p}\right) \tag{11.1}$$

where P_c = capillary pressure
γ = interfacial tension between oil and water
r_t = pore-throat radius of overlying cap rock.
r_p = pore radius of reservoir rock.

Equation (11.1) means that the capillary pressure, P_c, increases as the difference between r_t and r_p increases for a given value of interfacial tension, γ. This suggests that a sharp contact between reservoir and overlying rock, as is expected in the case of a prograding sandstone, makes for better capillary sealing.

Even if a petroleum accumulation is held by pressure seals, such a sharp contact between reservoir and overlying shale can improve the sealing efficiency, because the pressure seal is developed by a restricted fluid environment in the overlying shales.

In summary, the interpretation of sandstone body types such as prograding, aggrading and blocky helps in understanding petroleum accumulations as well as in understanding depositional environment. Such an interpretation can be made by a computer from digital well-log data; the least-squares method is used. A statistical method based on the sandstone-body type analysis produces percentages of each of these types in a given geological sequence. Mapping these statistical data of a given geological sequence in a sedimentary basin may be useful in interpreting the direction of sediment transportation and predicting the direction or area of significant sand deposition.

The technique discussed in this chapter uses digital gamma-log data. However, SP and resistivity (or conductivity) logs may also be used with minor revisions in the computer programs.

COMPACTION-CORRECTED GAMMA-RAY LOG

As gamma rays pass through sediments, they are absorbed and the energy levels decline. The amount of absorption increases as the formation density increases. Two formations having the same amount of radioactive material per unit volume but different bulk densities will show, on the gamma-ray log, different radioactivity levels; the less dense formations will appear more radioactive (Schlumberger, 1969). In a formation consisting of two materials, clay and nonclay, each with its own radioactivity level, the following relation exists (Schlumberger, 1969):

$$\mathrm{GR} \cdot \rho_b = A \cdot V_{\mathrm{clay}} + B \cdot V_{\mathrm{nonclay}} \tag{11.2}$$

where GR = gamma ray reading (API unit); ρ_b = formation bulk-density (g/cm^3); V_{clay} = volume fraction of clay; V_{nonclay} = volume fraction of nonclay; and A and B = coefficients.

The product of gamma ray and density ($\mathrm{GR} \cdot \rho_b$) is called 'normalized gamma ray' and is, according to Schlumberger, directly related to the volume fraction of clay or nonclay. The values A and B represent the $\mathrm{GR} \cdot \rho_b$ values in the zones of 100% clay and 100% nonclay respectively.

The normalized gamma-ray technique was applied experimentally to calculations of clay and nonclay contents in many wells in Canada. Cuttings and core samples of the same wells have been analysed by X-ray diffraction (Magara, 1978). Analysis by X-rays shows the volume fractions (or percentages) of different minerals such as montmorillonite, illite, chlorite, quartz, feldspar, calcite, etc., in rock samples. All clay minerals and all nonclay minerals were grouped separately, and the volume fraction (or percentage) of each group was calculated. This result was then compared with the log-calculated values mentioned above.

The trial, however, was not quite successful. Although the results for relatively deep or compacted intervals showed good agreement, in relatively shallow or less compacted intervals the log-calculated clay content was almost always less than that derived from X-ray analysis.

This discrepancy probably results from ignoring the effect of compaction on gamma-ray values. The assumption included in the previous equation is that the two formations have the same amount of radioactive material per unit volume, but different bulk densities. However, when the increase in bulk density is due mainly to compaction, this assumption is probably not valid. In other words, the amount of radioactivity per unit volume of rock should increase with compaction, or the increase of bulk density.

It is of interest that we commonly observe on the gamma-ray log a trend of shale base line that tends to increase in API units with depth or compaction. This increase is the result of shale compaction.

The gamma-ray reading can be corrected for compaction effect as follows. Suppose that a rock has a porosity of ϕ and a given mixture of clays and nonclays. The normalized gamma-ray, as shown in the left-hand side of eqn. (11.2), equals the total radioactivity of the rock material and the pore-water,

$$\mathrm{GR}\cdot\rho_b=(A\cdot V_{\mathrm{clay}}+B\cdot V_{\mathrm{nonclay}})\cdot(1-\phi)+W\cdot\phi \qquad (11.3)$$

where W = radioactivity of water.

The radioactivity of pore-water is usually very low, if it is compared with that of clays. If the pore-water radioactivity is assumed to be zero, eqn. (11.3) can be simplified to

$$\mathrm{GR}\cdot\rho_b=(A\cdot V_{\mathrm{clay}}+B\cdot V_{\mathrm{nonclay}})\cdot(1-\phi)$$

or

$$\frac{\mathrm{GR}\cdot\rho_b}{1-\phi}=A\cdot V_{\mathrm{clay}}+B\cdot V_{\mathrm{nonclay}} \qquad (11.4)$$

Equation (11.4) suggests that the value $(\mathrm{GR} \cdot \rho_b)/(1-\phi)$ can be directly related to the volume fraction of clays and nonclays at any compaction level. The porosity can be shown as follows:

$$\phi = \frac{\rho_m - \rho_b}{\rho_m - \rho_w} \tag{11.5}$$

where ρ_m = matrix density of the rock (g/cm^3)
ρ_w = density of pore-water (g/cm^3).

Introducing eqn. (11.5) into (11.4) and assuming that ρ_w equals 1 g/cm^3, we obtain

$$\frac{\mathrm{GR} \cdot \rho_b}{\rho_b - 1} = (A \cdot V_{\mathrm{clay}} + B \cdot V_{\mathrm{nonclay}}) \cdot \frac{1}{\rho_m - 1} \tag{11.6}$$

If new coefficients A' and B', which respectively equal $A/(\rho_m - 1)$ and $B/(\rho_m - 1)$, are defined, eqn. (11.6) will be simplified to

$$\frac{\mathrm{GR} \cdot \rho_b}{\rho_b - 1} = A' \cdot V_{\mathrm{clay}} + B' \cdot V_{\mathrm{nonclay}} \tag{11.7}$$

Equation (11.7) is simpler to use than eqn. (11.4), because its left-hand side contains only the two terms, GR and ρ_b, instead of GR, ρ_b and ϕ as in eqn. (11.4). The value $(\mathrm{GR} \cdot \rho_b)/(\rho_b - 1)$ or $(\mathrm{GR} \cdot \rho_b)/(1-\phi)$ in these equations may be called the compaction-corrected and normalized gamma ray, which is directly related to the proportions of clay and non-clay contents in sediment or rock.

The coefficients A' and B' were determined empirically in northern Canada by cross correlation of X-ray analysis data of cuttings and cores and the well-log data; they are about 200 and 50 API units (Magara, 1978). In a different basin, the values may differ significantly.

For application of this technique, a combination of gamma-ray and sonic logs is more practical than the gamma ray–density combination, since the sonic log is usually more available. The following empirical relationship, which is based on shales, may be used to obtain ρ_b from Δ_t (transit time, μs/ft):

$$\rho_b = 2{\cdot}99 - 0{\cdot}00616 \times \Delta_t \tag{11.8}$$

When good-quality formation-density and sonic logs are both available, the formation-density log is, of course, recommended to be used to provide the density values directly. An example of a compaction-corrected and normalized gamma-ray log is shown in Fig. 11.2; a

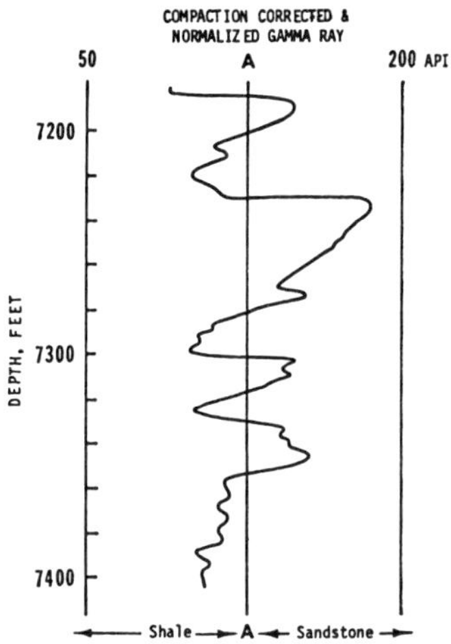

FIG. 11.2. Example of compaction-corrected and normalized gamma ray plot of a Beaufort well, Canada. (Reproduced from Magara (1979) by permission of the Plenum Publishing Corp.)

combination of gamma-ray and sonic is used because a formation-density log is not available for this well.

Since the plot shown in Fig. 11.2 is corrected for the effect of changing level of compaction, a single vertical line (A), which was determined empirically, can separate sandstones from shales. Corrections can also be made for erroneous log-calibrations and deteriorated bore-hole conditions.

IDENTIFICATION OF SANDSTONE-BODY TYPES BY COMPUTER ANALYSIS

Single Sandstone Body

Figure 11.3 shows three schematic composition profiles—they are all identical. By using a filter, sandstone bodies can be separated from shales. In the examples shown in Fig. 11.3 the interval between A and B is shown to be a sandstone by using the filter shown as F.

The computer first searches for the maximum nonclay percentage in the sandstone interval (A-B or L). This percentage is placed at the top of the sandstone (see point P) and two straight lines are drawn from P to

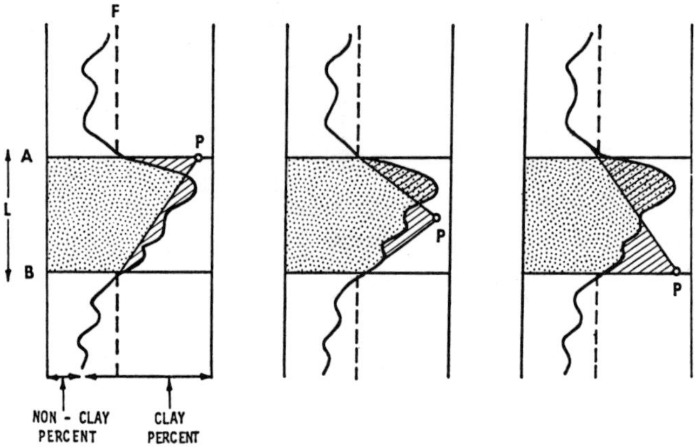

FIG. 11.3. Schematic diagram showing method of triangle fitting on an actual compositional plot, using least-squares method. (Reproduced from Magara (1979) by permission of the Plenum Publishing Corp.)

the upper and lower sand–shale boundaries as shown on the left of Fig. 11.3. Squares of the differences between the actual nonclay percentage and the percentage shown by these straight lines are summed in the sandstone interval and stored in the computer. Point P is then moved vertically downward a distance of 5 ft and a different triangle is drawn. The sum of the squares of the difference is again calculated and stored as in the previous case (see middle diagram of Fig. 11.3). Similar calculations at 5 ft intervals continue until point P reaches the base of the sandstone body (see right-hand diagram of Fig. 11.3).

The sums of the squares are then compared to find the least value among them. The position of P when this least-square value is obtained is schematically shown in Fig. 11.4. The vertical distance between point P and the upper sand–shale boundary is X and between P and the lower boundary $L-X$.

Although the value X or $L-X$ gives us an idea of sandstone-body type based on the triangle-fitting method as explained above, whether the sand is 'blocky' or one of the triangle shapes must be determined independently, as follows.

When the computer searches the maximum nonclay percentage in the body as mentioned previously, it also computes the average nonclay percentage of the sandstone body. A vertical line is drawn at this average nonclay value (Fig. 11.5), and the sum of the square of the difference between the value shown by this vertical line (or the average nonclay

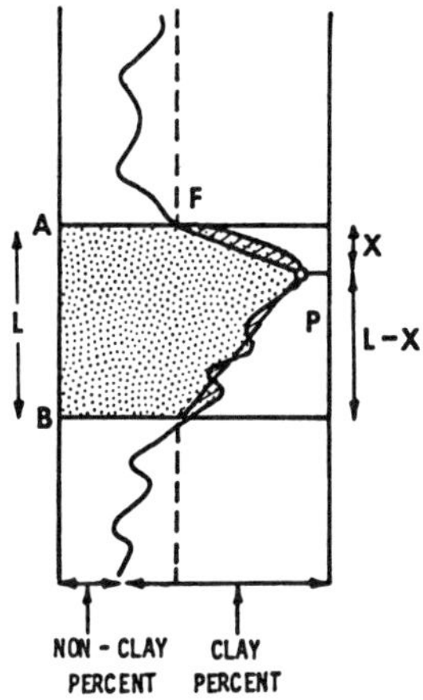

FIG. 11.4. Schematic diagram showing proportions of prograding and aggrading sandstone types by triangle fitting. (Reproduced from Magara (1979) by permission of the Plenum Publishing Corp.)

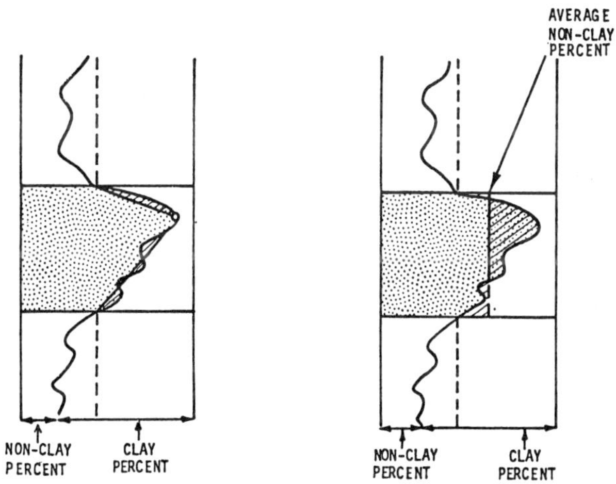

FIG. 11.5. Schematic diagram showing comparison between best triangle fitting and vertical line fitting. (Reproduced from Magara (1979) by permission of the Plenum Publishing Corp.)

percentage) and the calculated nonclay percentage at each depth is calculated.

If the sum of the squares based on the vertical-line fitting model is less than the least square previously obtained by the triangle fitting, the sandstone is called 'blocky' (Fig. 11.5).

When the least square of the triangle-fitting model is less than the sum of the vertical-line fitting, the sandstone is 'progradational' or 'aggradational'. In other words, if $L-X$ is more than 95% of the total thickness (or L), the sandstone is called 'progradational' in the current program. On the other hand, if X is more than 95% of the total, it is 'aggradational'. If $L-X$ or X is less than 95%, the percentage of each is printed by the computer.

Two Sandstone Bodies in Contact

If two sandstone bodies are in contact as shown in Fig. 11.6, use of a single-triangle model will not realistically predict the shape of the sandstone body. Two triangles must be drawn to find the best fit.

In order to make the double triangle fit, the minimum nonclay percentage as shown by M in Fig. 11.6 must first be determined. Then the maximum nonclay value in the upper zone, P_1 and in the lower, P_2, are searched out independently. The two average values in the upper and lower zones must also be calculated. The best-fit method in the upper or lower zone is essentially the same as that explained previously for the single sandstone body. However, the final answer is more complicated.

Figure 11.7 shows six choices of fitting in such a case. A and B are the cases for single fitting and all the others are for double fitting. The computer will find the best fit among these choices.

The number of sandstone bodies in contact can, however, be more than two, in which case the approach based on up to two sandstone

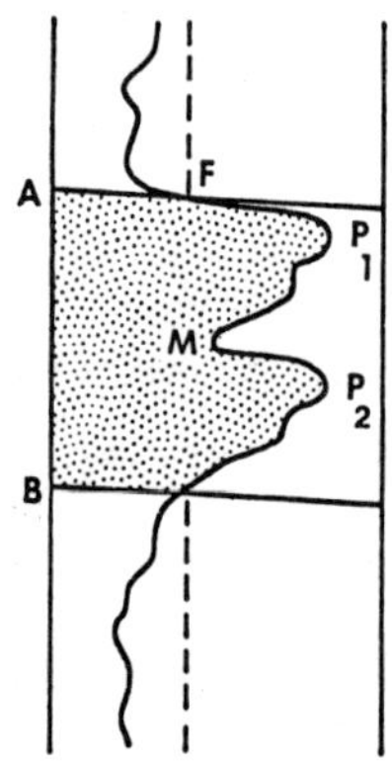

FIG. 11.6. Schematic compositional plot of two sandstone bodies in contact. (Reproduced from Magara (1979) by permission of the Plenum Publishing Corp.)

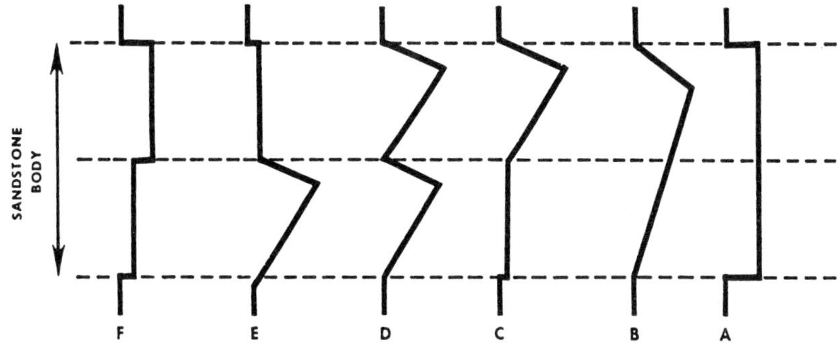

FIG. 11.7. Schematic diagram showing six choices of least-squares fitting for up to two sandstone bodies in contact. (Reproduced from Magara (1979) by permission of the Plenum Publishing Corp.)

bodies may not indicate the best fit. Because the computer program for these cases is so extremely complicated, a program for analysing up to two sandstone bodies has been written.

Sample Output of Computer Program to Analyse Sandstone-body Types

Figures 11.8 and 11.9 show examples of plots of compaction-corrected and normalized gamma ray on the left and the print-out of the computer-analysed sandstone-body types on the right, for a well in the Beaufort Basin. The print-out shows, from left to right, 'interval' of sandstone analysed, 'thickness' (if the sandstone is divided into two, the thicknesses of the upper and lower parts are shown), choice of 'aggrading' sandstone, of 'prograding' sandstone, of 'blocky' sandstone, of 'thin' sandstone (if the thickness of sandstone is equal to or less than 10 ft, no fitting trial is made and the sand is shown to be 'thin' sandstone).

The result of the computer analysis is shown by an asterisk, but in the case of the triangle-fitting, the percentage figures of aggrading and prograding sandstones are shown when one of these types does not exceed 95% of the total thickness. If the aggrading or prograding type is more than 95%, the same symbol (asterisk) is used to indicate that situation.

The sandstone between 11 438 and 11 508 ft (70 ft) is, for example, interpreted to be 20% aggrading and 80% prograding. The sandstone between 11 518 and 11 583 ft (65 ft) is divided into two bodies of 30 ft and 35 ft each. The upper sand is interpreted to be prograding, while the lower sand is 67% aggrading and 33% prograding.

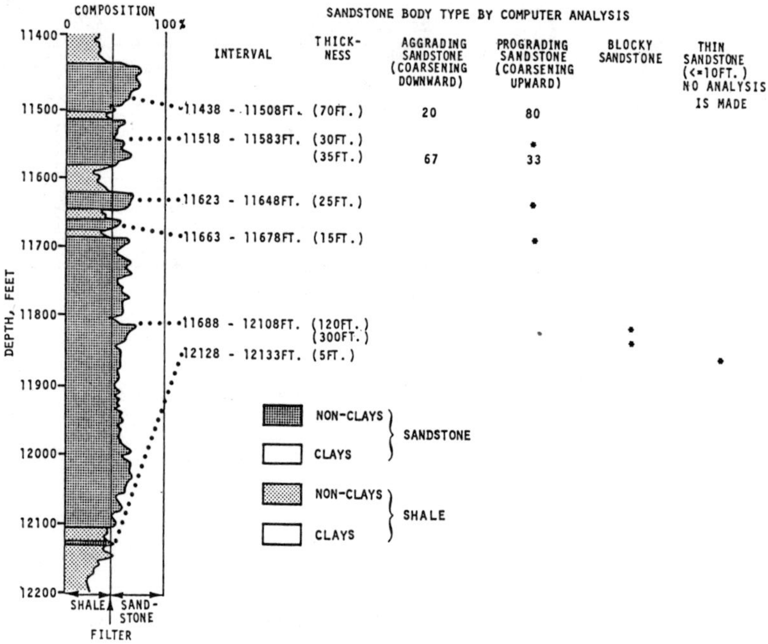

FIG. 11.8. Example of compositional plot (left) and print-out of computer-analysed sandstone body types (right) of a Beaufort well, Canada. (Reproduced from Magara (1979) by permission of the Plenum Publishing Corp.)

The computer-analysed sandstone-body types as shown in Figs. 11.8 and 11.9 are of interest to many geologists, but a similar analysis can also be done manually within a relatively short time.

However, if a geologist wished to have statistical data on the thickness or percentage of each sandstone type in a relatively thick (e.g., 2000 ft) geologic unit, a manual method would be quite time-consuming. In such an event a computer would be a very helpful tool.

Statistical Program to Compute Percentages of Three Sandstone-body Types in a Specified Geologic Unit

Figure 11.10 is a sample run of a statistical program to compute the percentages of three sandstone-body types between 3623 and 4623 ft in a Beaufort well. It shows the 'geologic zone' and 'depth' of the specified interval at the top. The program permits up to 50 geologic intervals in a well. The thickness (ft) and percentage figures of each lithology are

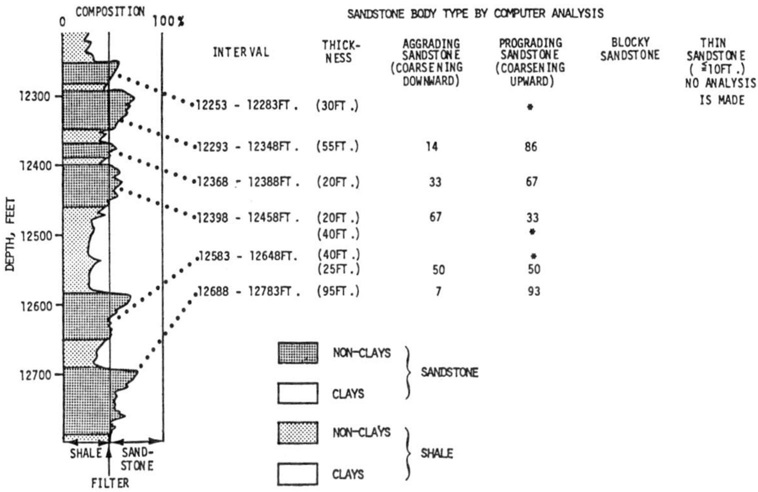

FIG. 11.9. Example of compositional plot (left) and print-out of computer-analysed sandstone body types (right) of a Beaufort well, Canada. (Reproduced from Magara (1979) by permission of the Plenum Publishing Corp.)

	Thickness (ft) (%)	*Prograd-ing %*	*Aggrad-ing %*	*Blocky %*
Sandstones	105 (11)	76	24	0
Shaly sands	255 (26)	49	39	12
Total sandstones	360 (36)	57	35	8
Sandy shales	600 (60)	39	38	23
Shales	40 (4)	0	100	0
Total shales	640 (64)	37	42	21

FIG. 11.10. Sample run of statistical program on sandstone and shale body types in a selected geologic section (No. 1 Geologic Zone: 3623–4623 ft). (Reproduced from Magara (1979) by permission of the Plenum Publishing Corp.)

shown on the left. The percentages of the three body types—prograding, aggrading and blocky—for each lithology are listed on the right.

The same program was run for a geologic interval (1) within the Tertiary section for 18 wells in the Beaufort area (Fig. 11.11). First of all, the percentages of the three sandstone types are plotted on a triangle diagram (Fig. 11.12). The numbers in Fig. 11.12 refer to the well

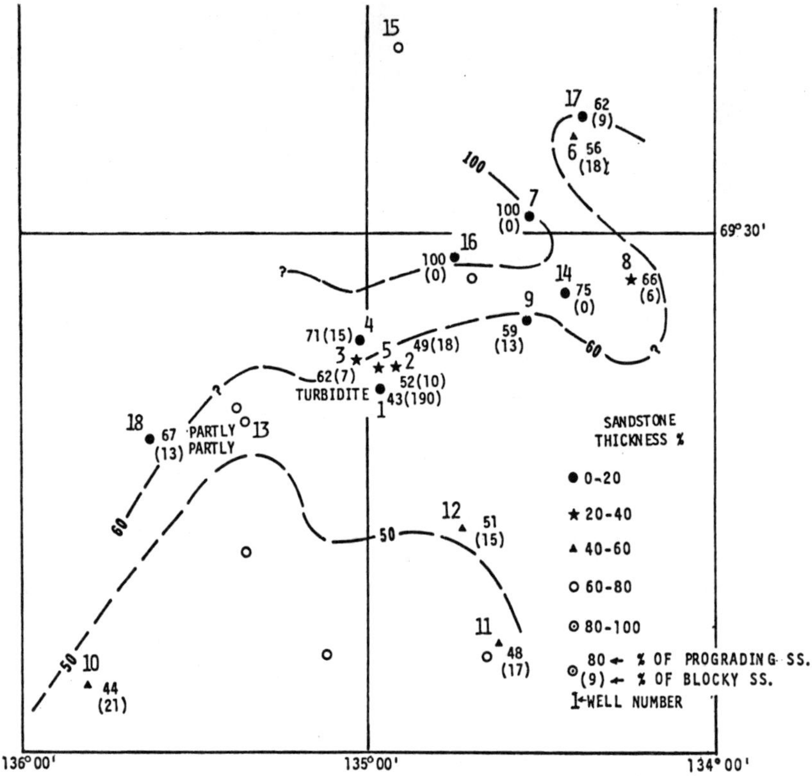

FIG. 11.11. Progradational percentage map of geologic interval 1 within Tertiary section, Beaufort Basin, Canada. (Reproduced from Magara (1979) by permission of the Plenum Publishing Corp.)

numbers shown in Fig. 11.11. Figure 11.12 also shows the sandstone thickness percentage for each well. There seems to be a general tendency for the percentage of aggradational and blocky types to decrease as the percentage of sandstone decreases (the percentage of progradational type increases). There is an exception at well 1, in which a relatively low percentage of sandstone is associated with a high percentage of aggradational and blocky types. These thin sandstones are interpreted to be turbidite. Such situations may be explained by the use of a schematic diagram (Fig. 11.13), in which a progradational sedimentation model is shown. On the left or in the landward area, where the sediments were mostly deposited above sea level, the sandstones are mainly of the

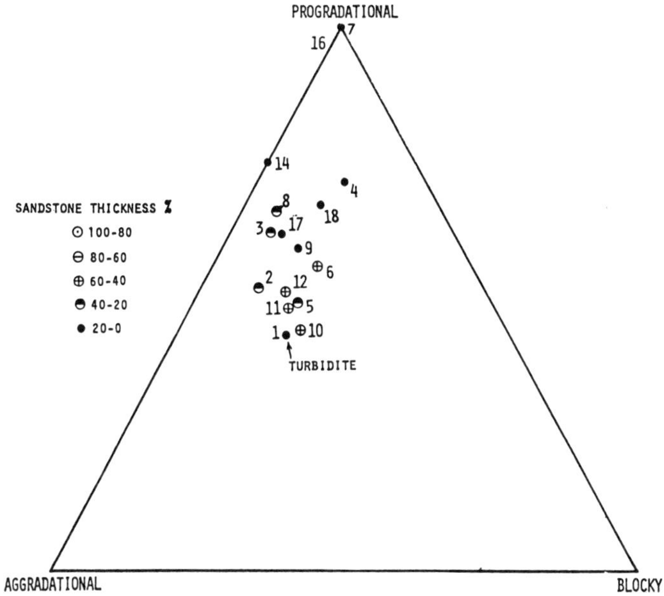

FIG. 11.12. Triangle diagram of progradational, aggradational and blocky sandstone percentages for geologic interval 1 in Beaufort Basin. Numbers refer to wells shown in Fig. 11.11. (Reproduced from Magara (1979) by permission of the Plenum Publishing Corp.)

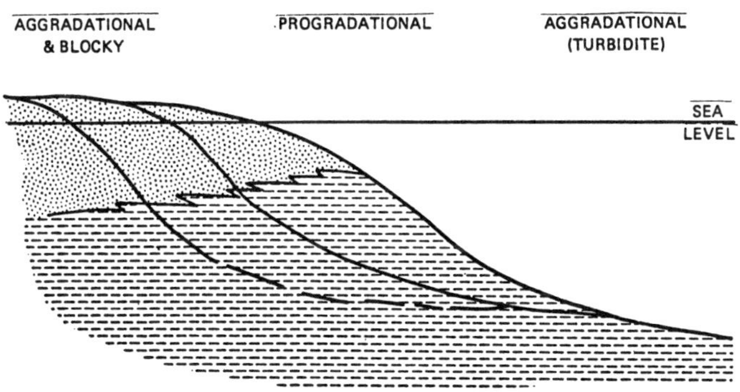

FIG. 11.13. Schematic diagram showing progradational sedimentation model. (Reproduced from Magara (1979) by permission of the Plenum Publishing Corp.)

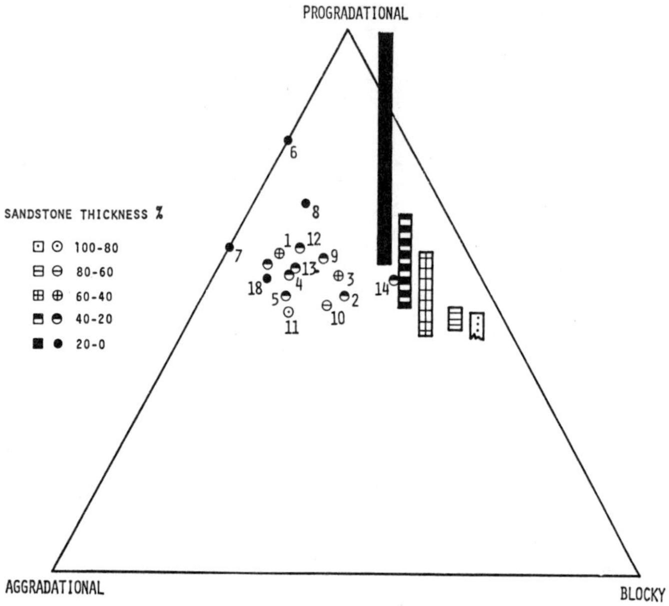

FIG. 11.14. Triangle diagram of progradational, aggradational and blocky sandstone percentages for geologic interval 2 in Beaufort Basin, Canada, Numbers refer to wells shown in Fig. 11.15. (Reproduced from Magara (1979) by permission of the Plenum Publishing Corp.)

aggradational and blocky types. This area is usually sand-rich. As we move seaward, we will find more progradational types and more shales (or fewer sandstones). If we move farther seaward, the percentage of shales will increase, but if we encounter turbidites they will be mainly of the aggradational type.

The areal distribution of the sandstone types is shown in Fig. 11.11. The contour lines are based on the percentage of the progradational type which increases northward. This direction is the same as the general direction of transportation of the sediments in the area. The percentage of sandstones generally decreases northward.

A similar study was also made for the interval below (geological interval 2) and the results are shown in Figs. 11.14 and 11.15. The triangle plot of Fig. 11.14 shows the general tendency for the percentage of prograding sandstone to increase as the percentage of sandstone in the

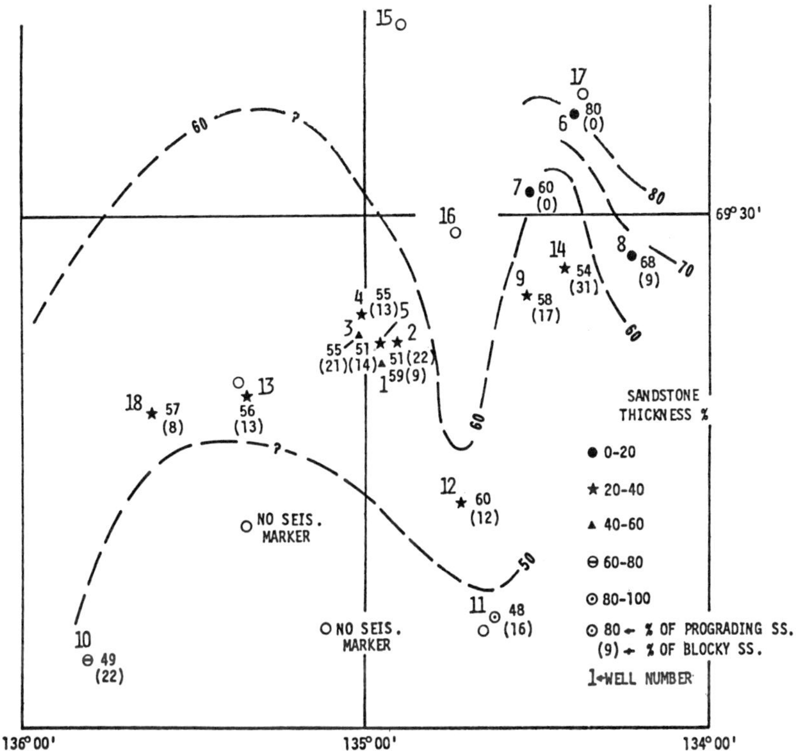

FIG. 11.15. Progradational percentage map of geologic interval 2 in Beaufort Basin, Canada. (Reproduced from Magara (1979) by permission of the Plenum Publishing Corp.)

interval decreases. The areal view of the progradational percentage is given in Fig. 11.15. The two delta lobes extending from south to north are clearly shown by this figure.

After sandstone-body types and sandstone thickness percentages had been studied for other Tertiary intervals in this area, the ranges of the progradational percentages for specified ranges of sandstone thickness percentage were obtained; they are shown by the thick vertical bars in Fig. 11.14. They demonstrate the general observation that the progradational percentage increases as the percentage of sandstone decreases. An exception, however, is the turbidite example at well 1.

REFERENCES

Berg, R. R. (1975) Capillary pressures in stratigraphic traps. *A.A.P.G.*, **59**, 939–56.

Connan, J. (1974) Time–temperature relations in oil genesis. *A.A.P.G.*, **58**, 2516–21.

Fisher, W. L. *et al.* (1969) Delta Systems in the Exploration for Oil and Gas, a Research Colloquium: Bureau of Economic Geology, University of Texas, Austin, Texas, 1969.

Magara, K. (1978) *Compaction and Fluid Migration—Practical Petroleum Geology*, Appendix 3-2, Elsevier, Amsterdam, New York, Oxford, 319 p.

Magara, K. (1979) Identification of sandstone body types by computer method. *International Association for Mathematical Geology*, **11**, 269–83.

Ruoff, W. A. (1976) A technique for interpreting depositional environments of sandstones from the SP log utilizing the computer. *Log Analyst*, **17**, no. 4.

Saitta, S. and Visher, G. S. (1968) Subsurface study of the southern portion of the Bluejacket–Bartlesville Sandstone, Oklahoma, Visher, G. S. (ed.), Oklahoma Geol. Soc., 1968.

Schlumberger (1969) *Log Interpretation Principles*, Schlumberger, New York.

Visher, G. S., Saitta, S. and Phares, R. S. (1971) Pennsylvanian delta patterns and petroleum occurrences in eastern Oklahoma. *Am. Assoc. Petroleum Geologists Bull.*, **55**, 1206–30.

Chapter 12

Litho-Statistical Study of Frio Formation in Texas Gulf Coast

with contributions by
W. E. GALLOWAY* and
D. K. HOBDAY†

INTRODUCTION

The Frio Formation, of Oligo–Miocene age, forms one of the principal progradational wedges of the Texas Gulf Coast Tertiary basin (Fig. 12.1). Several hundreds of feet of vertical aggradation of the ancient coastal plain by deposition of the stratigraphically updip equivalent Catahoula Formation was accompanied by construction of a continental platform wedge more than 15 000 ft thick and up to 50 miles in width. This wedge, which consists of interfingering marine and nonmarine sandstones and shales, has produced more than 16 billion barrels of hydrocarbons, making it one of the most important petroleum reservoirs of the Gulf Coastal basin. (Natural gas is converted to liquid equivalents on an approximate BTU basis of 1 bbl = 6000 ft^3.)

Major progradational wedges, such as the Frio Formation, consist of an updip and relatively shallow section of interbedded continental and marginal marine sandstone and mudstone underlain by several thousand feet of undercompacted slope and basinal mudstone. Resultant instability due to rapid loading of plastic, water-saturated sediment at the crest of a gently inclined subaqueous slope resulted in large-scale, syndepositional down-to-the-coast faulting and intrastratal deformation. Three structural provinces, which affect Frio deposition, can be defined on the basis of syndepositional structural style and nature of the diapiric material

* University of Texas at Austin, Department of Geological Sciences, Austin, Texas 78712-7909, USA.
† Bridge Oil Ltd., 60 Margaret Street, Sydney, New South Wales 2000, Australia.

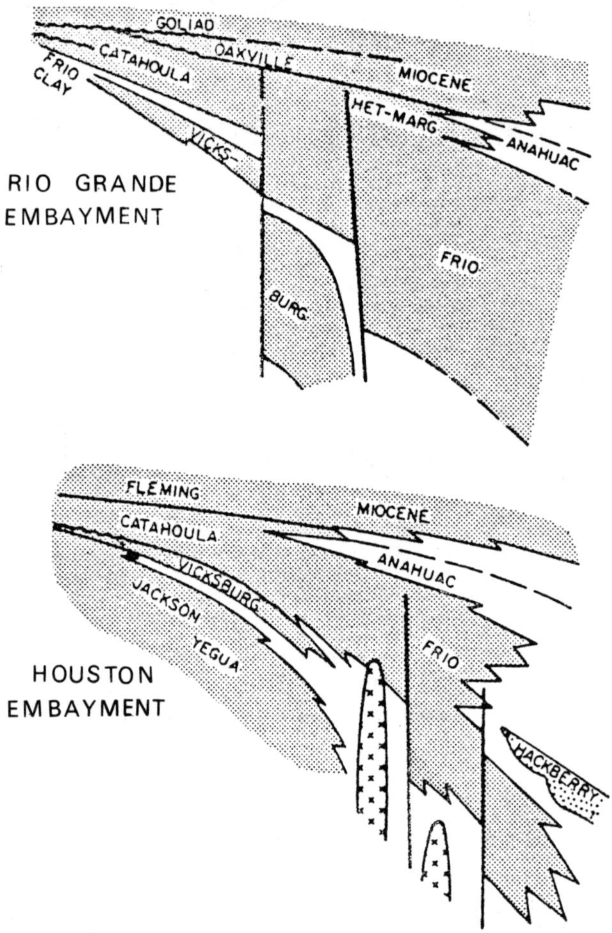

FIG. 12.1. Schematic Frio stratigraphic correlation sections for Houston and Rio Grande embayments, Texas Gulf Coastal plain, showing commonly applied nomenclature and principal bounding units (from Galloway *et al.*, 1982). (Reproduced by permission of the Bureau of Economic Geology, University of Texas at Austin.)

caught up in the deformational process. The broadly-defined Houston embayment of East Texas is characterized by salt diapirism and associated faulting and large salt withdrawal sub-basins (Gulf Coast Association of Geological Sciences, 1972; Bebout *et al.*, 1978). Across the San Marcos platform of Central Texas and southward toward the Rio

Grande embayment, underlying salt is thin or absent, and long, linear belts of growth faulting and associated shale ridges, massifs, and diapirs dominate the structural style (Bruce, 1973; Bishop, 1978). Within the Rio Grande embayment, large but more discontinuous belts of growth faulting and deep-seated shale ridges and massifs occur. In addition to the strike-parallel fault trends, subsidiary, oblique trends of deep, low-relief anticlinal structures are particularly prominent along the northern boundary of this province.

Division of the Frio Formation into a number of distinct but spatially related depositional systems (Galloway *et al.*, 1982) was accomplished primarily on the basis of sand isolith and percentage maps of the upper, middle, and lower Frio operational units (Figs. 12.2 and 12.3 for middle Frio) and facies maps (Fig. 12.4 for middle Frio) constructed from spontaneous potential (SP) and resistivity log patterns. No major progradational complexes, designated the Houston and Norias delta systems were centred on the Houston and Rio Grande embayments, respectively (Galloway *et al.*, 1983; also Fig. 12.5). The Houston delta system was characterized by numerous small, laterally displaced lobes, in contrast to the most clearly defined vertically stacked lobes of the considerably larger Norias delta system. These delta systems in East and South Texas were supplied by fluvial systems of different character. The East Texas Chita-Corrigan fluvial system carried a mixed load of sand, silt and clay, whereas the Gueydan system to the south was considerably coarser grained.

Separating these two major structurally-controlled fluvial/deltaic tracts was an area in Central Texas spanning the San Marcos arch, which was traversed by relatively minor coastal plain streams. Deposition of the Choke Canyon/Flatonia coastal lake and streamplain system in this area was dominated by silt and mud accumulation, with widely-spaced, minor belts of coarser grained stream deposits best developed in the western half. Shore-zone sediments of Central Texas were contributed largely along strike from adjacent deltaic depocentres, building up the substantial system of Frio coastal barriers and beach ridges first documented by Boyd and Dyer (1964). This trend is termed the Greta/Carancahua barrier/strand plain system (Fig. 12.5). The similar, but smaller, Buna strandplain/barrier system developed on the eastern flank of the Houston delta system and expands toward the Louisiana border. Updip from this zone of accreting beach ridges was an extension of the Chita/Corrigan fluvial system, which was progressively less active eastward where it incorporated muddy coastal lake and streamplain deposits.

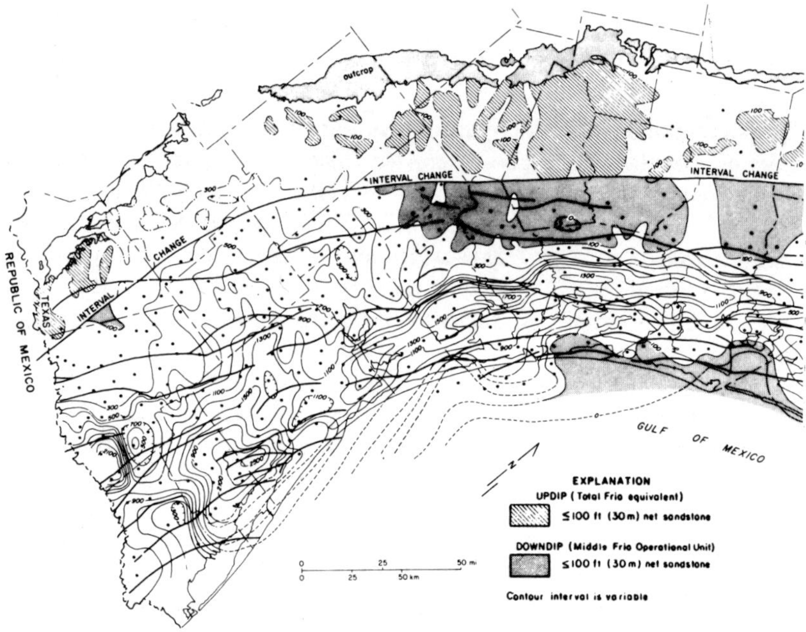

FIG. 12.2. Net sandstone isolith map of middle Frio operational unit. Updip, broken contours show sand content of undivided, Frio-equivalent Catahoula

FIG. 12.3. Sandstone percentage map of middle Frio operational unit (from

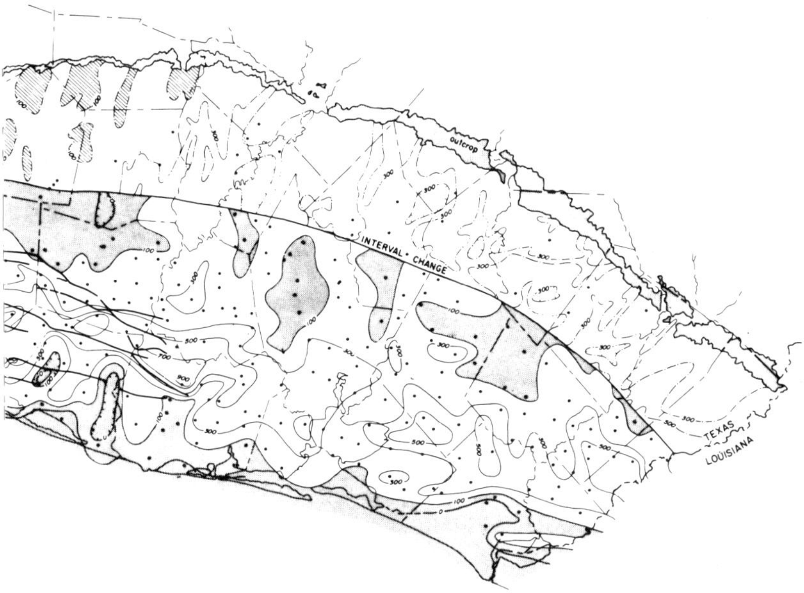

Formation (from Galloway *et al.*, 1982). (Reproduced by permission of the Bureau of Economic Geology, University of Texas at Austin.)

Galloway *et al.*, 1982). (Reproduced by permission of the Bureau of Economic Geology, University of Texas at Austin.)

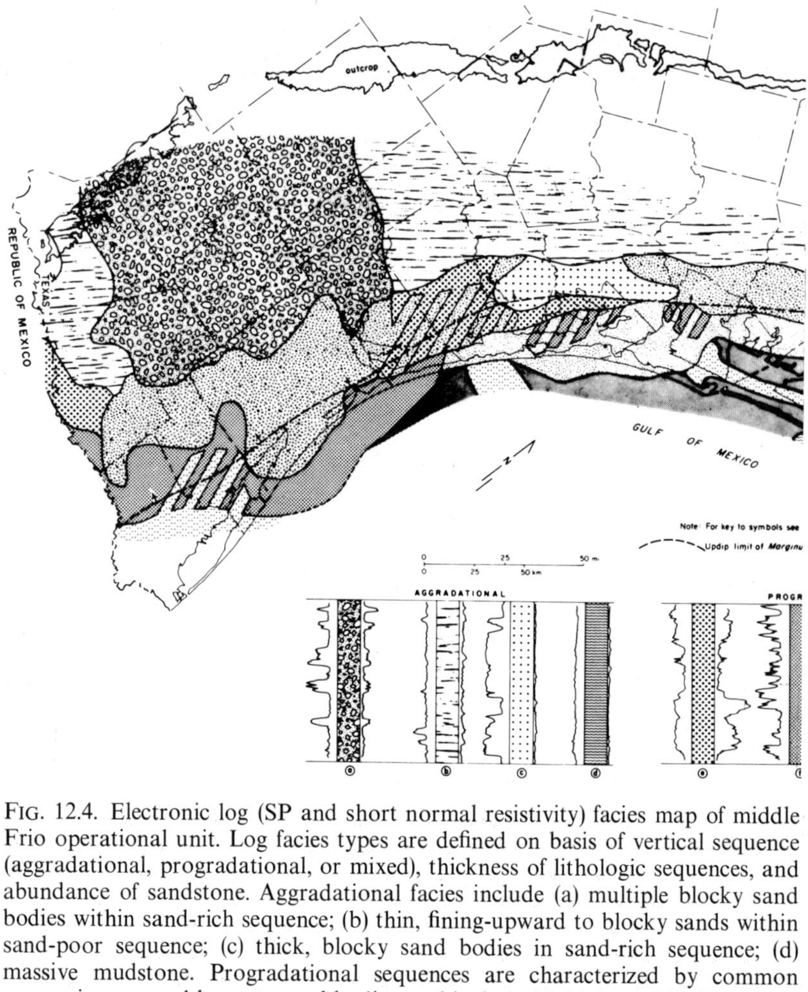

FIG. 12.4. Electronic log (SP and short normal resistivity) facies map of middle Frio operational unit. Log facies types are defined on basis of vertical sequence (aggradational, progradational, or mixed), thickness of lithologic sequences, and abundance of sandstone. Aggradational facies include (a) multiple blocky sand bodies within sand-rich sequence; (b) thin, fining-upward to blocky sands within sand-poor sequence; (c) thick, blocky sand bodies in sand-rich sequence; (d) massive mudstone. Progradational sequences are characterized by common coarsening-upward bases to sand bodies, and include (e) thick sand bodies within

Basinward Frio elements include a shelf platform inherited from Jackson (Eocene) and Vicksburg (Oligocene) deposition, and a broad unstable slope, which in East Texas was deeply incised and embayed. The extensive, deep-water Hackberry embayment of Louisiana continued into Orange and Jefferson Counties, and was possibly represented in other smaller slope canyons and embayments along the Texas coast (Stuckey, 1964). Although the 2000 ft thickness in Jefferson County

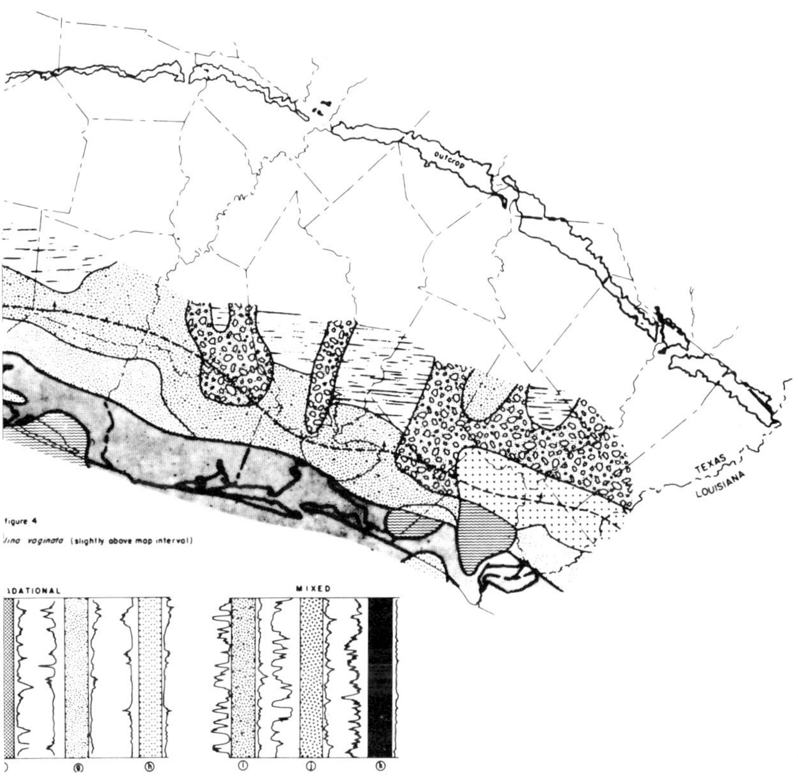

sand-rich sequence; (f) multiple thin to medium, erratically coarsening-upward sand units; (g) multiple thin sand units; (h) thick, well-developed coarsening-upward units within mud-rich section. Mixed log facies types include (i) composite sequences of coarsening- and fining-upward, thin to medium sand bodies; (j) blocky, thick sand bodies within sand-dominated sections; (k) thinly interbedded sand and mud forming thick packages that display both transitional and abrupt unit boundaries (from Galloway *et al.*, 1982). (Reproduced by permission of the Bureau of Economic Geology, University of Texas at Austin.)

represents a reduction from the maximum in Louisiana, both the 'Lower Hackberry sandstone' and 'Upper Hackberry Shale' are recognized. Together these correspond approximately to the middle Frio operational unit of the present study. Studies by Paine (1971) of Hackberry strata in Louisiana established a deep water origin; this has recently been confirmed by Berg and Powers (1980) in East Texas. In Louisiana, elongate slope channel sandstones with the typical internal sequence recording

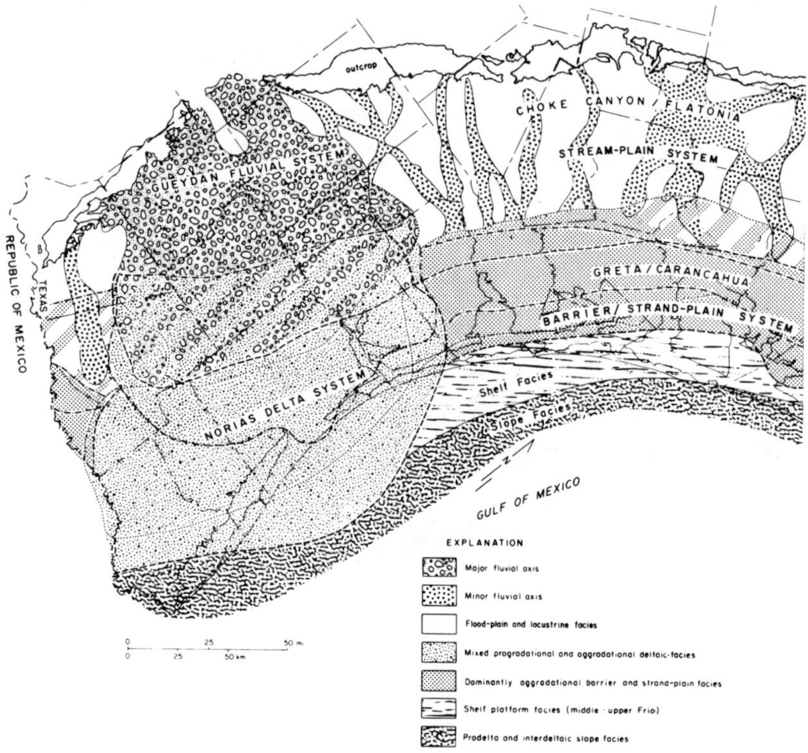

FIG. 12.5. Depositional systems of Frio Formation (from Galloway *et al.*, 1982). (Reproduced by permission of the Bureau of Economic Geology, University of Texas at Austin.)

waning flow conditions, merge basinward into blanket-like fan deposits which are thickest over topographic and structural depressions in the sub-Hackberry unconformity; these sandstones grade up into shales containing a deep-water fauna.

STATISTICAL CHARACTERIZATION OF SYSTEMS

In an attempt to establish objective characterization of the depositional systems, we compiled the statistical data on sandstones and shales of the Frio Major Stratigraphic Unit (MSU) on the basis of five environmental

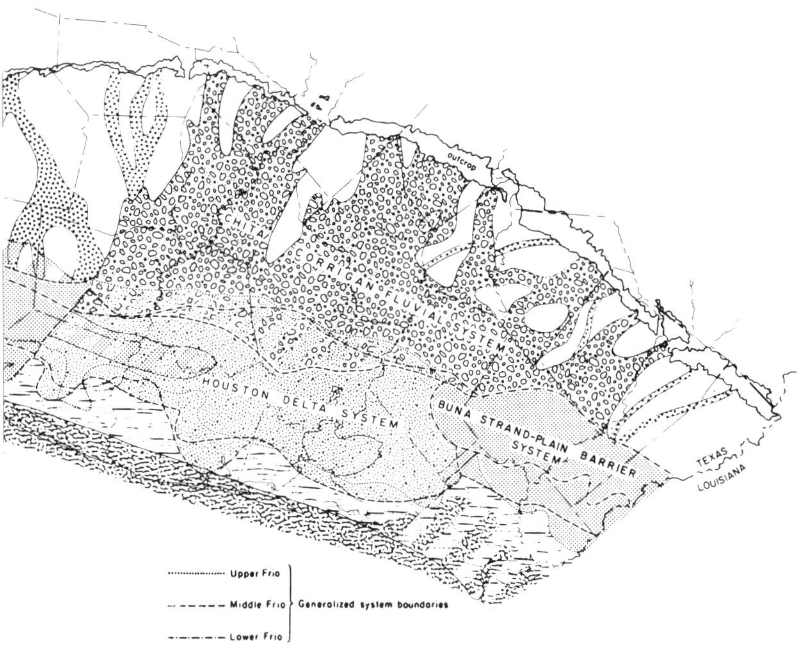

descriptions; streamplain, fluvial, delta, barrier, and shelf. The first four environmental sequences define major depositional systems. The shelf environment represents sand-poor sections that lie beneath or downdip of the sander deltaic or strandline facies. It is thus a component facies of each of the coastal systems as defined in this chapter. The statistical data extracted from electric logs of 860 wells include the thicknesses and the percentages of sandstones and shales and the frequency of interbedding of sandstones and shales.

In a given geologic unit, the number of sandstone bodies can be different from that of shale bodies even if the unit contains only the two rock types. This is due to the edge effect. The difference is, however, relatively small, in that if, for example, the number of sandstone bodies is 8, the number of shale bodies can be 7, 8, or 9, depending upon how the top and bottom of the unit are defined.

Because of the difficulties in defining the precise top and bottom boundaries of each operational unit and also in order to process this counting procedure more quickly, only the number of sandstone bodies were counted for the upper, middle, and lower Frio operational units in each of these 860 wells. Then, we assumed that the number of shale bodies is identical to that of sandstones.

Each operational unit represents a different total thickness; it is difficult to compare the numbers from the units of different thicknesses. Therefore, the number counted in each unit was normalized for a 1000 ft section for a better comparison.

FREQUENCY VS PERCENTAGE OF SANDSTONE UNITS

Figure 12.6 shows a cross plot of the number of sandstone bodies/1000 ft vs the percentage of sandstone for the five depositional environments as described above, of upper Frio. For each environment, the area which covers approximately 90% of the data points is indicated by a curved line with an appropriate environmental designation such as 'shelf'. The mean values of the number and the percentage of sandstones can be shown by the location of a square block for each depositional environment.

Three depositional environments, shelf, stream-plain, and fluvial, respectively, produce relatively vertical distribution patterns with changing average sandstone percentage (Fig. 12.6). The vertical distribution pattern of the plot means that, while the number of sandstone bodies increases, the sandstone percentage does not increase significantly within each depositional environment. This further suggests that, as the number of sandstones increases, the average thickness of sandstones decreases for the shelf, stream-plain, and fluvial environments.

On the contrary, the distribution patterns for the delta and barrier facies show roughly a 45° angle with wide ranges of the number and percentages of sandstones. This suggests that, within these environments, while the number of sandstone bodies increases, their average thickness remains relatively constant, thus increasing the percentage of sandstones in a given unit. In other words, within the delta and barrier facies, we would expect relatively uniform thickness of sandstones, even if the number is greater.

In the process of developing the basic environmental sequences from stream-plain, through fluvial and delta, toward barrier and shelf, the relative effect of fluvial and basin processes evolves. From stream-plain

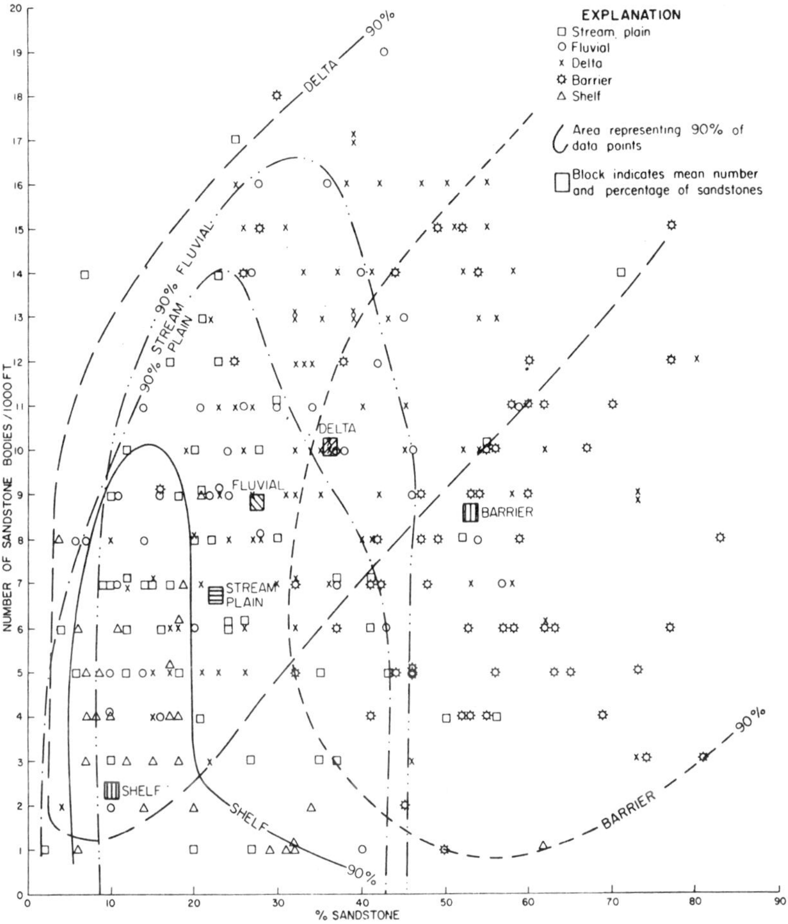

FIG. 12.6. Cross plot of percentage sandstone and number of sandstone bodies/1000 ft for five different environments of deposition of upper Frio.

to fluvial sequence, the average number of sandstone bodies gradually increases but the percentage does not seem to increase greatly. As we approach deltaic environments, both the number and the percentage of sandstones increase.

Because of the marine reworking, which tends to segregate the coarser grains from the finer grains more effectively, a clear separation between barrier (coarser) and shelf (finer) facies results. Although the points for

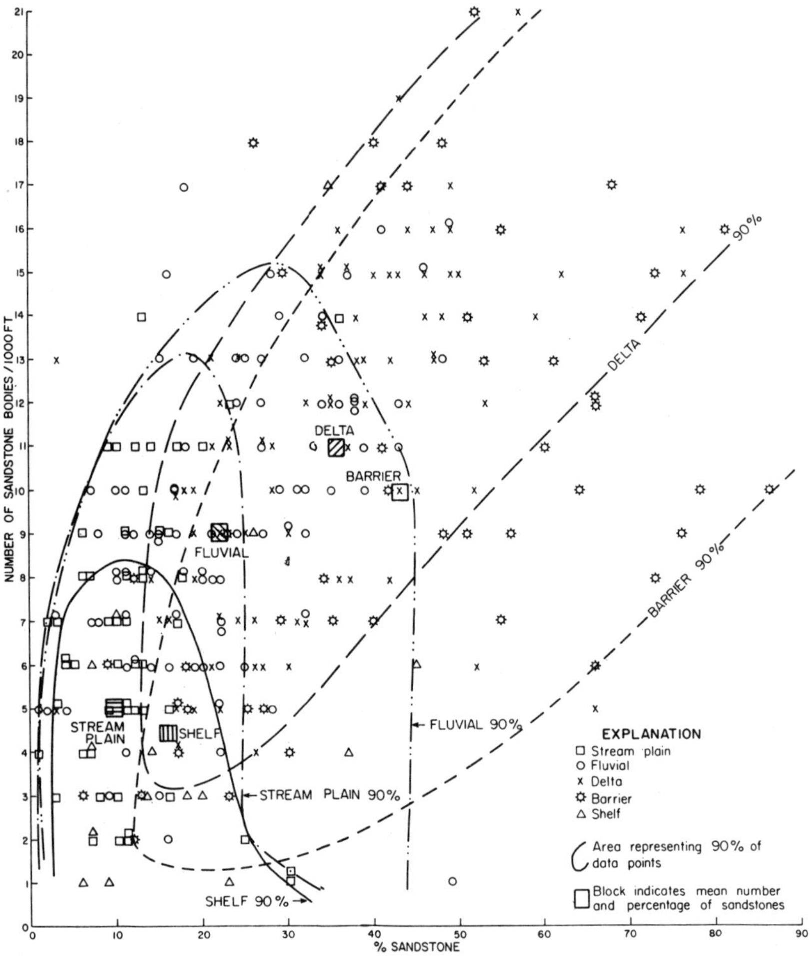

FIG. 12.7. Cross plot of percentage sandstone and number of sandstone bodies/1000 ft for five different environments of deposition of middle Frio.

the barrier facies are widely scattered, they show generally large numbers and percentages of sandstones. Shelf sequences may, however, be represented by two groups, respectively, of the relatively thin but numerous interbeds (plots in the upper left region within the shelf environment) and the relatively thick but less numerous sandstone development (plot in the lower right region).

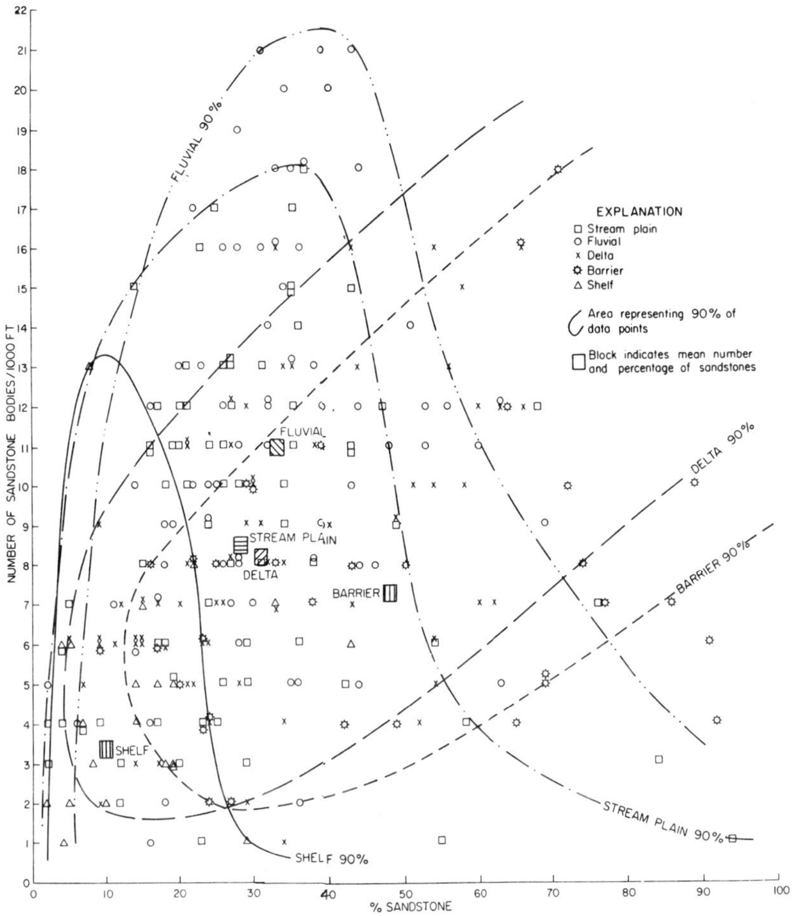

FIG. 12.8. Cross plot of percentage sandstone and number of sandstone bodies/1000 ft for five different environments of deposition of lower Frio.

Similar plots for the middle and lower Frio operational units are shown in Figs. 12.7 and 12.8, respectively. Although the shifts of the distribution patterns from those of Fig. 12.6 are observed, the relative locations and trends of the distribution stay almost the same. Therefore, we may conclude that a similar style of delta system sedimentation characterized middle and lower Frio.

Figure 12.9 summarizes the average number and percentage of sand-

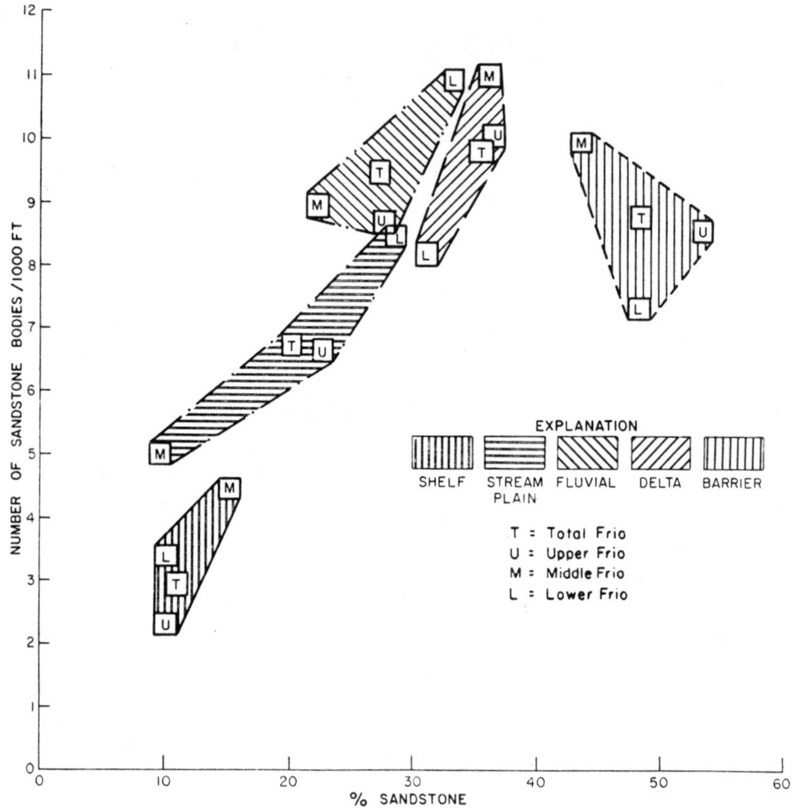

FIG. 12.9. Summary plot of percentage sandstone and number of sandstone bodies/1000 ft for five different environments of deposition of entire Frio.

stones for the upper, middle, and lower Frio, respectively, and for the total Frio MSU. This graph shows variation of the average values from the upper to lower Frio for each depositional environment, but each environment can be separated clearly from the other environments.

THICKNESS OF SANDSTONE AND SHALE UNITS

Distribution of number of sandstones vs thickness for the five depositional environments of the total Frio are shown in Fig. 12.10. Note

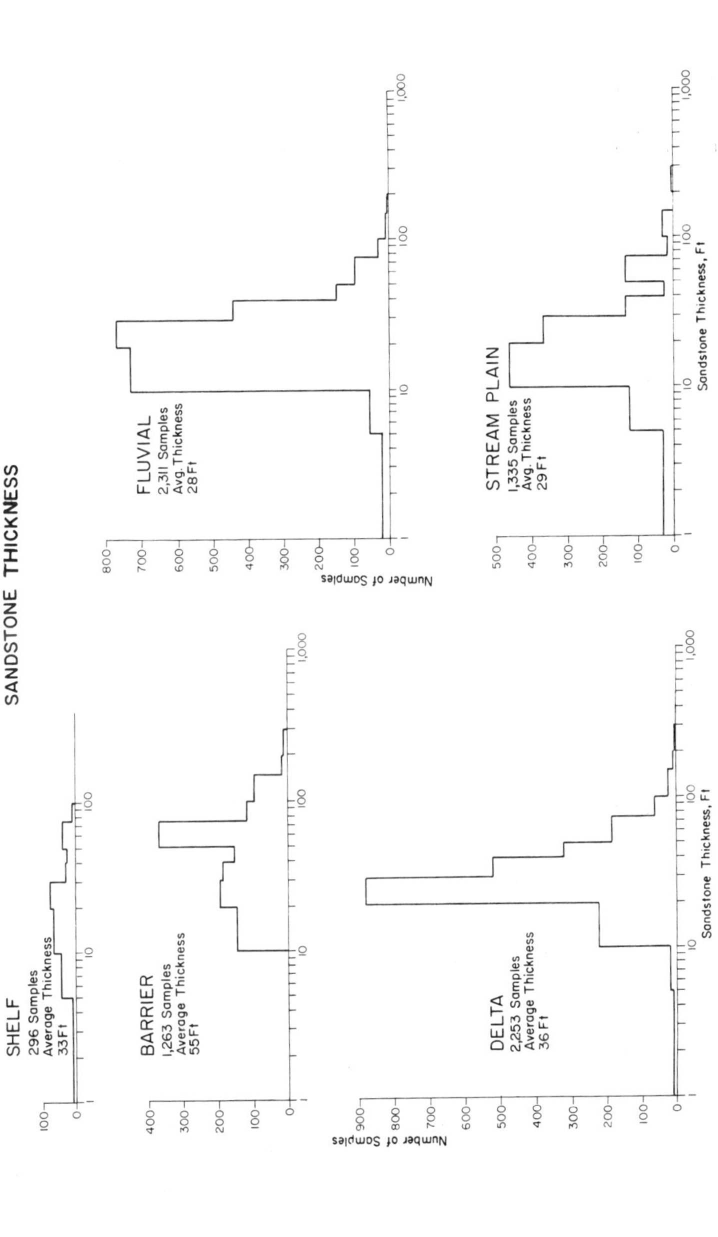

FIG. 12.10. Frequency plot of sandstone thickness for each of five different environments of deposition of entire Frio.

that the x axis of each graph in this figure has a logarithmic scale. The plots apparently show log-normal distributions. Number of samples used and the average thickness of sandstones for each of the five depositional environments are also shown in Fig. 12.10.

To test log-normality of the thickness distribution of sandstones, the percentage of the cumulative number of sandstones was plotted against the sandstone thickness for the five depositional environments on log-probability paper (Fig. 12.11). Note that a log-normal population would be shown by a straight line on log-probability paper. The distribution of sandstone thickness for each of the five environments can be approximated by a straight line, suggesting that there is a log-normal distribution of samples. Except for the shelf environment, all the other environments indicate relatively parallel distribution patterns with the lines shifted from left to right. This shows that the thickness of sandstones varies from environment to environment.

The distribution pattern for the shelf environment is different from those of the others; it shows a more inclined trend, suggesting that the sandstone thickness is more widely distributed for this environment. Thus, the standard deviation of sandstone thickness for the shelf environment is greater than those of the other environments.

Using the cumulative percentage plots such as those shown in Fig. 12.11, one would be able to determine graphically the median (mid-point value in the thickness distribution) and the standard deviation (how widely the distribution is made). The median sandstone thickness can be estimated directly from the plot by reading the thickness at the 50 percentile point for each environment. The approximate standard deviation, on the other hand, can be read from the difference between the median sandstone thickness and the thickness at either 16 or 84 percentile point on the y axis. The thickness corresponding to 16 percentile may be considered as the statistical minimum, and that to 84 percentile the statistical maximum.

The values estimated from Fig. 12.11 are shown in Table 12.1, along with the average sandstone thickness estimated earlier. Note that, in all cases, the median sandstone thickness is significantly less than the average thickness. This is primarily due to their log-normal distribution patterns. The statistical maximum and minimum in Table 12.1 may be considered as the statistically determined upper and lower limits of sandstone thickness, respectively, for the purpose of assessment of the sandstone thickness for each environment. Such an assessment is essen-

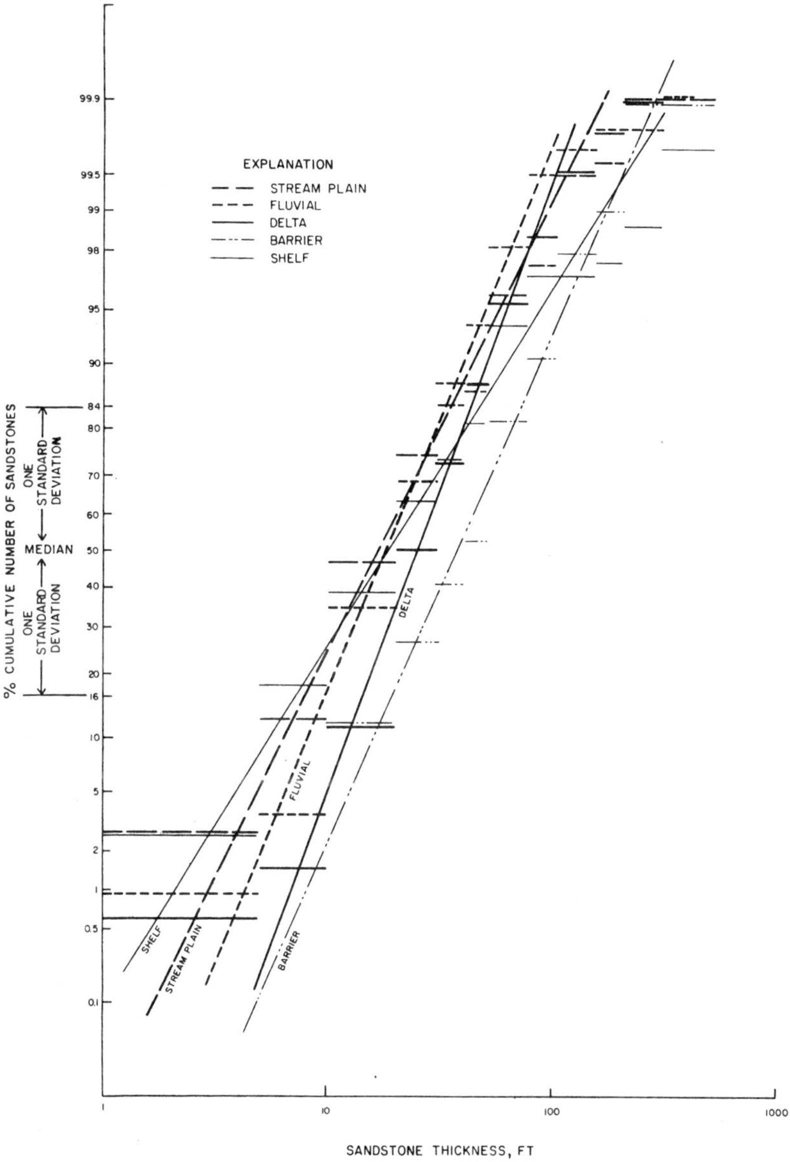

FIG. 12.11. Cumulative frequency plots of sandstone thickness on log-probability paper for five different environments of deposition of entire Frio.

TABLE 12.1
Statistical summary of sandstone and shale thicknesses

Sandstone thickness (ft)

Environment	*Minimum**	*Median*	*Maximum†*	*Average*
Stream-plain	8	17	35	29
Fluvial	10	18	33	28
Delta	14	25	42	36
Barrier	19	37	72	55
Shelf	7	18	47	33

Shale thickness (ft)

Environment	*Minimum*	*Median*	*Maximum*	*Average*
Stream-plain	46	84	150	118
Fluvial	32	56	98	77
Delta	26	47	87	66
Barrier	17	37	78	59
Shelf	70	150	320	300

*Median – one standard deviation.
†Median + one standard deviation.

tial for an application of Monte Carlo random simulation for the petroleum resource evaluation.

The fact that the sandstones of the Frio MSU are distributed log-normally may also be important in evaluating how most sandstones pinch out in a certain direction. It suggests that the pinch-out would occur logarithmically rather than arithmetically. In other words, when contour lines of the sandstone thickness around a pinch-out area are drawn, an equal spacing of the contours, which means a uniform or normal change of the sandstone thickness, is unjustified. There must be more abrupt changes from a certain thickness to near-zero thickness. Therefore, the statistical data shown above can improve the style of sandstone isopach mapping.

Figure 12.12 shows the plots of number of samples vs shale thickness of the total Frio MSU for the five depositional environments. The distribution pattern of each environment for shales can also be approximated by a straight line (Fig. 12.13), indicating a log-normal distribution. The trend lines of the stream-plain, fluvial, and delta are nearly parallel with slight horizontal shifts of the trends. The lines representing both the

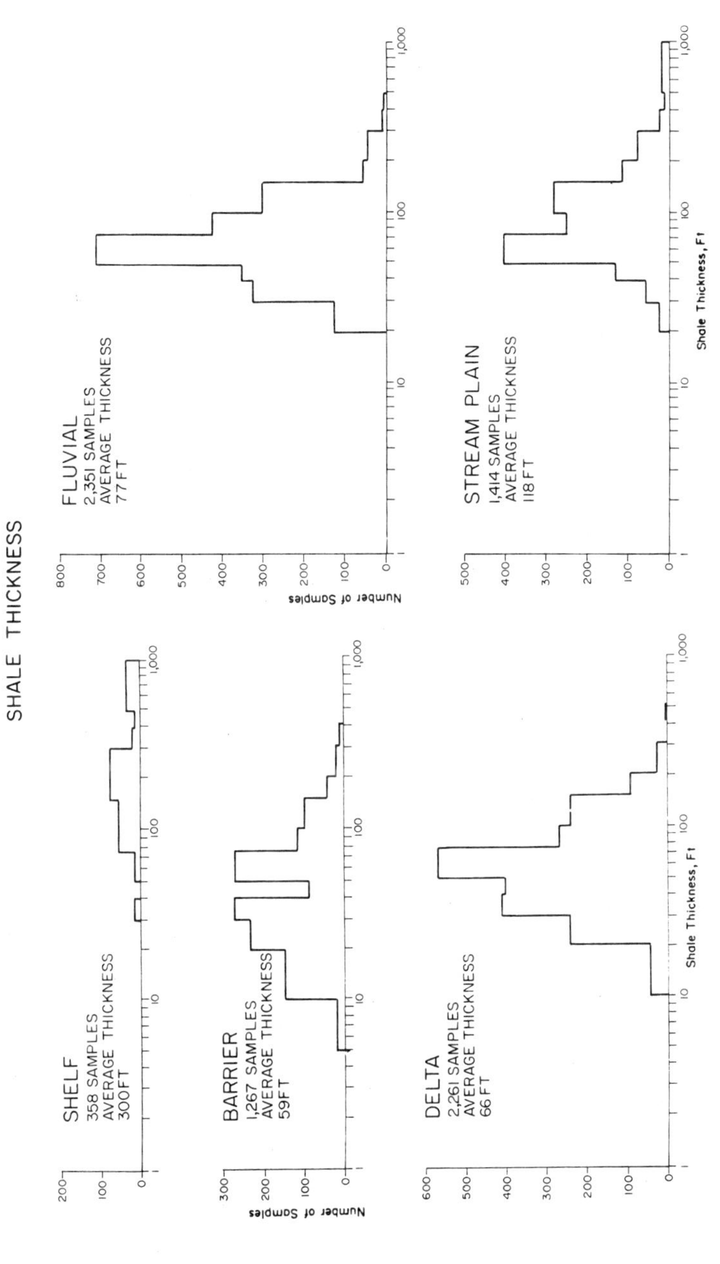

FIG. 12.12. Frequency plot of shale thickness for each of five different environments of deposition of entire Frio.

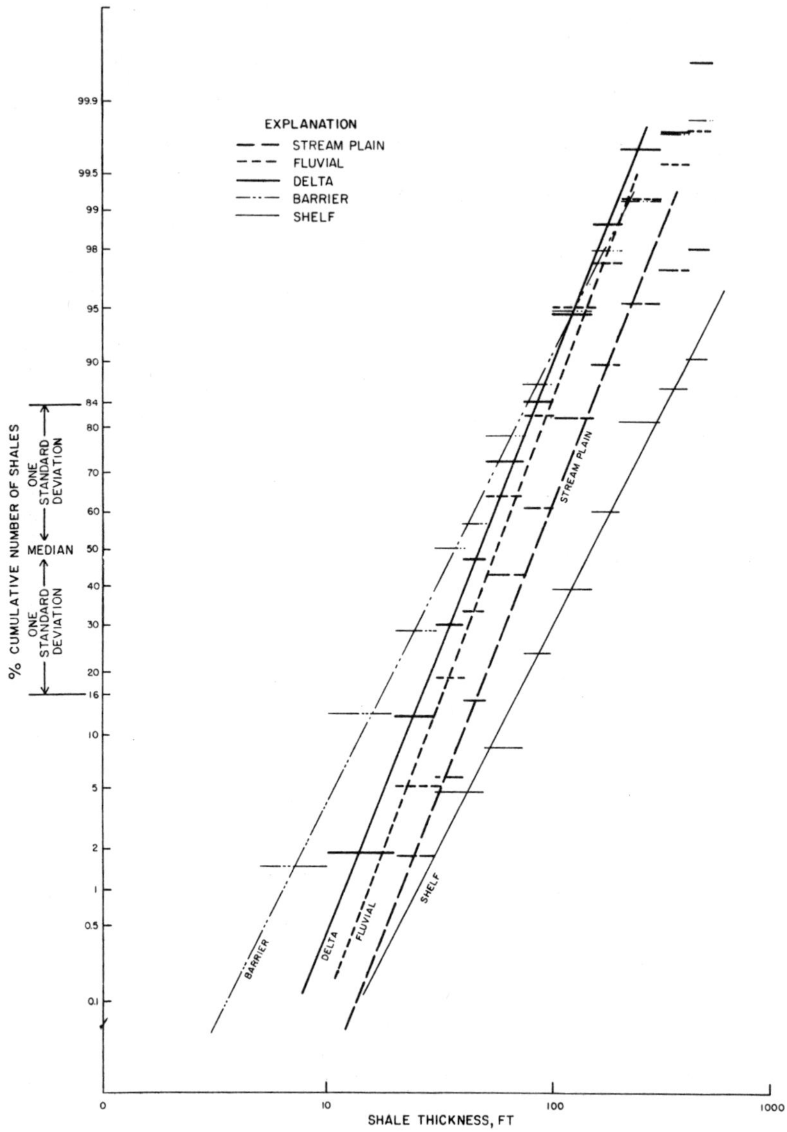

FIG. 12.13. Cumulative frequency plots of shale thickness on log-probability paper for five different environments of deposition of entire Frio.

barrier and shelf environments are flatter than those of the other environments, suggesting that the populations are more widely distributed. The standard deviations for the barrier and shelf, both of which reflect greater marine influence than the others, are thus greater. The statistical summary for shales of the five different environments is included in Table 12.1.

NUMBER OF SANDSTONE AND SHALE UNITS

Figure 12.14 depicts the plot of number of sandstones (or shales)/1000 ft vs number of locations for each of the five depositional environments. Note, in this case, that a normal graph paper is used, on the basis of the assumption that the distribution pattern of the sandstone–shale interbedding would be normal (not log-normal).

This assumption seems to be reasonable, because the plot of the cumulative frequency of the sandstone (or shale) number/1000 ft on the normal probability paper can be approximated by a near-straight line for each environment (Fig. 12.15). Note that, toward the upper end of each plot, the trend becomes flatter, suggesting that such a normal distribution does not exist where the number of sandstones (or shales) is relatively large (more than 10 sandstones or shales/1000 ft for shelf and more than 20 sandstones or shales/1000 ft for most of the other environments).

The statistical figures estimated directly from Fig. 12.15 are summarized in Table 12.2. Here the median figures are relatively close to the average values, because the distribution patterns are almost normal (not log-normal).

SANDSTONE PERCENTAGE

Sandstone percentage in a geologic unit is a product of the number and the thickness of the sandstones. Due to the fact that the distribution pattern of the number is normal whereas that of the thickness is log-normal, one would anticipate the distribution of sandstone percentage to display an intermediate pattern between normal and log-normal.

Figure 12.16 shows the frequency distribution of the sandstone percentage on normal graph paper, and Figs. 12.17 and 12.18 depict the percent cumulative frequency of such on the normal and log-normal

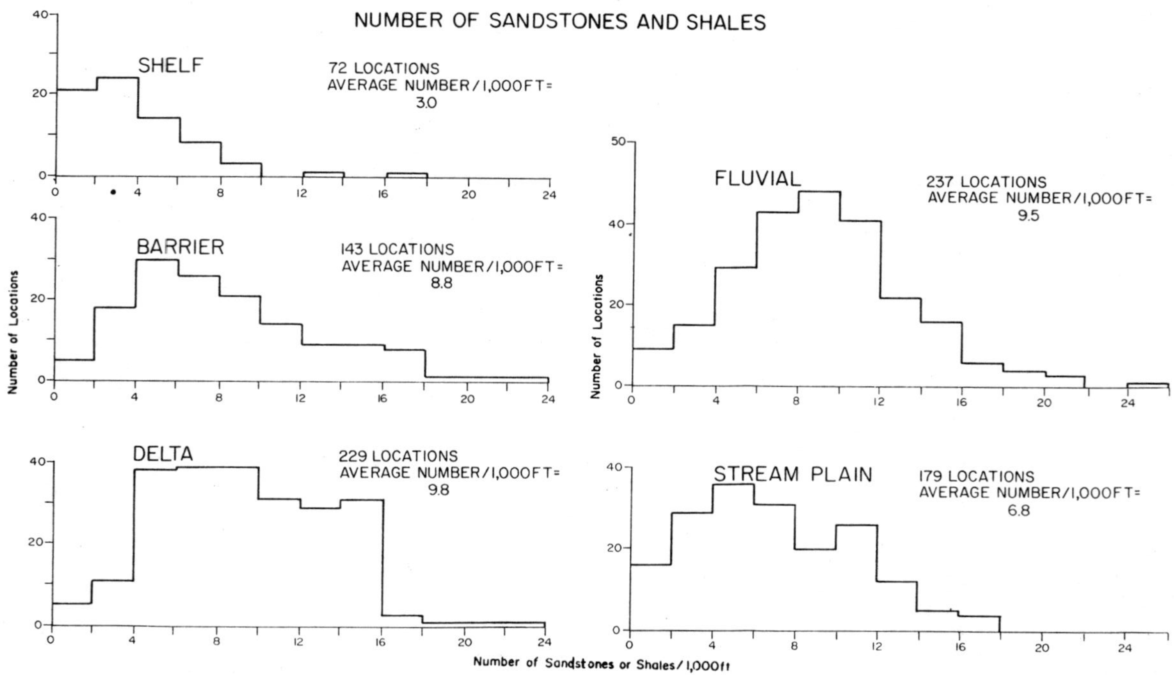

FIG. 12.14. Frequency plot of number of sandstone (or shale) bodies for each of five different environments of deposition of entire Frio.

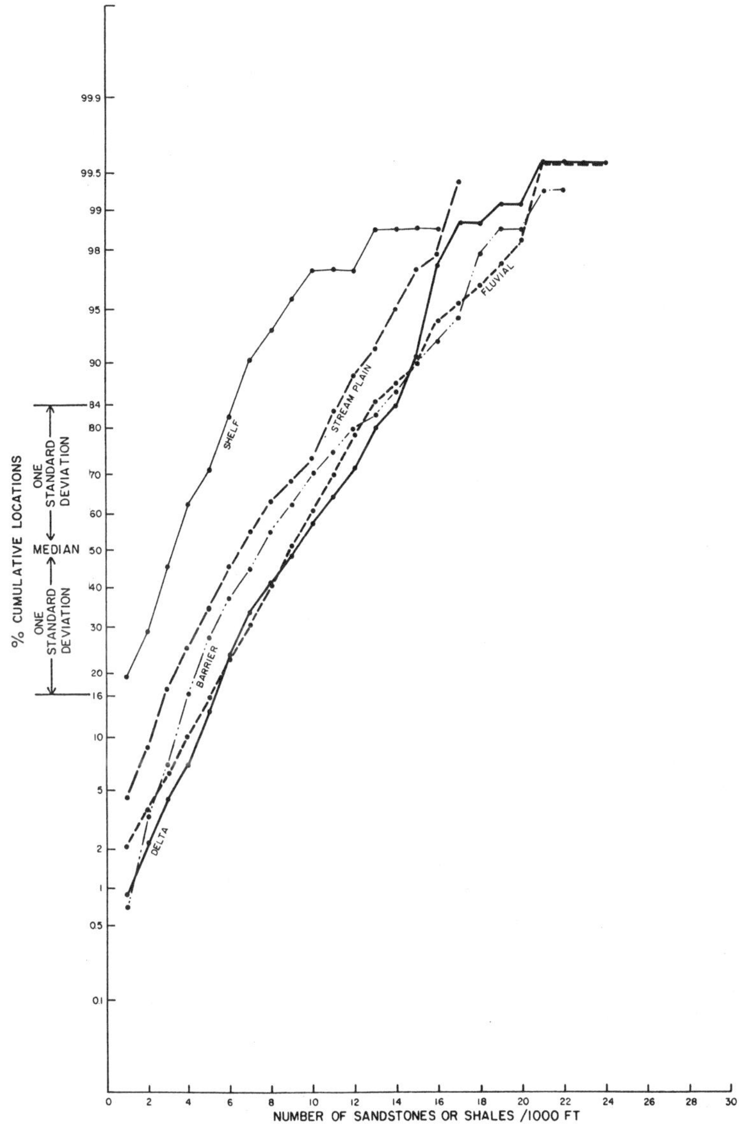

FIG. 12.15. Cumulative frequency plots of number of sandstone (or shale) bodies on probability paper for five different environments of deposition of entire Frio.

TABLE 12.2
Statistical summary of frequency of interbedding (Sandstone or shale) (number/1000 ft)

Environment	*Minimum**	*Median*	*Maximum†*	*Average*
Stream-plain	2·8	6·5	11·2	6·8
Fluvial	5·0	9·0	13·0	9·5
Delta	5·3	9·2	14·0	9·8
Barrier	4·0	7·5	13·6	8·8
Shelf	0·5	3·3	6·2	3·0

*Median – one standard deviation.
†Median + one standard deviation.

probability papers, respectively. It is seen from Figs. 12.17 and 12.18 that straight-line approximations cannot be justified in any of these two papers, confirming that the distribution patterns of the sandstone percentage lie between normal and log-normal. Table 12.3 summarizes the sandstone percentage statistics determined from Fig. 12.17 (normal probability paper).

SUMMARY

The statistical summary of the five depositional facies sequences of the Frio MSU is shown in Fig. 12.19. From left to right, sandstone percentage, number of sandstones (or shales) per 1000 ft, sandstone thickness, and shale thickness are depicted for the five depositional facies. The block of each facies show the statistically determined minimum (left end of each block) and maximum (right end of each block) values. A thick vertical line in each block indicates the median and a dashed line shows the mean (or average) value.

The thickness and percentage of sandstones in the barrier facies are greatest among the five depositional facies, resulting in the thinnest average thickness of the interbedded shales. One would, therefore, anticipate good reservoir development in this facies, but the hydrocarbon sealing condition may be relatively poor.

The delta facies has the largest number of sandstone and shale beds in a given total thickness (1000 ft). Both the percentage and thickness of sandstones are reduced in the delta environment, and the thickness of shales increases. Therefore, there may be improved sealing conditions in this facies.

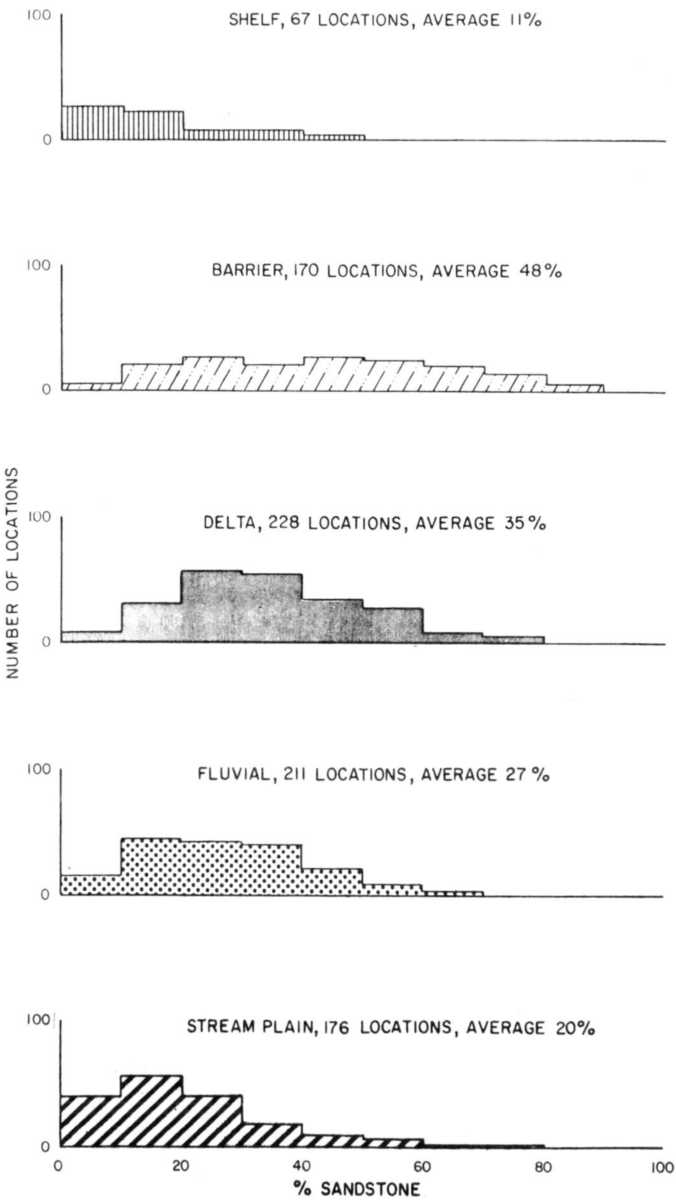

FIG. 12.16. Frequency plot of percentage sandstone for each of five different environments of deposition of entire Frio.

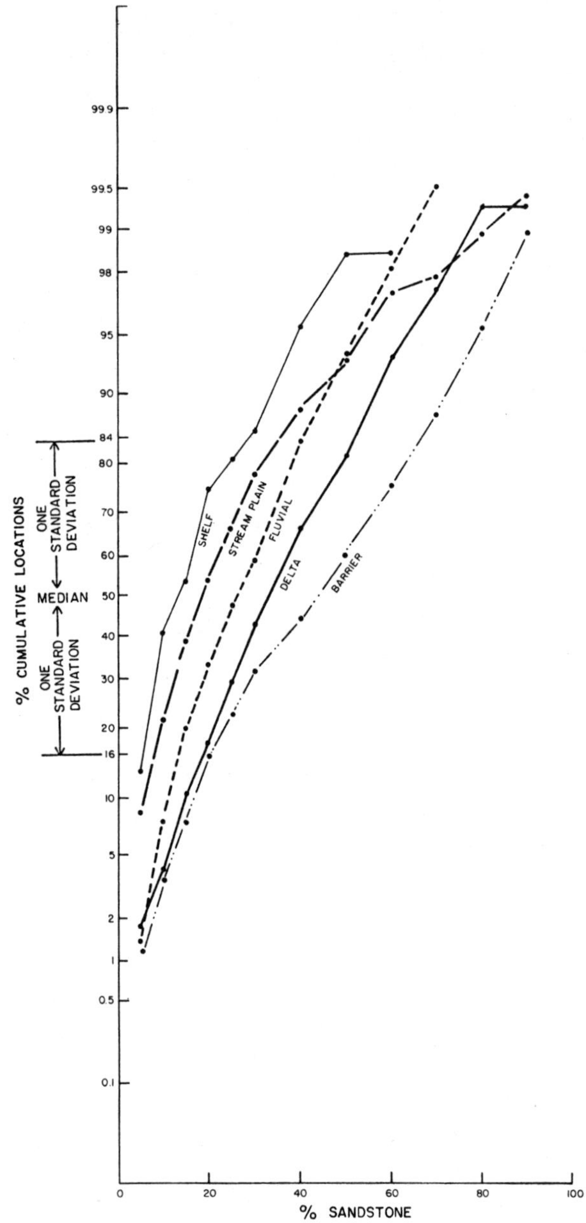

FIG. 12.17. Cumulative frequency plots of percentage sandstone on probability paper for five different environments of deposition of entire Frio.

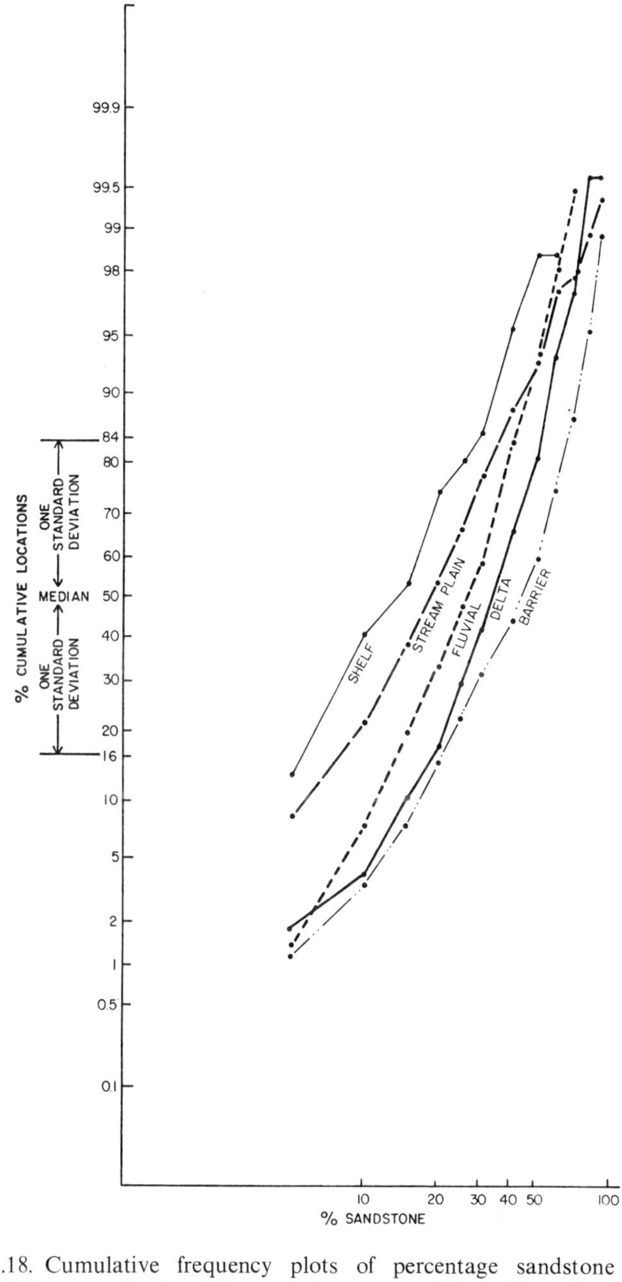

FIG. 12.18. Cumulative frequency plots of percentage sandstone on log-probability paper for five different environments of deposition of entire Frio.

TABLE 12.3
Statistical summary of sandstone percentage

Environment	*Minimum**	*Median*	*Maximum†*	*Average*
Stream-plain	8	19	36	20
Fluvial	14	26	41	27
Delta	19	33	52	35
Barrier	20	43	67	48
Shelf	6	14	29	11

*Median − one standard deviation.
†Median + one standard deviation.

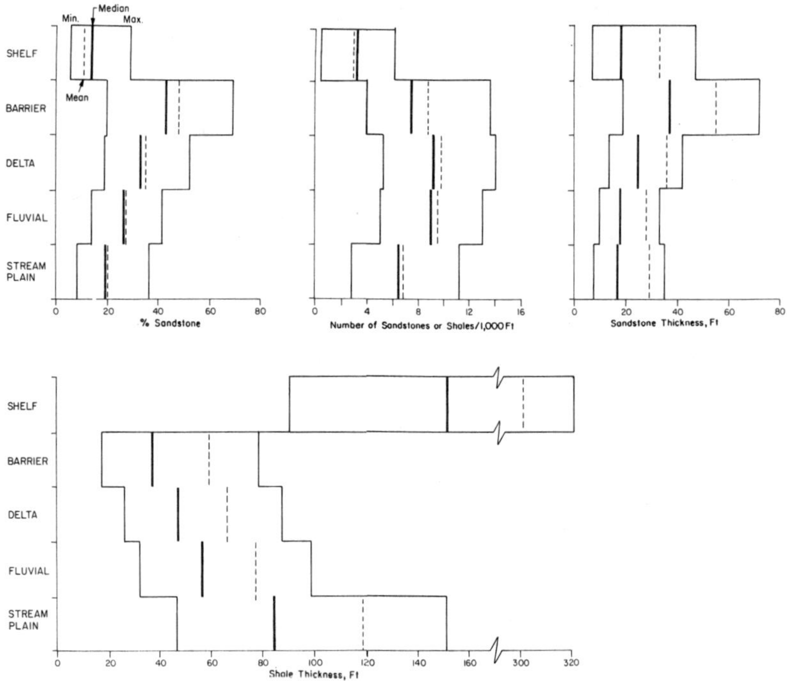

FIG. 12.19. Summary plots of various litho-statistical data of entire Frio.

The progressive decrease of both the sandstone percentage and number per 1000 ft characterizes the fluvial and stream-plain environments. The shale thickness, accordingly, increases. In these environments, the reservoir quality and quantity could be a problem, but in general, the sealing condition is excellent. However, the quality of organic matter and level of petroleum generation of these facies may present additional problems in creating significant oil (and maybe gas) accumulations in these systems, unless large scale vertical or lateral migration of petroleum from the zones of better organic facies is involved.

The shelf facies, on the other hand, is characterized by low sandstone percentage and low frequency of interbedding. The average thickness of sandstones is not extremely thin but that of shales is quite thick. Within this facies, the reservoir development is the most critical factor for oil and gas accumulations.

Importantly, the differing depositional systems, which consist of different proportions and combinations of these basic five facies sequences, are statistically different in terms of reservoir-source-seal distributions and abundance.

REFERENCES

Bebout, D. G., Loucks, R. G. and Gregory, A. R. (1978) Frio sandstone reservoirs in the deep subsurface along the Texas Gulf Coast, Report of Investigations No. 91, The University of Texas at Austin, Bureau of Economic Geology, 92 p.

Berg, R. R. and Powers, B. K. (1980) Morphology of turbidite-channel reservoirs, Lower Hackberry (Oligocene), southeast Texas, Gulf Coast Association of Geological Societies Trans., Vol. 30, pp. 41–48.

Bishop, R. S. (1978) Mechanism for emplacement of piercement diapirs. *A.A.P.G.*, **62**, 1561–83.

Boyd, D. R. and Dyer, B. F. (1964) Frio barrier bar system of South Texas, Gulf Coast Association of Geological Societies Trans., Vol. 14, pp. 309–22.

Bruce, D. H. (1973) Pressured shale and sediment deformation: Mechanism for development of regional contemporaneous faults. *A.A.P.G.*, **57**, 878–86.

Galloway, W. E., Hobday, D. K. and Magara, K. (1982) Frio Formation of the Texas Gulf Coastal Plain—Depositional systems, structural framework, and hydrocarbon origin, migration, distribution and exploration potential, The University of Texas at Austin, Bureau of Economic Geology, Report No. 122, 78 pp.

Gulf Coast Association of Geological Societies (1972) Tectonic Map of the Gulf Coast Region, U.S.A., *A.A.P.G.*, Tulsa, Ok.

Paine, W. R. (1971) Petrology and sedimentation of the Hackberry sequence of southwest Louisiana, Gulf Coast Association of Geological Societies Trans., Vol. 21, pp. 37–55.

Stuckey, C. W. (1964) The stratigraphic relationship of the Hackberry, Abbeville, and Harang faunal assemblages, Gulf Coast Association of Geological Societies Trans., Vol. 14, pp. 209–16.

Chapter 13

Litho-Statistical Study of Type Sections of Jurassic and Cretaceous Formations in Saudi Arabia

INTRODUCTION

In forming reservoir, source, and cap rocks for petroleum accumulations, a sedimentary sequence must have a set of physical, chemical, and geological characters which help to generate, migrate, and trap hydrocarbons. Such characters can be expressed in terms of porosity, permeability, organic content and type, organic maturity, capillary pressure, etc. They are directly or indirectly controlled by the basic lithological characteristics of the sequence, such as lithology and thickness.

Knowing a certain set of lithologic characteristics would not guarantee those of good reservoir, source, and cap rocks, but may give us a clue in guessing a possibility of having it. Comparison of such possibilities in different areas would be valuable in assessing exploration prospects in the region.

In this chapter, the data of lithology and thickness of the Jurassic and Cretaceous formations at the type localities in Saudi Arabia (Fig. 13.1) defined by Powers *et al.* (1966) are summarized and analysed statistically. The statistical results were considered in the light of either reservoir, source, or cap rock capacity. However, because there are significant differences in both litho-facies and depositional environment between the Eastern Province of Saudi Arabia where most oil fields occur and the Central Province where most type localities except that of the Arab Formation (principal reservoir of Jurassic age) exist, the result of the litho-statistical study may only be used as a clue to understand or infer the probable characteristics of reservoir, source, and cap rock in this region.

It is strongly recommended that in the future study of the subsurface data of the Eastern Province is undertaken to understand true statistical

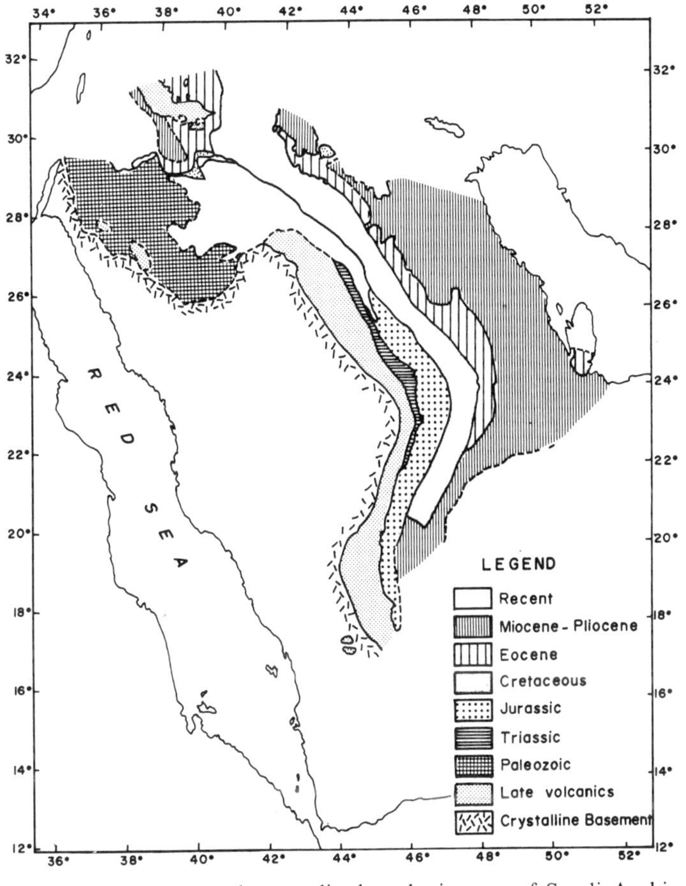

FIG. 13.1. Location map and generalized geologic map of Saudi Arabia (from Powers *et al*., 1966). (Reproduced by permission of the U.S. Geological Survey, United States Department of the Interior.)

relationships with the occurrences of oil there. However, at present we are unable to conduct such research because of unavailability of data due to their confidential nature.

The following guidelines were employed for interpreting the litho-statistical data.

1. Source Rock

A typical source rock is defined as a fine-grained rock which has generated and expelled hydrocarbons in a sufficient amount.

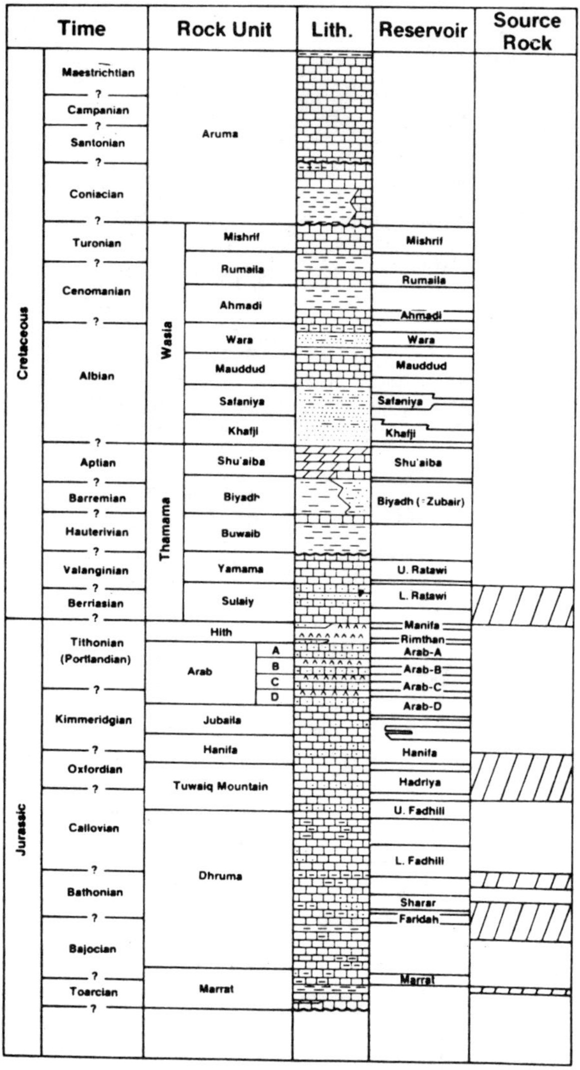

FIG. 13.2. Generalized Cretaceous and Jurassic stratigraphic section showing lithology, reservoir units, and source rocks (from Ayres *et al.*, 1982). (Reproduced by permission of the American Association of Petroleum Geologists.)

Consequently, the evaluation of source rock depends on the study of the richness of organic carbon (about 0·4% or more), type of organic matter, thermal maturity, and expulsion efficiency.

Ayres *et al.* (1982) concluded that the organic rich carbonate rocks have been identified in the Middle Jurassic and Lower Cretaceous formations in the northeastern part of Saudi Arabia as having an average organic carbon content of 3–5% (Fig. 13.2). To generate significant amounts of hydrocarbons, however, source rocks must have a certain minimum thickness, which is arbitrarily considered to be 4 m in this chapter. In short, formations which have relatively high frequency values of limestones or shales of 4 m or more are considered to be potential source rocks. To become true and effective source rocks, they of course must retain sufficient amounts of organic matter and enough maturity.

2. Reservoir

A reservoir is a hydrocarbon container which is made of relatively continuous porous media. A typical reservoir may be either calcarenite, fractured carbonate, sandstone, fissured crystalline rock, dolomite, or any other porous rock. Two major types of reservoirs have been described in the northeastern part of Saudi Arabia; calcarenite associated with dolomite, and sandstone. The result of a litho-statistical study of the Arab Formation, the most important petroleum reservoir in this region, suggests that thinly interbedded (< 1 m) calcarenite, dolomite, and aphanitic limestone made a good reservoir. For sandstones, however, thick and homogeneous reservoirs may be suitable for trapping large quantities of oil.

3. Seal

A seal is a very low permeable layer which completely covers the reservoir and forms a vertical discontinuity against upward hydrocarbon migration. In Saudi Arabia, both thick anhydrite and shale are the main seals.

INTERPRETATION OF SOURCE, RESERVOIR, AND SEAL ROCKS OF JURASSIC AGE

From the litho-statistical interpretation of the Jurassic formations, potentials of each formation as a source, reservoir, and seal may be estimated. Based on the geochemical study by Ayres *et al.* (1982), the

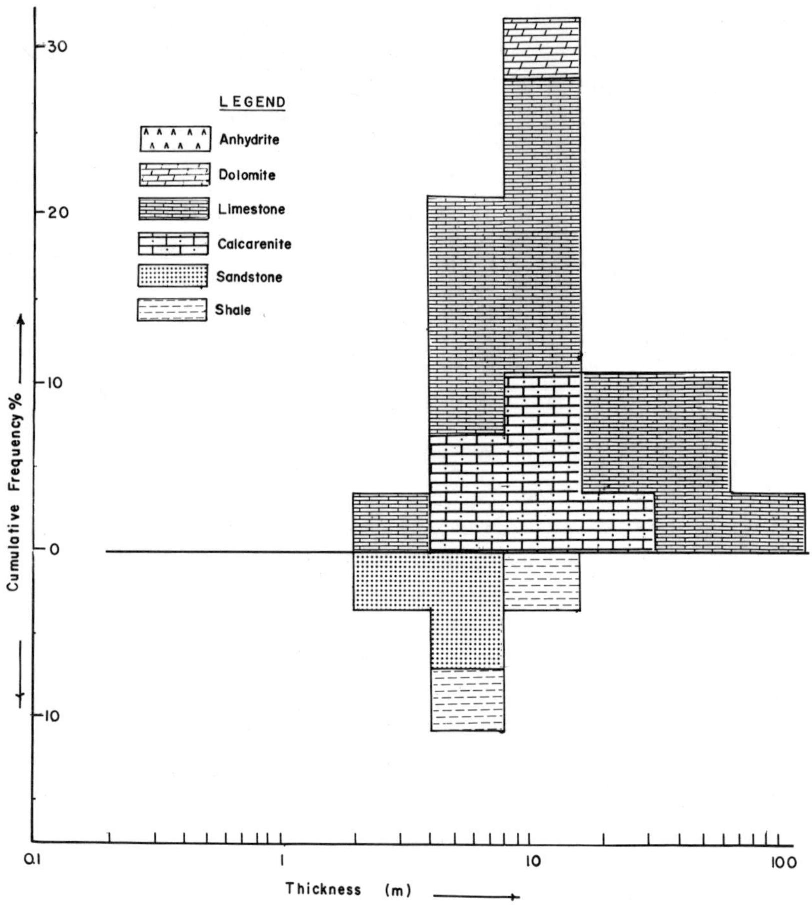

FIG. 13.3. Lithological frequency plot of Tuwaiq Mountain Formations.

Tuwaiq Mountain limestone is considered as a typical source rock (Fig. 13.2). The litho-statistics shows that relatively high percentages of aphanitic limestone occur between the thicknesses of 4 and 64 m (Fig. 13.3). Tuwaiq Mountain would thus be a good source rock, if it contains a relatively large quantity of organic matter as described by Ayres *et al.* (1982).

The Arab Formation, which consists of thin and cyclic calcarenite beds associated with dolomite and aphanitic limestone, is the most important

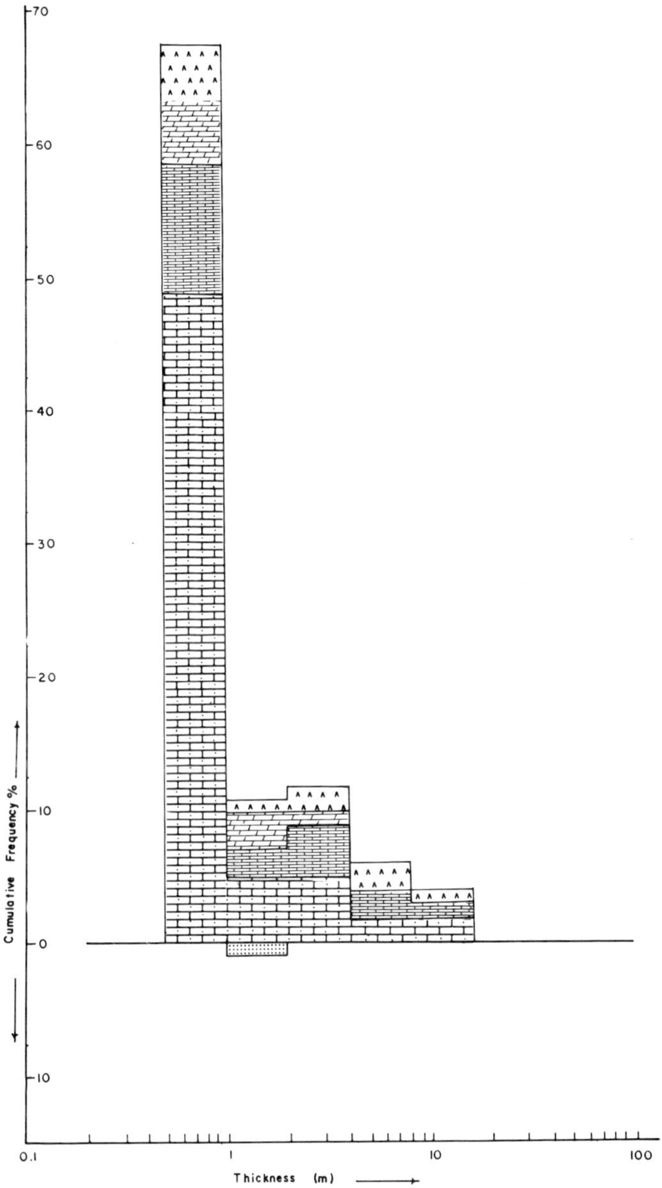

FIG. 13.4. Lithological frequency plot of Arab Formation.

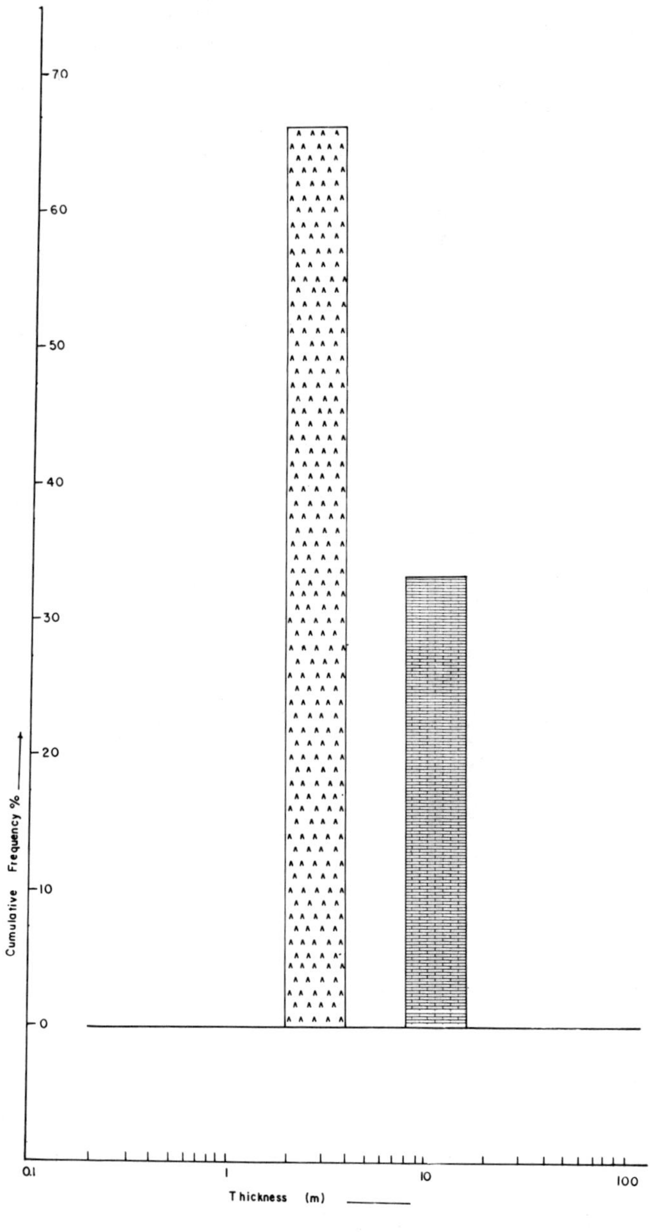

FIG. 13.5. Lithological frequency plot of Hith Formation.

petroleum reservoir in this region (Fig. 13.4). It has a very high calcarenite frequency value of about 49% within the thickness interval of 0.5–1 m. Anhydrite of Arab shows frequency values of 1–4% within the thickness interval of 0.5–16 m (Fig. 13.4).

The Hith Formation which consists of massive and compact anhydrite with occasional layers of fine-grained pure limestones is considered here as a typical seal. A high frequency value of about 67% occurs in the thickness interval of 2–4 m (Fig. 13.5). The aphanitic and pure limestone has a frequency value of about 33% within the thickness interval of 8–16 m.

OTHER JURASSIC FORMATIONS

The Jurassic sedimentary sequence of Saudi Arabia consists of the Marrat, Dhruma, Tuwaiq Mountain, Hanifa, Jubaila, Arab, and Hith Formations (Fig. 13.2). The litho-statistical data of the formations other than the previous three formations (Tuwaiq Mountain, Arab, and Hith) will be made as follows.

Marrat shows relatively high frequency percentages of both aphanitic limestone and shale in a wide thickness range (Fig. 13.6). This suggests that the Marrat Formation has a good chance of becoming a source rock if it is mature with sufficient organic matter. It may also show an excellent sealing capacity because of its shales. Relatively low frequency values of calcarenite and sandstone suggest that Marrat has a minor chance of being a reservoir for significant amounts of hydrocarbons (Fig. 13.6).

Figure 13.7 shows the Dhruma lithological frequency plot. There are relatively high frequency percentages of both fine-grained limestone and shale in wide ranges of thickness intervals, suggesting an excellent chance of source rocks. Relatively high frequencies of calcarenite occur between thicknesses of 0·5 and 4 m. This suggests that Dhruma can be a good reservoir, if it is sealed by shales (Fig. 13.7).

The frequency plot of Hanifa shows relatively high values of aphanitic limestone (1–64 m thickness) and shale (4–16 m), suggesting an excellent chance of source rocks if they contain sufficient amounts of organic matter (Fig. 13.8). The chances of Hanifa being a reservoir are also high, because of its relatively high frequencies for both calcarenite and dolomite.

The plot of the Jubaila Formation is shown in Fig. 13.9. Jubaila can be good source rocks if it contains high organic contents, because of

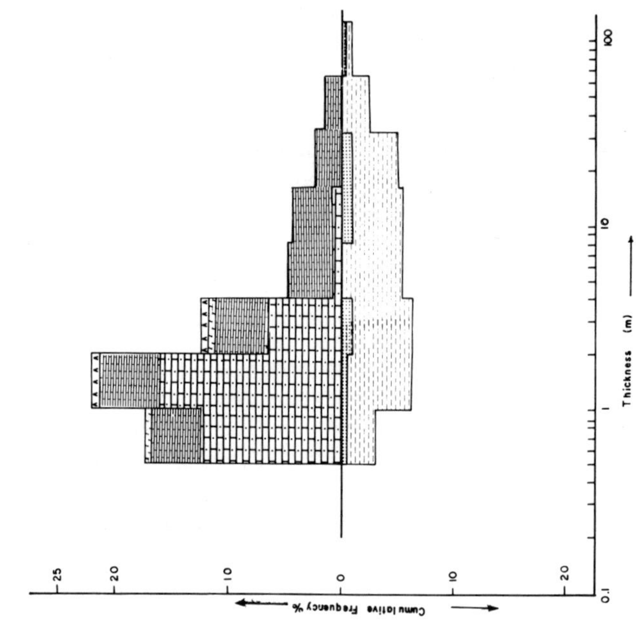

FIG. 13.7. Lithological frequency plot of Dhruma Formation.

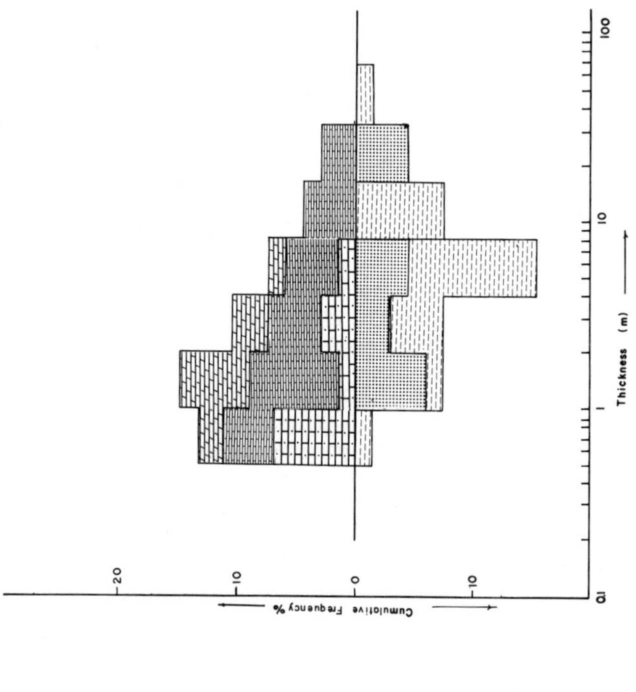

FIG. 13.6. Lithological frequency plot of Marrat Formation.

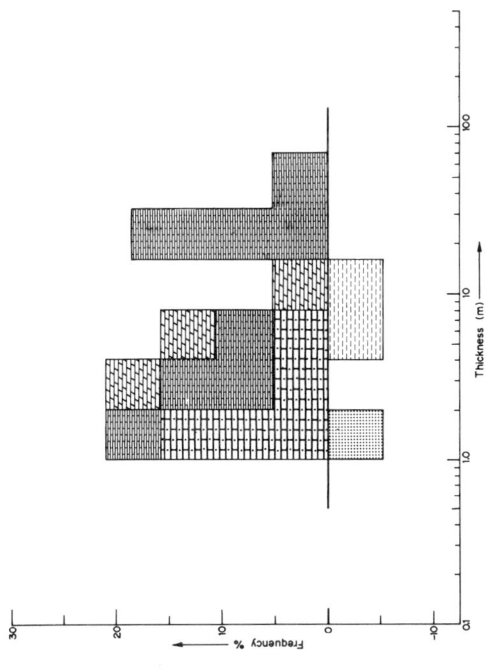

FIG. 13.8. Lithological frequency plot of Hanifa Formation.

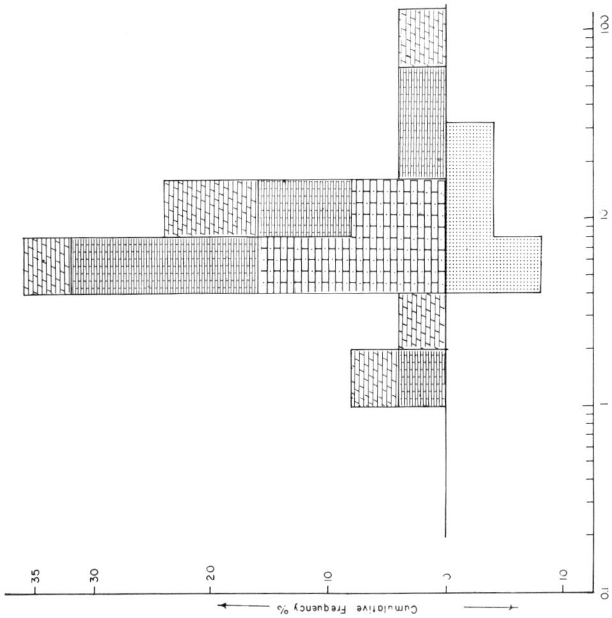

FIG. 13.9. Lithological frequency plot of Jubaila Formation.

relatively high frequencies of aphanitic limestone. There are also significant developments of porous sections (potential reservoirs) by both calcarenite and dolomite. Anhydrite and shale are both absent, suggesting that Jubaila has little chance of being a seal.

In summary, the Tuwaiq Mountain Formation is considered as a typical source rock, whereas the Hanifa, Jubaila, Marrat, and Dhruma Formations indicate very good chances of becoming source rocks from a purely litho-statistical standpoint. The Arab Formation, which consists of thin and cyclic calcarenite beds, is believed to be the best reservoir among the Jurassic formations. Hanifa, Jubaila, Marrat, and Dhruma also show good potential for being a reservoir.

For seals, the Hith anhydrite is considered as a typical seal, and Arab also has good sealing zones. Other Jurassic formations show relatively poor sealing capacities or chances, except the shales in the Marrat and Dhruma Formations.

CRETACEOUS FORMATIONS

Litho-statistical plots were made for the Sulaiy, Yamama, Buwaib, Biyadh, Wasia, and Aruma Formations.

The frequency plot of the Sulaiy Formation, as shown in Fig. 13.10, depicts only two lithologies of aphanitic limestone and calcarenite, suggesting good potentials for being source and reservoir rocks. To become a source rock, it must contain sufficient amounts of organic matter and attain sufficient maturity, of course.

The plot of Yamama shows relatively high frequencies of both calcarenite and aphanitic limestone in the thickness interval of 0·5–4 m (Fig. 13.11). Although there is a high reservoir possibility with calcarenite, good source sections may not be present because of their thickness (<4 m). Anhydrite and shale are completely absent in Yamama (Fig. 13.11), suggesting virtually no chance of being a seal.

The litho-statistical plot of Buwaib, shown in Fig. 13.12, suggests good possibilities for both reservoir and source rocks (if they are mature).

Both Biyadh and Wasia at the type localities consist mainly of clastic rocks (sandstone and shale), possibly acting as reservoir, source, and cap rocks (Figs. 13.13 and 13.14). To become a source rock, of course, it must contain a sufficient amount of organic matter and be of sufficient maturity.

The plot for the Aruma Formation shows aphanitic limestone, calca-

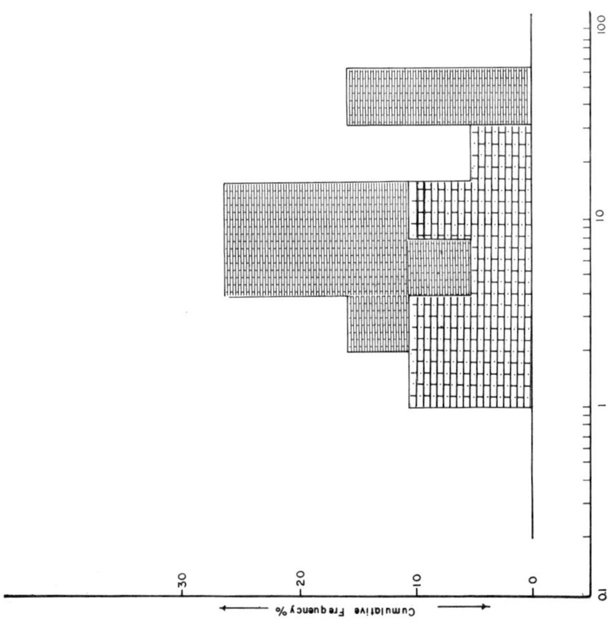

FIG. 13.10. Lithological frequency plot of Sulaiy Formation.

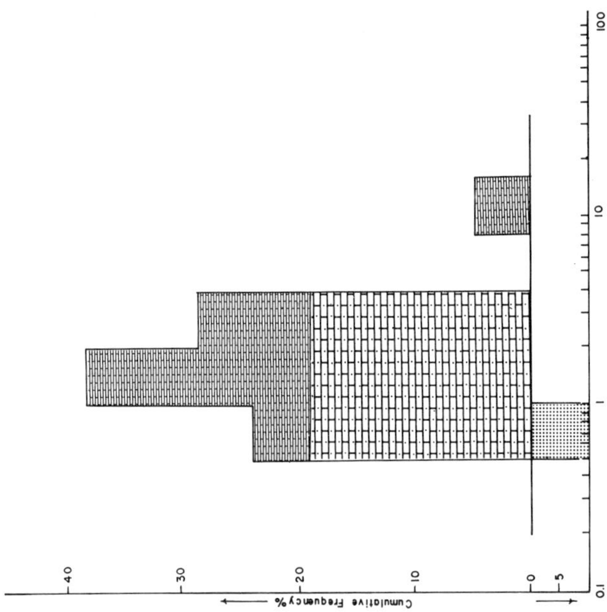

FIG. 13.11. Lithological frequency plot of Yammama Formation.

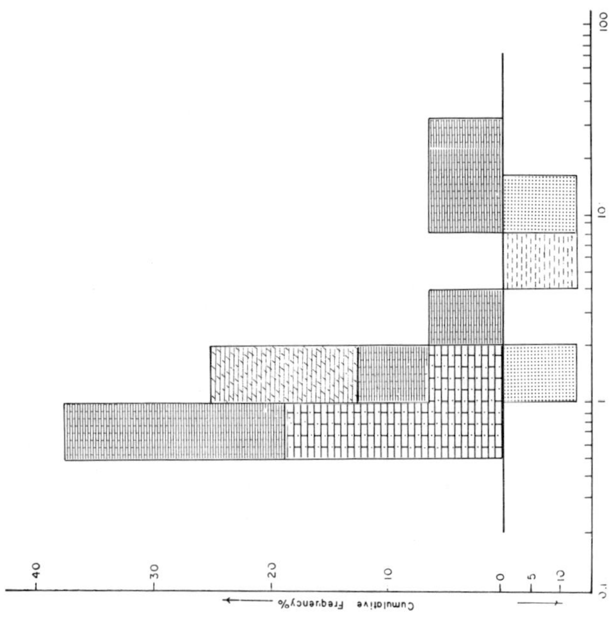

FIG. 13.12. Lithological frequency plot of Buwaib Formation.

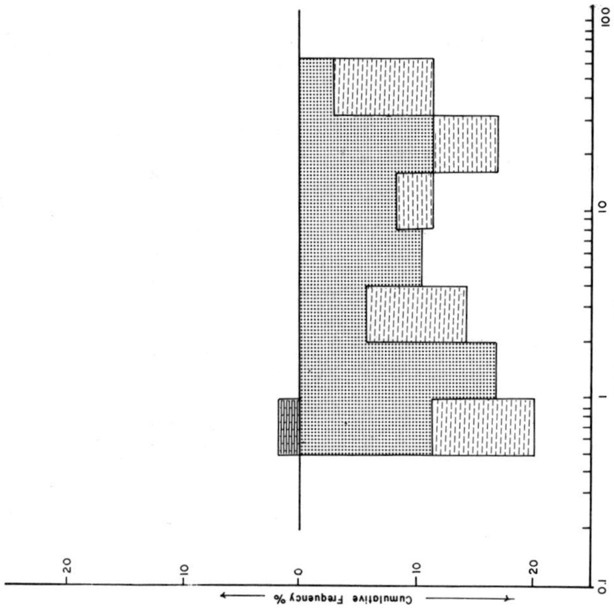

FIG. 13.13. Lithological frequency plot of Biyadh Formation.

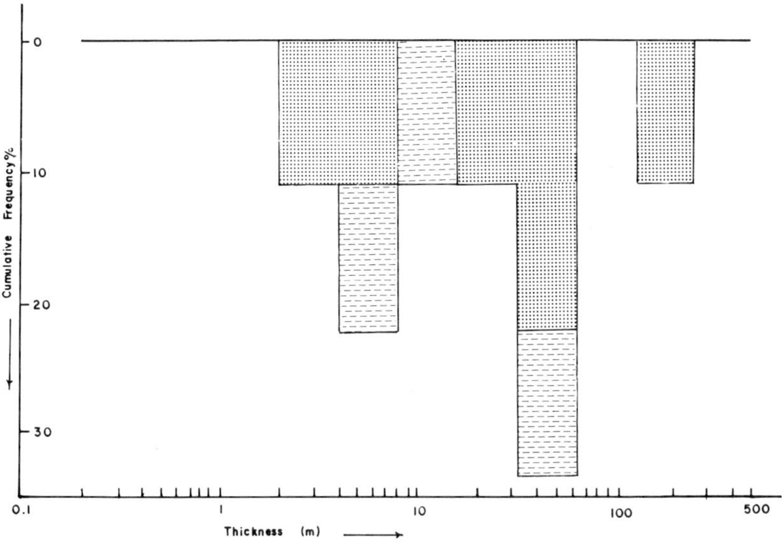

FIG. 13.14. Lithological frequency plot of Wasia Formation.

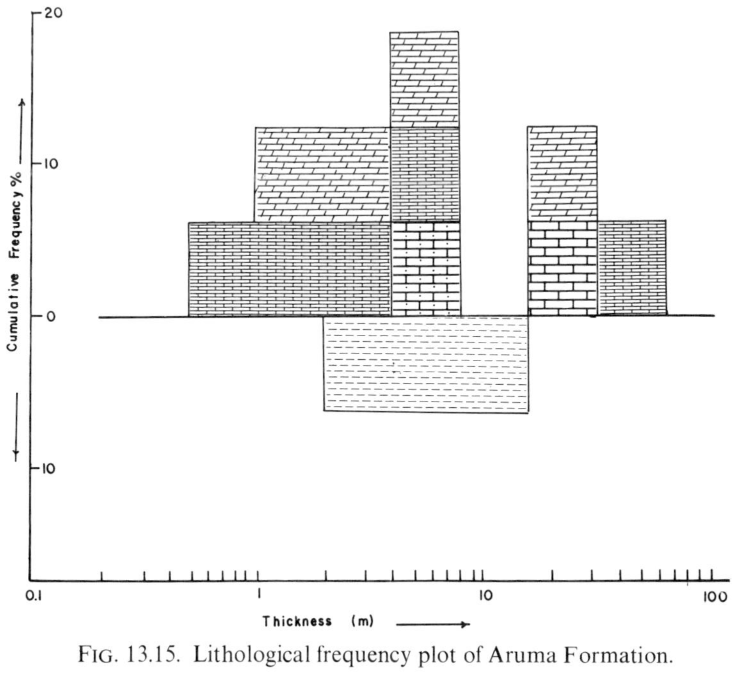

FIG. 13.15. Lithological frequency plot of Aruma Formation.

renite, dolomite, and shale (Fig. 13.15). Although the potentials for a reservoir and cap rock exist, this formation cannot be good source rock mainly because of its low maturity.

In summary, the Sulaiy Formation presents the best chance to be the main source rocks, from the litho-statistical standpoint. The shales of both Biyadh and Wasia may also be source rocks, if they are sufficiently mature.

For the reservoir potential, the Wasia sandstones have chances to be the best reservoir rocks in the Cretaceous sequence, followed by the Yamama calcarenite. The Biyadh sandstones may also be reservoirs.

Although anhydrite is generally absent in the Cretaceous sections, shales can act as excellent seals.

SUMMARY PLOTS

The summary plot for the entire Jurassic formations is shown in Fig. 13.16, and that for Cretaceous in Fig. 13.17.

The cumulative frequency plots for both Jurassic and Cretaceous on log-probability graphs are shown in Figs. 13.18 and 13.19, respectively. The statistical values, such as minimum, most likely, and maximum thicknesses were also estimated graphically, using the method described in Chapter 12, and are shown in Table 13.1.

COMPARISON WITH OIL RESERVOIR STATISTICS

The above-mentioned litho-statistical results may be compared with the frequency of the main oil reservoirs in this region (Fig. 13.20). The highest reservoir frequency percentage occurs in the Arab Formation with about 30%, whereas Jubaila, Hanifa, and Dhruma have frequency values of about 20, 11, and 9%, respectively (Fig. 13.20). Within the Cretaceous sequence, only Wasia has a percentage figure of about 9, whereas the others have less than 5%.

LITHOLOGICAL TRIANGLE PLOTS OF JURASSIC AND CRETACEOUS FORMATIONS

Based on the lithological data of the type sections defined by Powers *et al.* (1966), triangle plots have been made by computer for both the

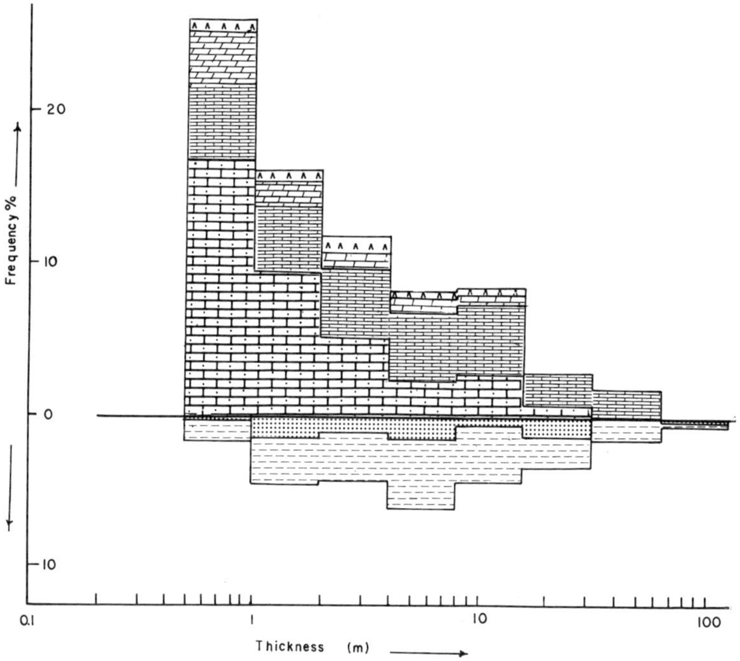

FIG. 13.16. Lithological frequency plot of total Jurassic formations.

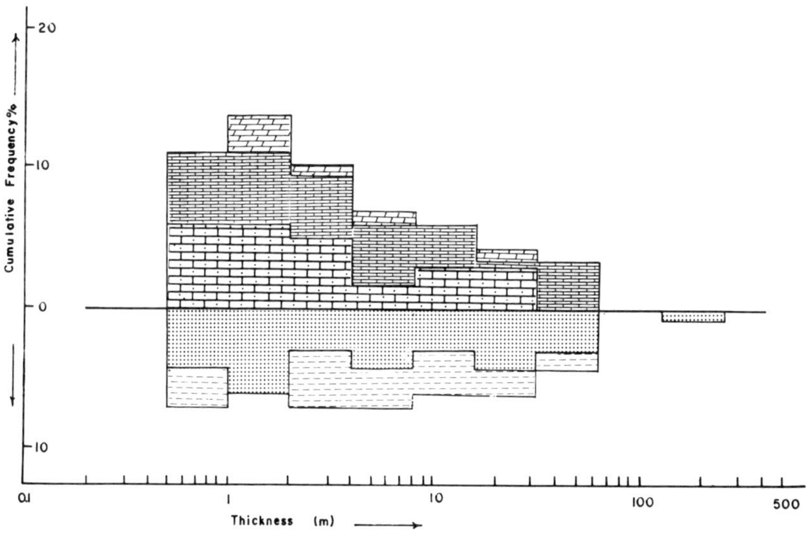

FIG. 13.17. Lithological frequency plot of total Cretaceous formations.

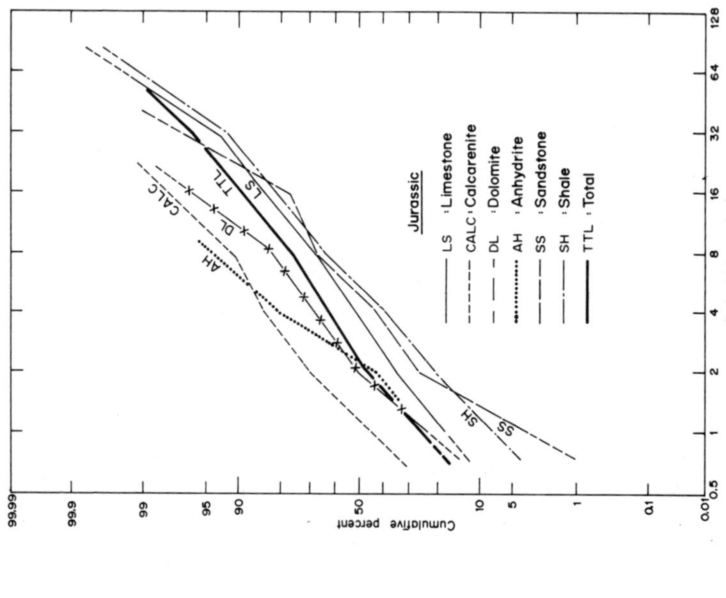

FIG. 13.18. Cumulative frequency plot of total Jurassic formations, on log-probability paper.

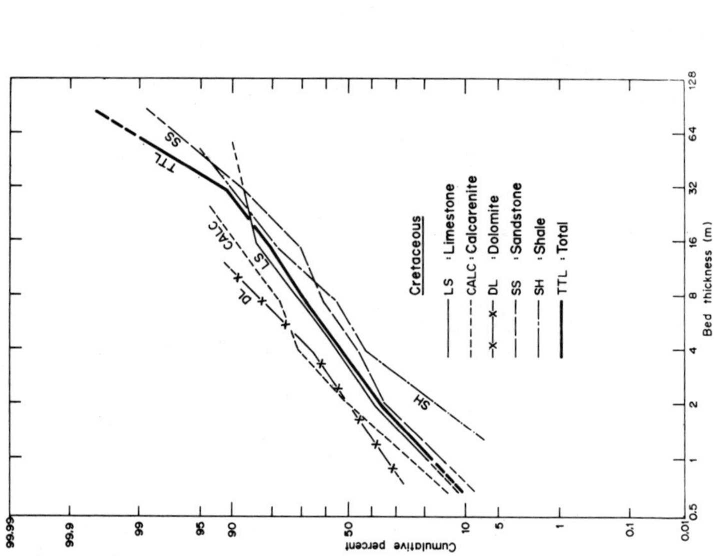

FIG. 13.19. Cumulative frequency plot of total Cretaceous formations, on log-probability paper.

TABLE 13.1
Statistical summary of Jurassic and Cretaceous formations (m)

Jurassic formations

Rock type	*Minimum*	*Most likely**	*Maximum*
Limestone	—	3·9	16·0
Calcarenite	—	1·2	4·0
Sandstone	1·7	5·2	19·8
Shale	1·7	6·0	19·8
Dolomite	—	2·0	11.0
Anhydrite	—	2.3	5.0
Total Jurassic	—	2.4	12.0

Cretaceous formations

Rock type	*Minimum*	*Most likely**	*Maximum*
Limestone	0·7	3·3	16·0
Calcarenite	0·5	2·0	11·3
Sandstone	1·1	5·0	28·0
Shale	2·0	6·0	24·0
Dolomite	—	2·0	8.3
Total Cretaceous	0·8	3·6	18·5

*Same as Median.

Jurassic and Cretaceous formations. Figure 13.21 shows the plot for the Jurassic formations (Upper, Middle, Lower, and entire Jurassic), and Figure 13.22 shows the plot for the Cretaceous formations (Upper, Lower, and entire Cretaceous). For the input data of lithological composition, and for the definition of the Upper, Middle, etc. of Jurassic and Cretaceous formations, refer to Table 13.2.

This program also calculates the D-function and relative entropy function described by Forgotson (1960). The D-function can be defined as

$$D = 100\ (1 - [(\Delta p)_m - (\Delta p)_{vm}]) \tag{13.1}$$

where $\Sigma_{i=0}^{n} p_i = 1$, and $(\Delta p)_m$ is the maximum of all positive differences of adjacent p's and $(\Delta p)_{vm}$ the next largest difference.

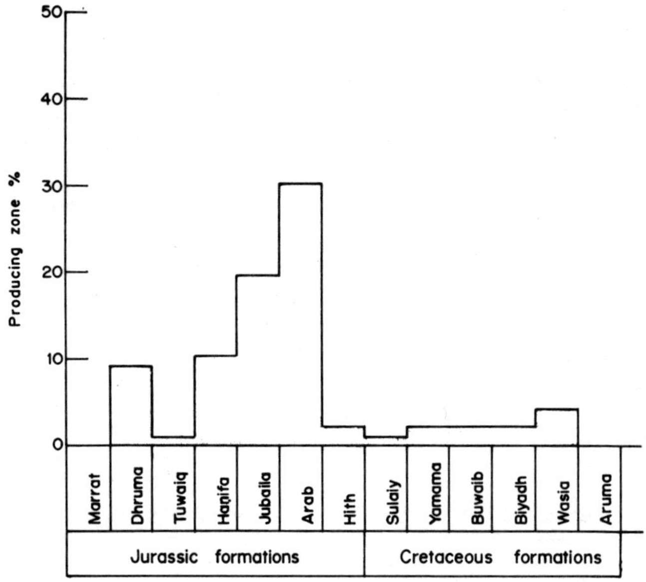

FIG. 13.20. Frequency plot of oil-producing zones in Saudi Arabia, compiled by Sail *et al.* (1986).

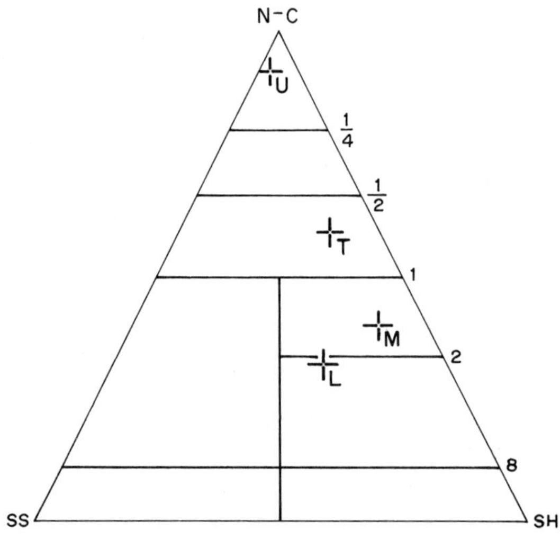

FIG. 13.21. Lithological triangle plot of Jurassic formations based on data of type sections in Saudi Arabia, defined by Powers *et al.* (1966). U = Upper, M = Middle, L = Lower and T = total Jurassic, N–C = non clastics, SS = sandstone and SH = shale.

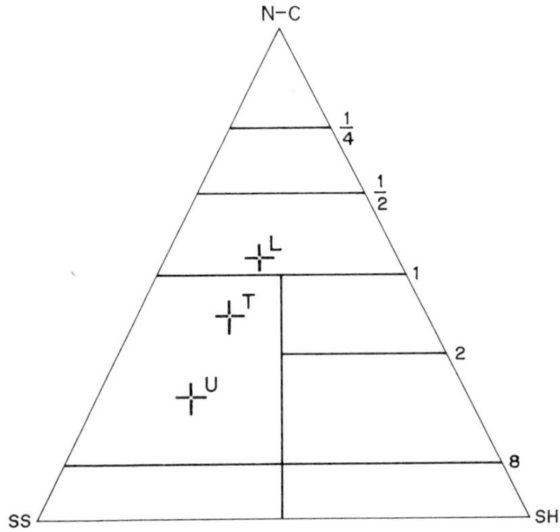

FIG. 13.22. Lithological triangle plot of Cretaceous formations based on data of type sections in Saudi Arabia, defined by Powers *et al.* (1966). U = Upper, L = Lower and T = total Cretaceous, N–C = non clastics, SS = sandstone and SH = shale

The relative entropy function (Hr) is expressed as

$$100\ Hr = \frac{-100\ \Sigma_i p_i \log p_i}{Hm} \qquad (13.2)$$

The entropy of a set of probabilities $p_1, p_2, \ldots, p_n$ is given as $H = -\Sigma_i\ p_i \log p_i$, where p_i is the probability of occurrence of each component. In a three-component system, Hm equals $-3(0{\cdot}33 \log 0{\cdot}33) = 1{\cdot}0986$. The results of calculation of the D- and relative entropy functions are shown in Table 13.2. A value of the D-function closer to 100 (or of the relative entropy function closer to 1) suggests relatively an even mixture of the three lithological components (non-clastics, sandstone, and shale in these examples), whereas a smaller value means that the composition belongs to more or less a single component.

According to the results shown in Table 13.2, the values of both the D- and relative entropy functions of the Upper Jurassic formations are significantly less than those of other formations, because of the distinctively higher percentage of non-clastic (carbonate) rocks in the Upper Jurassic. The Upper Jurassic, which includes the Arab formation, is the prime petroleum reservoir section in Saudi Arabia.

TABLE 13.2
Lithological composition, calculated D-function and relative entropy function of type sections of Jurassic and Cretaceous ages in Saudi Arabia. Original data derived from Powers *et al.* (1966).

Interval	*Non-classic (%)*	*Sand-stone (%)*	*Shale (%)*	*D-function*	*Relative entropy function*
Upper[a] Cretaceous	25	56	19	88	0·898
Lower[b] Cretaceous	53	28	19	94	0·918
Total Cretaceous	41	40	19	98	0·954
Upper[c] Jurassic	92	6	2	18	0·295
Middle[d] Jurassic	40	10	50	80	0·859
Lower[e] Jurassic	32	25	43	86	0·978
Total Jurassic	59	10	31	93	0·823

[a] Aruma and Wasia Formations
[b] Biyadh, Buwaib, Yamama, and Sulaiy Formations
[c] Hith, Arab, Jubaila, Hanifa, and Tuwaiq Mountain Formations
[d] Dhruma Formation
[e] Marrat Formation

CONCLUDING STATEMENT

As described at the beginning of this chapter, the source, reservoir, or sealing capacity of a rock can be controlled by a set of physical, chemical, and geological factors included in the rock. The litho-statistical study may provide only a possibility of interpretation. Such an interpretation could be misleading if the areas of statistical sampling and of oil production are not the same, as is the case described in this chapter.

Therefore, the main purpose of this chapter is to present the methods of analysis for potential future applications. It is recommended that similar research is continued using better data basis in the future, in order to improve our understanding on source, reservoir, and cap rocks in this region.

REFERENCES

Arabian American Oil Company Staff (1959) Ghawar Oil Field, Saudi Arabia. *AAPG Bull.*, **43**, 434–54.

Ayres, M. G., Bital, M., Jones, R. W., Slentz, L. W., Tarter, M. and Wilson, A. O. (1982) Hydrocarbon habitat in main producing areas, Saudi Arabia. *AAPG Bull.*, **66**, No. 1, 1–9.

Beydoun, Z. K. and Dunnington, H.V. (1975) *Petroleum Geology and Resources of the Middle East*, Scientific Press Ltd., Beaconsfield, Bucks, UK, 99 p.

Law, J. (1957) Reasons for Persian Gulf Oil abundance. *AAPG Bull.*, **41**, No. 1, 51–69.

Forgotson, Jr., J. M. (1960) Review and classification of quantitative mapping techniques. *AAPG Bull.*, **44**, 83–100.

Levorsen, A. I. (1954, 1967) *Geology of Petroleum*, W. H. Freeman, San Francisco.

Murris, R. J. (1980) Middle East: Stratigraphic evolution and oil habitat. *AAPG Bull.*, **64**, 597–618.

Powers, R. W. (1962) Arabian Upper Jurassic Carbonate rocks; in W. E. Ham, ed., Classification of Carbonate rocks. AAPG Mem. 1, pp. 122–92.

Powers, R. W., Ramirez, L. P., Redmond, C. D. and Elberg Jr., E. L. (1966) Geology of the Arabian Peninsula, Sedimentary Geology of Saudi Arabia, U.S. Geol. Survey Prof. Paper, 560–D, D 127 p.

Sail, I. A., Abdulla, H. M. and Magara, K. (1986) Litho-statistical study of both the Jurassic and Cretaceous type sections in Saudi Arabia, King Abdulaziz University, Jeddah, Saudi Arabia. (In press).

Chapter 14

Porosity of Sandstone, Shale and Limestone

INTRODUCTION

In causing compaction or porosity reduction of shales and sandstones, many geological, physical, chemical, and mineralogical factors are considered to have been effective. Research activities in evaluating the significance of each of these factors and in further predicting the porosity values under a given set of geological, physical, chemical, and mineralogical conditions have been conducted (e.g. Maxwell, 1964).

So far, the understanding of the internal features of shales and sandstones has been greatly improved by modern geochemical methods and by studies using X-ray analysis, thin sections, and scanning electron microscopes.

In the first half of this chapter, I intend to discuss the difference between shale porosity–depth and sandstone porosity–depth profiles of many sedimentary basins, in evaluating the most important factors causing the porosity reduction of shales and sandstones. The approach is based on the difference between the natures of the porosity trends on normal graph paper: the shale porosity–depth profile tends to be a curved line (an exponential function), while the sandstone porosity–depth profile is shown as a relatively straight line if there was no significant development of secondary porosity, or if there was no significant retardation of sediment diagenesis owing to the presence of hydrocarbons and of abnormally high fluid pressure. In short, the first part of this chapter deals mostly with the primary porosity.

The second half of this chapter, on the other hand, describes possible causes of creating secondary porosity by leaching calcite grains and cements.

SHALE POROSITY–DEPTH RELATIONSHIP

A summary of shale porosity–depth relationships in different parts of the world is shown in Fig. 14.1. Shale porosity decreases with increase of depth. The rate of porosity decrease is rapid at shallow depths and slows at greater depth of burial. This can be shown by a curved line on normal graph paper.

The shale porosity can also be influenced by subsurface fluid pressure; the higher the pressure the greater the porosity at a given burial depth or under a given overburden pressure. This suggests that many of the porosity–depth curves in Fig. 14.1, especially the ones indicating relatively high porosity values at depth, may be the result of the above-normal hydrostatic pressure.

If the area has experienced significant uplift and erosion, the shale porosity–depth trend is shifted in the direction of increased compaction at any specific depth, in comparison with the trend in an area of no erosion.

Rubey and Hubbert (1959) proposed an exponential function, which is based on Athy's (1930) curve derived from the Pennsylvanian and Permian shales in northern Oklahoma, as follows:

$$\phi = \phi_0 \exp(-cZ) \tag{14.1}$$

where

ϕ = value of the shale porosity at depth Z
ϕ_0 = porosity at the surface ($Z = 0$)
and c = constant of dimension (length^{-1}).

Rubey and Hubbert considered that because of the elapsed time since deposition, Athy's relation is the one closest to the compaction-equilibrium conditions under which fluid pressure reached normal hydrostatic pressure. Even in relatively young sedimentary basins, the shale porosity–depth relationship can be approximated by an exponential function for a section of normal hydrostatic pressure where the compaction equilibrium has been attained. Magara (1971) reported such an exponential function above 7000 ft of the Gulf Coast area, based on Dickinson's (1953) shale porosity–depth relationship.

The exponential function of shale porosity–depth proposed by Rubey and Hubbert may be explained as follows. At an earlier stage of compaction, the shale retains higher permeability so that more pore-fluid

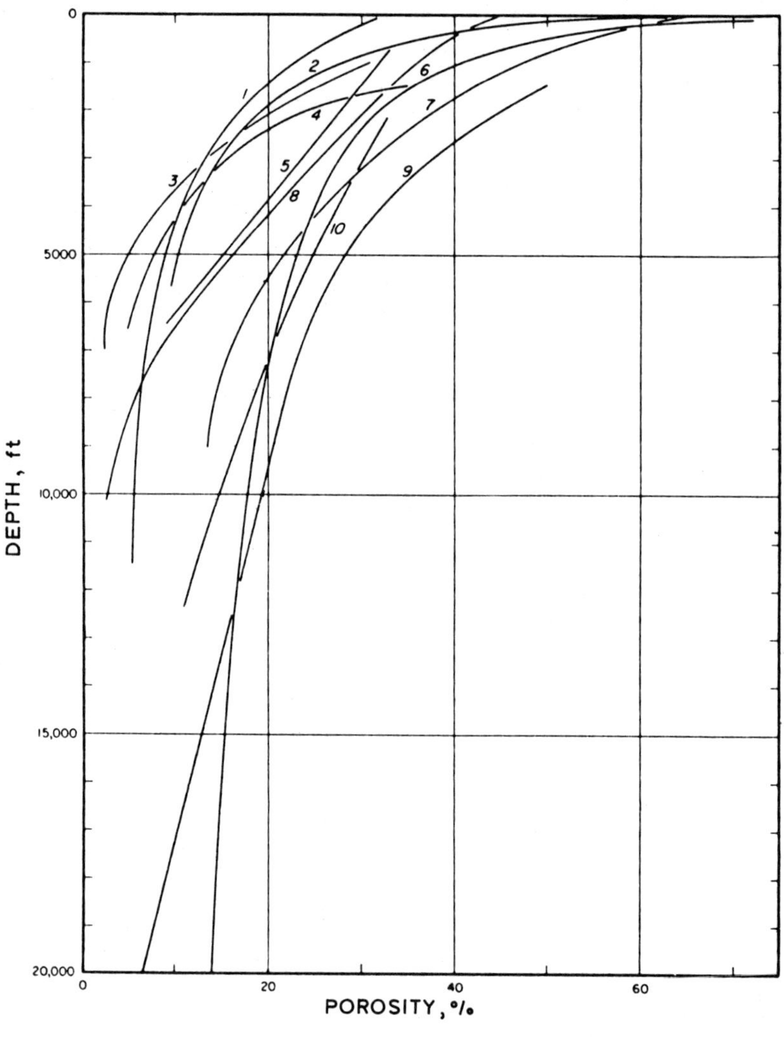

FIG. 14.1. Relationship between porosity and depth for shales (from Rieke and Chilingarian, 1974) 1 = Proshlyakov (1960); 2 = Meade (1966); 3 = Athy (1930); 4 = Hosoi (1963); 5 = Hedberg (1936); 6 = Dickinson (1953); 7 = Magara (1968); 8 = Weller (1959); 9 = Ham (1966); 10 = Foster and Whalen (1966). (Reproduced by permission of Elsevier Science Publishers, Amsterdam.)

can be squeezed from it for a given increase in burial depth or overburden pressure. The result is a faster rate of porosity reduction at shallower depths.

As more fluid is expelled from the shale, its porosity is reduced. Permeability is also reduced with the reduction in porosity. Therefore, at deeper burial depths, the amount of fluid expulsion would be much less than at shallower depths, for a given increase in burial depth or overburden pressure. Such changing rates of fluid expulsion or porosity reduction can be shown as an exponential function such as eqn. (14.1).

Low (1976) studied the viscosity of water in montmorillonite, shown by the dashed line in Fig. 14.2. The viscosity increases from about 1 to 8 cp toward the clay surfaces.

Because Low's viscosity value is plotted against the g H_2O/g clay which can be converted to porosity, his relationship can be explained in terms of changing porosity; the viscosity of water in montmorillonite tends to increase as the porosity decreases or as compaction progresses.

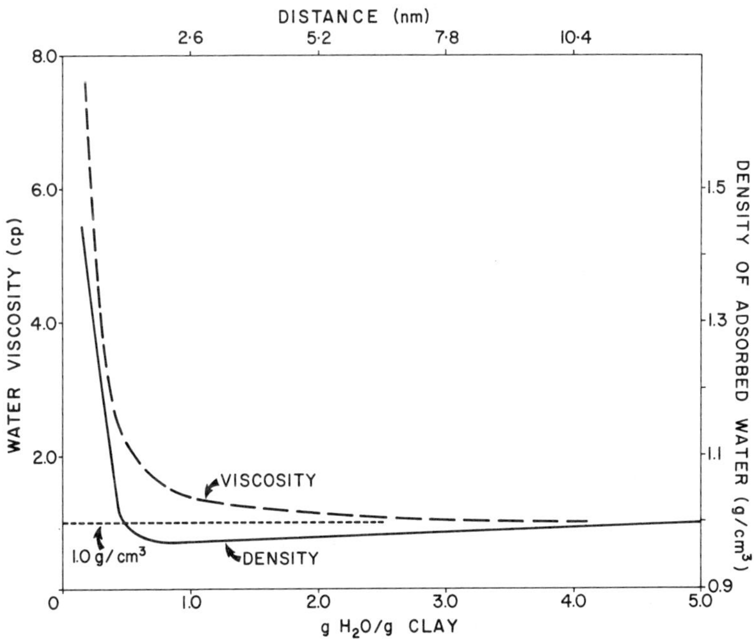

FIG. 14.2. Viscosity and density of adsorbed water. Data derived from Low (1976) and Martin (1962).

The water density values from Martin (1962) are also included in Fig. 14.2.

Because of the worldwide evidence of the tendency of shale porosity to reduce with increasing depth (Fig. 14.1), one may be able to conclude that even this viscous water can be expelled from shales as they are buried deeply. Relatively high temperatures in the subsurface could have helped the shales to move the water by reducing viscosity slightly.

The rate of expulsion of water, however, slows with depth, partly because the water viscosity of the shale increases continuously as compaction increases, even though the higher temperature in the subsurface tends to reduce the water viscosity slightly.

In summary, then, the reduced rate of shale compaction at depth, as shown by most curves in Fig. 14.1 or shown by the exponential function of eqn. (14.1), may be the result of the combined effect of decreasing shale permeability and of increasing water viscosity with burial and compaction.

The change of shale compaction rate with depth may also be explained on the basis of the concept of grain-to-grain contact of shale. As depicted in Fig. 14.3, there are relatively light grain-to-grain contacts at a

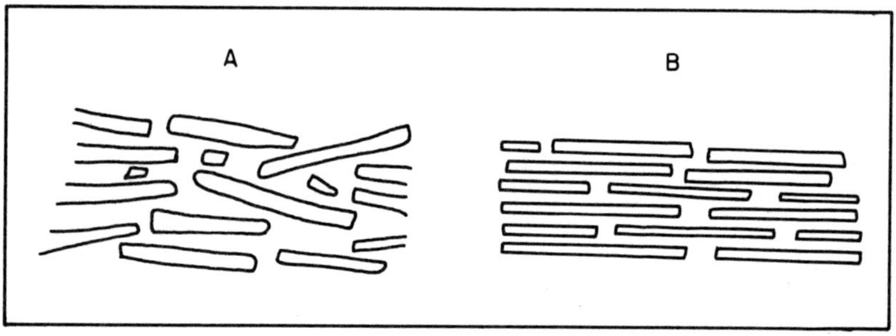

FIG. 14.3. Schematic diagrams showing mechanism of shale compaction.

shallower burial depth (A). The area of contact between grains per given cross-sectional area or given volume of shale is relatively small at this stage. If a certain amount of overburden pressure is added at this stage, the shale will compact at a relatively high rate because the pressure applied per given contact area is relatively large. (The amount of pressure increase (P_i) is constant in this case, but the contact area (A_c) is relatively small at the shallow burial stage. Hence, the pressure increase per contact area (P_i/A_c) is relatively large.)

At a deeper burial stage (Fig. 14.3B), on the other hand, for a given increase of the overburden pressure, the increase of pressure applied to a given contact area will be relatively small because of a larger grain-to-grain contact area per given volume of shale. The rate of compaction or of porosity reduction would accordingly be smaller at this later stage. The change of rate of shale compaction can therefore be expressed by an exponential function such as eqn. (14.1).

SANDSTONE POROSITY–DEPTH RELATIONSHIP

By comparison, the porosity–depth relationship of sandstone is generally closer to a straight line on arithmetic (or normal) graph paper, as are the plots in Fig. 14.4. Here, Galloway (1974) has shown two porosity–depth relationships of sandstones associated with two different geothermal gradients in the Northeast Pacific arc. Figure 14.5 shows other examples of relatively straight porosity–depth relationships of sandstones (with and without matrix) on normal graph paper (Loucks *et al.*, 1979). The data included in Fig. 14.5 are derived from lower Tertiary sandstones in the Texas Gulf Coast. If there is significant development of secondary sandstone porosity at deep burial by leaching the grains, by fracturing, and perhaps by the presence of abnormal fluid pressure, the porosity may increase with depth. If such leaching or fracturing occurs, the porosity at depth should be greater than that shown in Figs. 14.4 and 14.5.

In summary, the porosity–depth relationship for sandstones tends to be expressed by a straight line on normal graph paper if there was no significant development of secondary porosity. This means that the rate of sandstone porosity reduction is almost the same whether the sandstone is at a shallow or deep burial depth.

As is in the case of shales, the grain-to-grain contact area of the sandstones will increase as the porosity decreases. Therefore, for a given increase of overburden pressure, the increase of pressure per contact area will decrease with burial. Yet, for sandstone, the rate of porosity reduction at a deep burial depth is almost the same as that at a shallow depth.

The absolute permeability of most sandstones also decreases with decreasing porosity. An example of such a porosity–permeability relationship is shown in Fig. 14.6 (from Bebout *et al.*, 1978). Archie (1950) also reported porosity–permeability relationships of sandstone, lime-

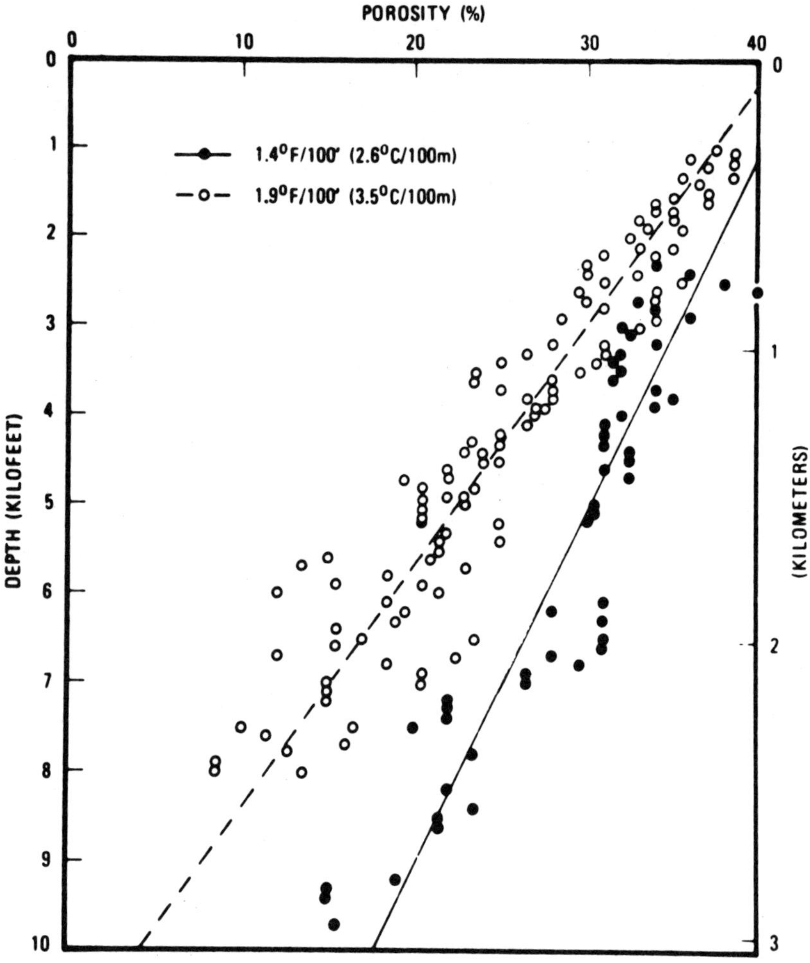

FIG. 14.4. Porosity–depth relationships of sandstones. From Stephenson (1977) after Galloway (1974). (Reproduced by permission of the American Association of Petroleum Geologists.)

stones, and muddy sands, such as shown in Fig. 14.7. According to Archie's relationship, an increase in porosity of about 3% produces a tenfold increase in permeability. If, therefore, the mobility of pore-water, which is influenced by the absolute permeability, has a principal control on the rate of reduction of sandstone porosity, the reduction rate would

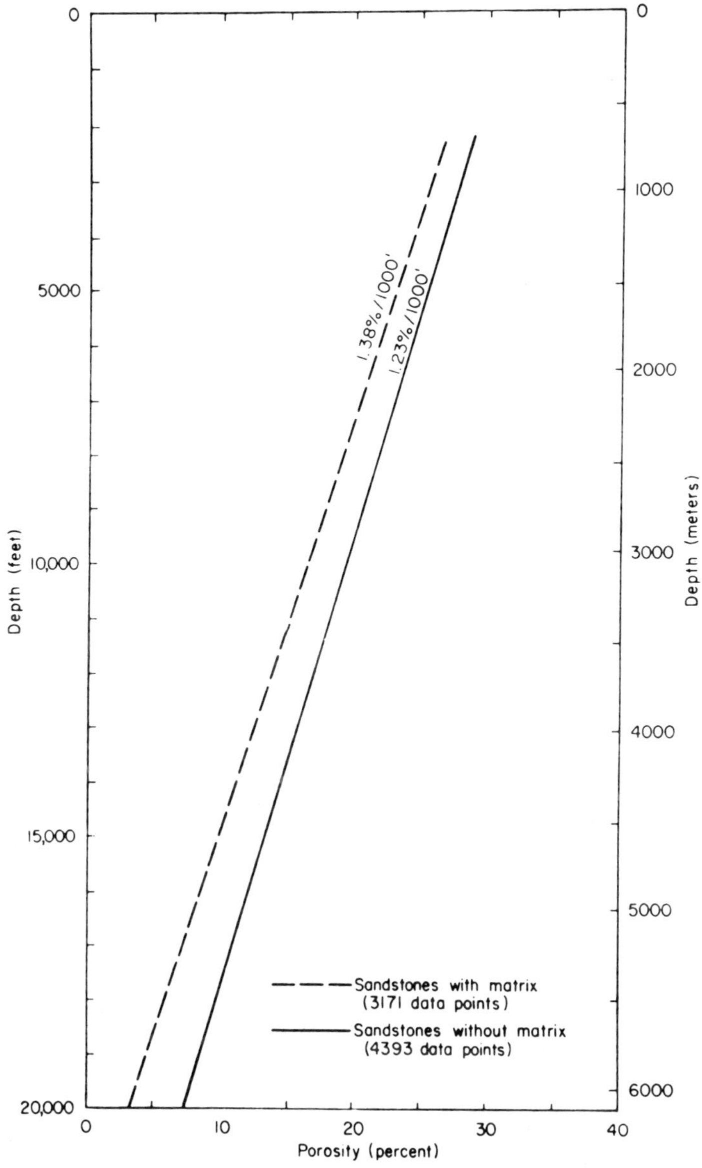

FIG. 14.5. Mean sandstone porosity versus depth relationships for lower Tertiary sections with and without clay matrix of Texas Gulf Coast area. From Loucks *et al.* (1979). (Reproduced by permission of the Bureau of Economic Geology, University of Texas at Austin.)

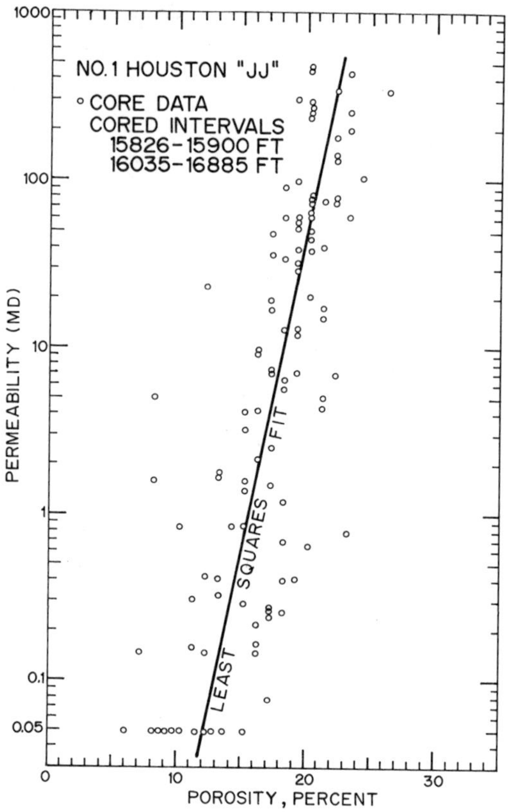

FIG. 14.6. Porosity–permeability relationship of sandstones, Texas (from Bebout *et al.*, 1978). (Reproduced by permission of the Bureau of Economic Geology, University of Texas at Austin.)

slow with burial. This means that, to continue a diagenetic process in sandstones, there may be certain minimum requirements for continuous water supply which can be controlled by the permeability of the sandstones. However, the subsurface data suggests that the rate of sandstone porosity reduction is almost constant during continuous burial, although the mobility of water continuously decreases as the permeability decreases with burial.

In other words, the mobility does not seem to have had a significant effect on the rate of reduction of porosity as far as the absolute

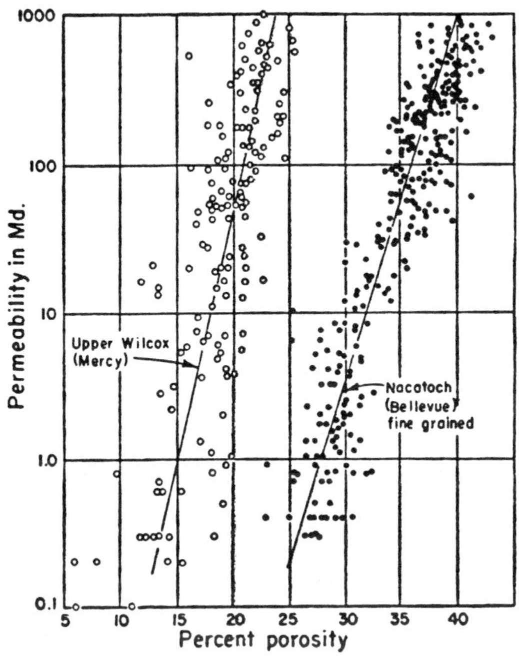

FIG. 14.7. Porosity–permeability relationships of Upper Wilcox(Eocene) and Nacatoch (U. Cretaceous) sandstones of Gulf Coast area (from Archie, 1950). (Reproduced by permission of the American Association of Petroleum Geologists.)

permeability of most sandstones in drillable depth ranges is concerned. Sandstones are primarily water-saturated in this case.

However, if the movement of pore-water has been inhibited by an early accumulation of hydrocarbons, the rate of porosity reduction may have been greatly reduced. Such examples in the Gifhorn Trough were reported by Philipp *et al.* (1963). In hydrocarbon-saturated sections of a reservoir, the relative permeability for water is virtually zero, so that the water cannot move even if the reservoir itself has some absolute permeability. The supply of formation water and ions may be inhibited in this case, reducing the rate of sandstone diagenesis.

Another approach to evaluating the effect of pressure on sandstone porosity reduction is concerned with the porosity values of sandstones having different grain sizes.

Füchtbauer (1974) showed porosity–depth relationships of micaceous silty sandstones of Upper Carboniferous age. The sandstone porosity is higher at a given depth within the section between 8000 and 20 000 ft if the median grain size is larger (coarser grains).

A sandstone of a coarser grain size will have fewer grain-to-grain contacts per given volume at a given burial depth than another sandstone of a finer grain size. For a given incremental increase in overburden pressure, the incremental pressure increase per given contact area must be greater for the coarser grained sandstone because of fewer contacts. Yet, the rate of porosity reduction for the coarser grained sandstone is less than that for the finer grained sandstone. This observation also suggests that the pressure applied per given contact area is not the principal factor in reducing sandstone porosity. Some other factors, especially chemical and mineralogical ones, must have controlled the reduction of sandstone porosity.

Regarding the mobility of water in sandstones, the permeability tends to increase with increase of grain size (decrease of specific surface area) for a given porosity level (refer to the Kozeny–Carman relationship). Therefore, if the mobility of pore-water and the supply of additional ions were the most critical factors in controlling the rate of porosity reduction, the coarser grained sandstone must have lost porosity more quickly, unless the sandstone is saturated with hydrocarbons. The subsurface observation does not seem to support this possibility.

In summary, neither the concept of the pressure applied per given contact area, nor of the mobility of pore-water as long as sandstones are saturated with water would satisfactorily explain the relatively constant rate of porosity reduction for the sandstones throughout the burial history.

COMPARISON OF SANDSTONE POROSITY–DEPTH AND SHALE POROSITY–DEPTH RELATIONSHIPS

The observations made in the preceding section may lead us to conclude that both the pressure applied on grain-to-grain contact and the expulsion or movement of pore-water are the principal causes for the reduction of shale porosity. This suggests that the shale compaction is primarily controlled by physical forces of sedimentary rock, although other chemical and mineralogical factors can also influence the rate of shale compaction. This is particularly true for shales of relatively younger geologic ages because of the shorter duration of time involved in this case.

The above interpretation may be used to explain why the shale properties, such as porosity, density, transit time, and resistivity, are used effectively for detecting subsurface fluid pressure, which is a physical measure, and why the sandstone properties are not used effectively. Because the shale compaction has been controlled primarily by physical forces, the shale compaction can be used to interpret the physical properties of sedimentary rocks. The thickness of sedimentary rocks removed by erosion, which also influences the physical measure of sedimentary rocks, can be estimated from shale compaction data as well (Magara, 1976).

On the other hand, in the case of sandstones, the rate of porosity reduction seems to be controlled neither by the pressure or stress applied per given contact area nor by the mobility of pore-water if the sandstone is saturated by water. (If a sandstone has a very low porosity (say less than 5%), thus having extremely low permeability, the rate of diagenetic process may be reduced even though the sandstone is saturated with water.) If the sandstone is saturated by hydrocarbons, the rate of primary

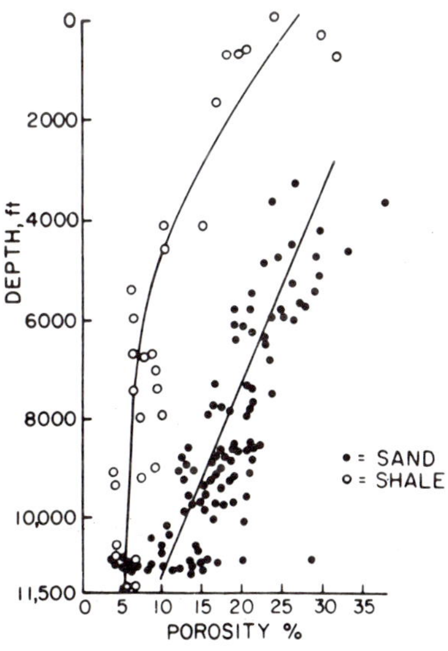

FIG. 14.8. Porosity–depth relationships of sandstones and shales of Jurassic and lower Cretaceous sections (from Proshlyakov, 1960).

porosity reduction may be reduced because of the restricted water movement in such a section. It may be concluded that, as long as sandstones have some grain-to-grain contacts and their pore-fluid contains some ions, there seems to be chances of precipitating minerals, thus causing the reduction of porosity. This suggests that the reduction of sandstone porosity is primarily controlled by chemical and mineralogical agents, rather than by physical agents. This may explain why such factors as types of grains, properties of formation water, temperature, and geologic ages are believed to be of prime importance in controlling sandstone porosity by many investigators. Temperature and the duration of time (geologic age) must be critical factors for any chemical reaction under a given chemical environment (properties of formation water and grains).

This may further explain why the sandstone porosity cannot be used to evaluate either pore-pressure or thickness of erosion, both of which are the physical measures of sedimentary rocks.

Figure 14.8 shows an interesting comparison of the two porosity trends of sandstones (straight line) and shales (curved line) reported by Proshlyakov (1960).

CAUSES OF LEACHING CALCITE GRAINS AND CEMENTS

It has been proposed that carbonic acid, which was formed by carbon dioxide (CO_2) in solution, originated from organic matter undergoing thermal maturation and directly from the atmosphere, can be a cause of the leached minerals (Schmidt and McDonald, 1979). The thermal generation of carbon dioxide may be explained by the cross plot of the hydrogen index (mg hydrocarbons/g organic carbon) and the oxygen index (mg CO_2/g organic carbon), such as shown in Fig. 14.9 (Tissot, 1977). This plot shows the changing hydrogen and oxygen indices for three types of organic matter (I: algal, II: herbaceous, and III: woody).

Type III organic matter (woody), for example, shows that, with thermal maturation, the oxygen index decreases significantly but the hydrogen index does not. This means that a significant amount of carbon dioxide will be generated and lost from the woody organic matter with a small loss of hydrocarbons. This may further suggest that the woody organic matter is a poor source for hydrocarbons but a good source for carbon dioxide.

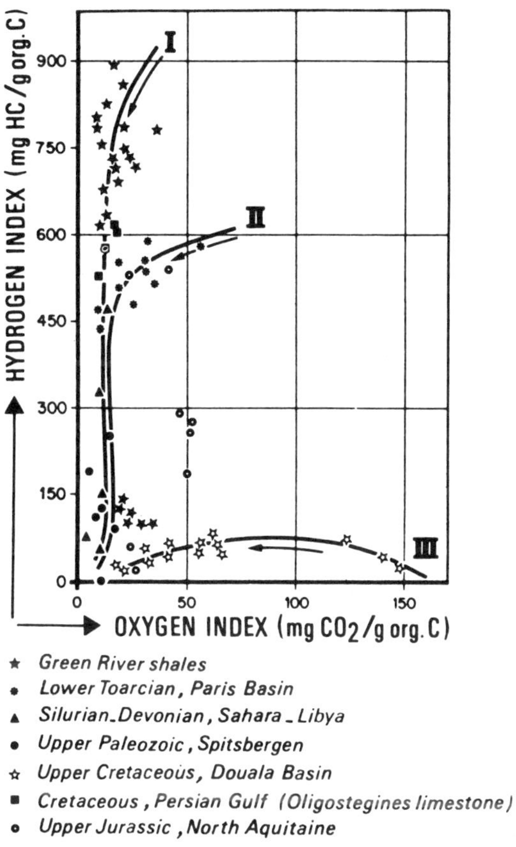

FIG. 14.9. Plots of hydrogen and oxygen indices for three types of organic matter (I, II, and III) from different parts of the world (from Tissot, 1977).

Type I organic matter (algal), on the other hand, will produce by the thermal maturation process a large amount of hydrocarbons with a small amount of carbon dioxide. Type II organic matter (herbaceous) may present a situation between those of these two types (I and III).

In summary then, to produce a significant quantity of hydrocarbons by thermal maturation, the algal organic matter would be the best type. However, for the generation of carbon dioxide, which may cause significant development of secondary porosity by leaching, the woody organic matter would produce the best result. In the case of formation of deep oil or gas accumulations, therefore, a certain mixture of these two types of organic matter may provide the best result. Another possibility is that these two types of organic matter occur in different lithological units (e.g. algal type in shales and woody type in sandstones), but the fluids from these units may eventually be mixed after thermal maturation.

Regarding the timings of generation of these chemical products through maturation, one may prefer to have carbon dioxide, first, and hydrocarbons, second, for a deep reservoir because with this order, there is a chance of creating the secondary leaching porosity before the significant hydrocarbon generation and migration. Fortunately, the natural order of generation of these products is the same as mentioned above (Tissot, 1977), so that one may generally expect a good timing relationship of secondary porosity and hydrocarbon generation and migration.

Figure 14.10 shows a graph of the solubilities of calcite, amorphous silica, and quartz versus pH. The calcite solubility increases as pH decreases (acidic condition).

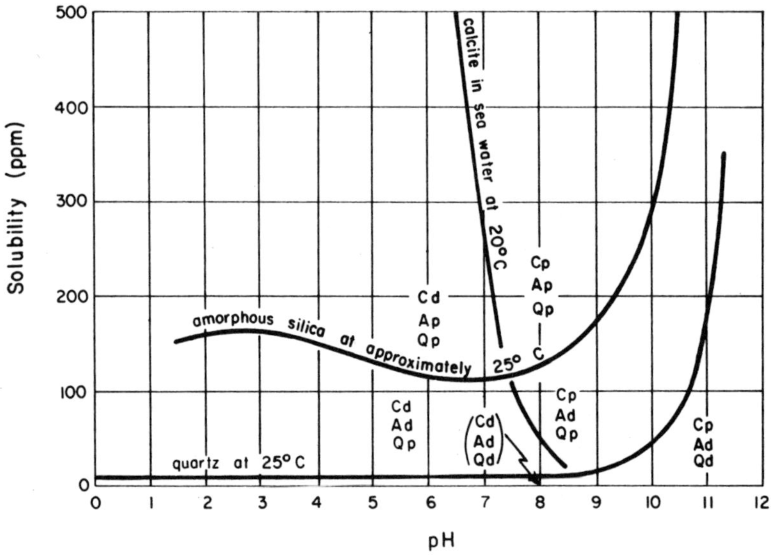

FIG. 14.10. Relationships between pH and solubility of calcite, amorphous silica, and quartz. C, calcite; A, amorphous silica; Q, quartz; p, precipitates; d, dissolves (from Blatt, 1966). (Reproduced by permission of the Wyoming Geological Association.)

Schmidt and McDonald (1979) cited Savkevic's (1971) concept that both the interlayer water released from montmorillonite and the carbon dioxide generated from organic matter could cause subsurface brines to become so undersaturated with respect to carbonate that the latter could be dissolved. From the study of the Smackover sandstone of Mississippi, Parker (1974) mentioned that carbon dioxide derived from a plutonic intrusion was the prime cause for leaching the carbonaceous material. However, this may not be applicable to many other areas.

Feldspar grains may also be leached out by carbonic acid generated from the aqueous solution of carbon dioxide.

As suggested by Schmidt and McDonald (1979), leaching of calcite grains and cement would have taken place in the geopressured interval where the temperature is high enough to cause CO_2 generation and petroleum maturation. Dissolution of CO_2 gas in water would have caused the acidic environment (low pH) that facilitated calcite reaching. Figure 14.11 shows a plot of pH of formation water vs depth in the Gulf Coast. The points in this figure are scattered widely, but show a general decreasing tendency of pH below about 10 000 ft (3 km) where geopressure exists and organic materials are geochemically more matured than in the shallower hydropressure zone (Dow, 1978).

The compaction fluid, which contains a significant quantity of dissolved ions, moved upward to the shallower interval and was mixed with the water there. The pH value would increase because of mixing with other brines of more normal pH. Precipitation of calcite can be expected there.

There are also other causes that possibly change the solubility of calcite, such as pressure and temperature. Figure 14.12 shows the relationships between temperature, partial pressure, p_{CO_2}, and solubility of calcite, constructed from the data by Wedepohl (1976). This figure shows that decreasing temperature increases the solubility, but decreasing pressure decreases the solubility. When a hot and high-pressured brine moves from deep to shallow, both temperature and pressure decline, but their effects on calcite solubility are different.

Let us discuss a hypothetical example. Let us assume that a hot and high-pressured brine exists at depth, where the pressure gradient is 0·8 psi/ft. This brine moved upward for a distance of 3000 ft (1000 m) where the pressure gradient is hydrostatic (0·465 psi/ft). The drop in fluid pressure is about 1000 psi (69 bars). If the geothermal gradient is 1·4°F/100 ft (2·55°C/100 m), which is an average figure for the Gulf Coast, the drop in temperature is 42°F (23°C). Therefore, the ratio of the

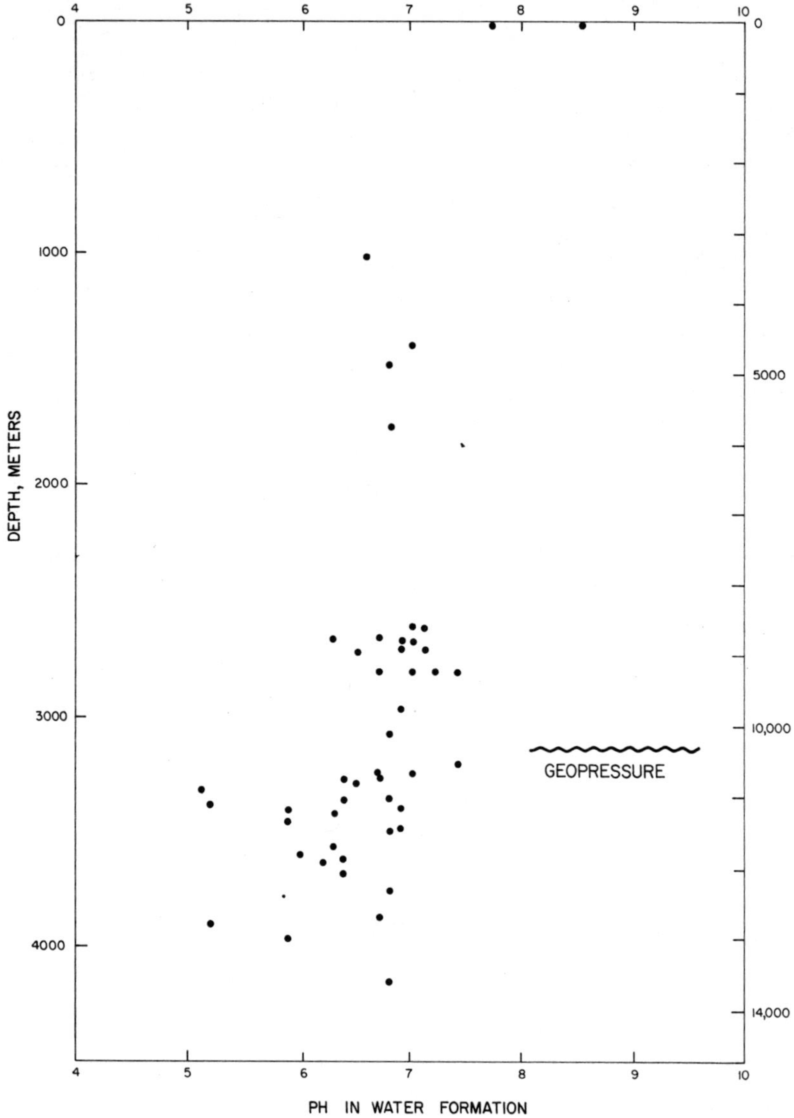

FIG. 14.11. Plot showing pH of formation water versus depth in Texas Gulf Coast. Data derived from Kharaka *et al.* (1977) (from Magara, 1981). (Reproduced by permission of the American Association of Petroleum Geologists.)

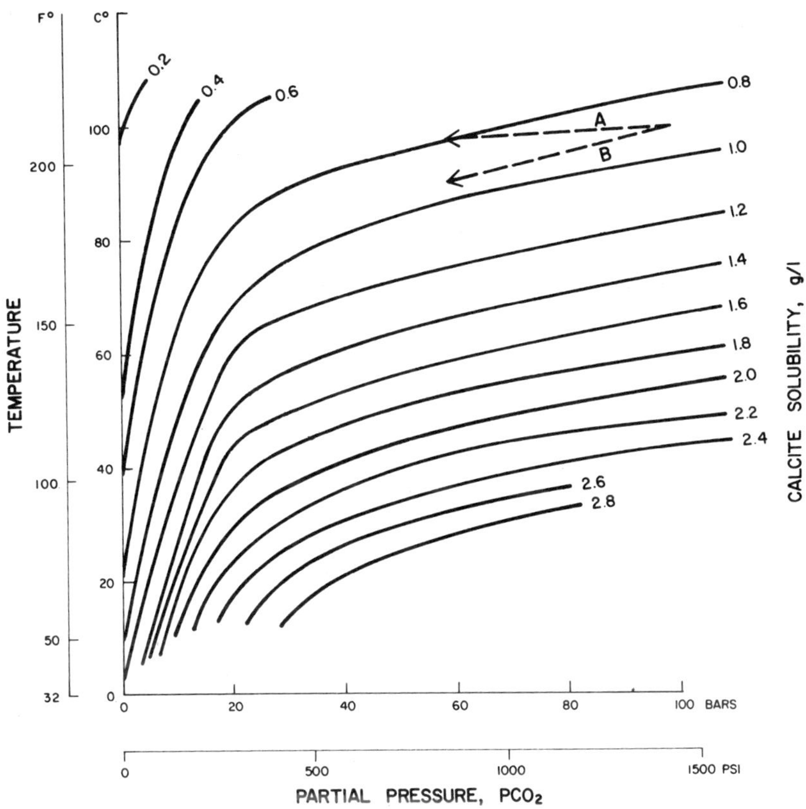

FIG. 14.12. Relations among temperature, partial pressure (p_{CO_2}), and solubility of calcite. A and B show hypothetical cases of changing temperature and pressure from geopressured interval to hydropressured interval (A) and within hydropressured interval (B). From Magara (1981). (Reproduced by permission of the American Association of Petroleum Geologists.)

drop in temperature over the drop in pressure by the movement of the brine from high- to normal-pressure environment for a distance of 3000 ft (1000 m) is about 0·33°C/bar, which is shown by the dashed line A in Fig. 14.12. The solubility decreases after the movement of the brine, suggesting a possibility of precipitating calcite at the shallower depth.

If, on the other hand, the brine moved from deep to shallow within a hydrostatic pressure zone, the drop in fluid pressure for a given temperature drop is much less than that in the previous high-pressure case. The ratio of the drop in temperature over that of pressure in this case is greater than that of the previous case. This second case is shown by another dashed line B, which is parallel to the solubility lines in Fig. 14.12, suggesting that, if the brine moves from deep to shallow within a hydrostatically pressured interval, calcite will not be precipitated significantly by the effects of changing pressure and temperature.

The last factor one may have to consider is the difference in concentration of other ions in formation water between the deep geopressured and the shallow hydrostatically pressured intervals. It is known that the concentrations of other ions is abnormally low in an undercompacted and geopressured interval (Overton and Timko, 1969, and Magara, 1974). As suggested by Folk and Land (1975), the solubility of a mineral ion generally decreases with concentration of other ions. Therefore, it is most likely that precipitation of calcite takes place after the movement of brines from a deep and relatively fresh geopressured interval to a shallow and more saline hydropressured interval.

SECONDARY POROSITY OF LIMESTONE IN IRAN

Limestone porosity is usually more difficult to predict. This is particularly true if most limestone porosity is of a secondary nature.

The Asmari limestone reservoir (Oligocene–Miocene age) of the Zagros Folded Belt in Iran is known for its high productivity due to its secondary porosity and permeability. Figure 14.13 shows a limestone porosity–depth plot of both the Asmari and the Bangestan (U. Cretaceous) limestone reservoirs. Below about 6000 ft, there is significant development of secondary porosity.

This area experienced a period of extremely fast sedimentation prior to the strong tectonic movements associated with the development of the Zagros thrust fault. Refer to Fig. 10.5 in Chapter 10 which shows relatively high rates of the excess fluid-pressure-increase due to fast sedimentation during the Tertiary period at location B (see Fig. 10.6 for location) in Iran. The rate of sedimentation or burial in this area was particularly fast during the Tertiary period.

At present, high fluid pressures are still preserved in many reservoirs as shown in Fig. 14.14. Such high fluid pressures in the deeper sections may have charged vertically the shallower sections with relatively low original pressures, and possibly caused natural hydrofracturing in the shallower sections (Chapter 9).

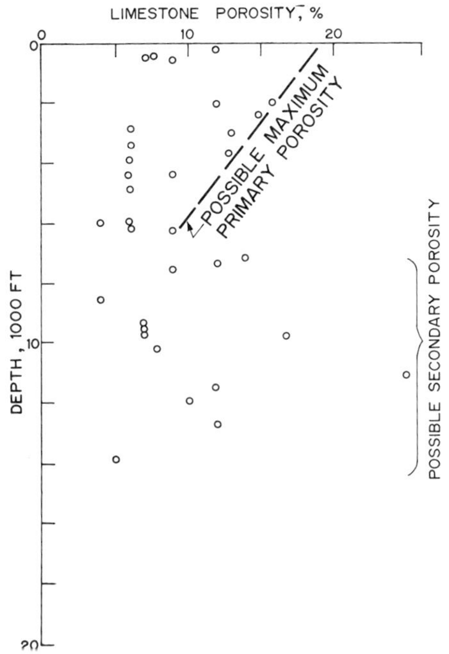

FIG. 14.13. Limestone porosity–depth plot of Asmari (Oligo–Miocene) and Bangestan (Upper Cretaceous) reservoirs in Zagros folded belt, Iran (data derived from Petroconsultants, 1984).

During the period of the Zagros tectonic movements, the fractures, which were previously formed, would have propagated further throughout the reservoir sections. During, and after the significant tectonic movements, the uppermost part of the sedimentary column in relatively uplifted areas would have been removed by erosion, possibly causing the late-stage fracturing (Chapter 9). The removal of the sedimentary rocks itself may also have caused fractures by the reduced subsurface stress and the physical relaxation of the rocks.

Although most of the methods of estimating thickness of erosion described in Chapter 7 cannot be applied because of lack of data to hand, a plot of temperature versus depth of this area may provide a clue for guessing an erosional effect. Figure 14.15 shows such a plot for the Asmari and Bangestan reservoirs. Because the temperature values measured in the bore holes are usually lower than the true natural values due to the circulation effect of drilling fluid, a trend line was drawn through

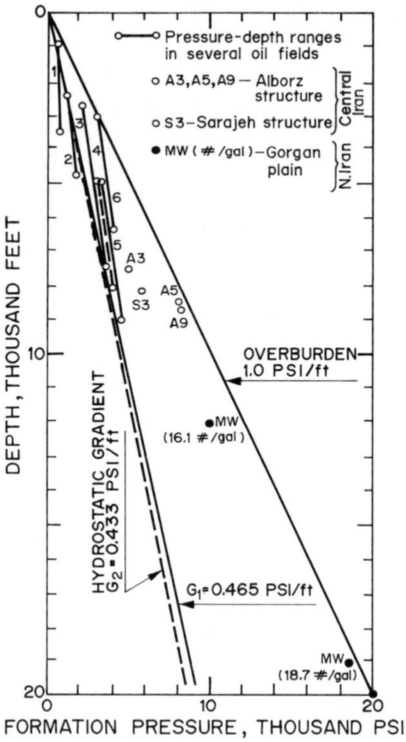

FIG. 14.14. Fluid pressure–depth plot of reservoirs in Iran (from Fertl, 1976). (Reproduced by permission of Elsevier Science Publishers, Amsterdam.)

the measured highest temperature values. This trend line indicates a geothermal gradient of about 1·1°F/100 ft (2·0°C/100 m) and is extrapolated to about 135°F (57°C) at the present surface (after erosion), which is significantly higher than the mean annual surface temperature of 80–90°F (27–32°C) in this area.

This surface temperature difference suggests a possibility of significant erosion in this area; the paleo subsurface temperature before erosion may still be preserved in the present subsurface.

Maturation of organic matter, particularly of terrestrial origin, and generation of CO_2 gas and formation of the carbonic acid as described previously may be an additional cause of creating secondary porosity by dissolution of calcite grains and cement.

In summary, the secondary porosity of the limestone reservoirs in Iran may have been generated by the following mechanisms.

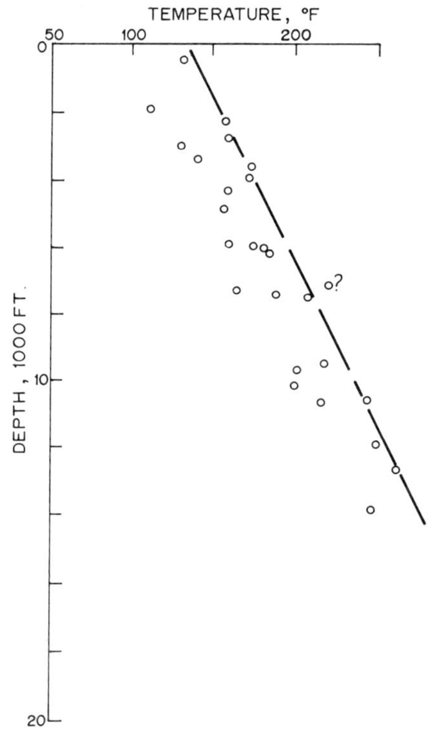

FIG. 14.15. Temperature–depth plot of Asmari (Oligo–Miocene) and Bangestan (Upper Cretaceous) formations in Zagros folded belt, Iran (data derived from Petroconsultants, 1984).

1. Charging of highly pressured fluid to cause natural hydrofractures.
2. Tectonic forces to cause fracture porosity.
3. Uplift and erosion to create more fractures due to the late-stage fluid charging and the relaxation of rocks.
4. Organic maturation and generation of CO_2 and acidic fluid to open the existing fractures by the dissolution of calcite.

REFERENCES

Archie, G. E. (1950) Introduction to petrophysics of reservoir rocks. *A.A.P.G.*, **34**, 943–61.

Athy, L. F. (1930) Density, porosity and compaction of sedimentary rocks. *A.A.P.G.*, **14**, 1–24.

Bebout, D. G., Loucks, R. G. and Gregory, A. R. (1978) Frio sandstone reservoirs in the deep subsurface along the Texas Gulf Coast, Bureau of Economic Geology, University of Texas, no. 91, 93 p.

Blatt, H. (1966) Diagenesis of sandstones: processes and problems, *Wyoming Geol. Assoc. 20th Ann. Conf. Proc.*, 63–5.

Dickinson, G. (1953) Geological aspects of abnormal reservoir pressures in Gulf Coast Louisiana. *A.A.P.G.*, **37**, 410–32.

Dow, W. G. (1978) Petroleum source beds on continental slopes and rises; *AAPG Bull.*, **62**, 1584–606.

Fertl, W. H. (1976) *Abnormal Formation Pressures*, Elsevier, Amsterdam, 382 p.

Folk, R. L., and Land, L. S. (1975) Mg/Ca ratio and salinity: two controls over crystallization of dolomite, *AAPG Bull.*, **59**, 60–8.

Foster, J. B. and Whalen, H. E. (1966) Estimation of formation pressures from electrical surveys—offshore Louisiana. *J. Pet. Technol.*, **18**, 165–71.

Füchtbauer, H. (1974) *Sediments and Sedimentary Rocks, 1: Part II*, Halsted Press, New York, 464 p.

Galloway, W. E. (1974) Deposition and diagenetic alteration of sandstone in northeast Pacific arc-related basins: Implications for graywacke genesis. *Geol. Soc. Amer. Bull.*, **85**, 379–90.

Ham, H. H. (1966) New charts help estimate formation pressures. *Oil Gas J.*, **64**, 58–63.

Hedberg, H. D. (1936) Gravitational compaction of clays and shales. *Am. J. Sci.*, **31**, 241–87.

Hosoi, H. (1963) First migration of petroleum in Akita and Yamagata Prefectures. *Jpn. Assoc. Mineral., Petrol. Econ. Geol. J.*, **49**, 43–55, 101–14.

Kharaka, Y. K., Callender, E. and Carothers, W. W. (1977) Geochemistry of geopressured geothermal waters from the Texas Gulf Coast, in *Proceedings, 3d Geopressured geothermal energy conference* (J. Meriwether (ed.)), Univ. Southwestern Louisiana, **1**, GI-121–GI-166.

Loucks, R. G., Dodge, M. M. and Galloway, W. E. (1979) Sandstone consolidation analysis to delineate areas of high-quality reservoirs suitable for production of geopressured geothermal energy along the Texas Gulf Coast, Bureau of Economic Geology, University of Texas, 98 p.

Low, P. F. (1976) Viscosity of interlayer water in montmorillonite. *Soil Sci. Soc. Amer. Proc.*, **40**, 500–5.

Magara, K. (1968) Compaction and migration of Fluids in Miocene mudstone, Nagaoka Plain, Japan, *A.A.P.G.*, **52**, 2466–501.

Magara, K. (1971) Permeability considerations in generation of abnormal pressures. *Soc. Pet. Engng. J.*, **11**, 236–42.

Magara, K. (1974) Compaction, ion-filtration, and osmosis in shales and their significance in primary migration, *AAPG Bull.*, **58**, 283–90.

Magara, K. (1976) Thickness of removed sedimentary rocks, paleopore pressure, and paleotemperature, southwestern part of Western Canada Basin. *A.A.P.G.*, **60**, 554–65.

Magara, K. (1981) Fluid dynamics for cap rock formation in Gulf Coast, *A.A.P.G.*, **65**, 1334–43.

Martin, R. T. (1962) Adsorbed water on clay: a review, *Clays Clay Miner.*, 9

(*Proc. 9th Natl. Conf. Clays and Clay Minerals*, 1960), Pergamon Press, New York, pp. 28–270.

Maxwell, J. C. (1964) Influence of depth, temperature, and geologic age on porosity of quartzose sandstone. *A.A.P.G.*, **48**, 697–709.

Meade, R. H. (1966) Factors influencing the early stages of compaction of clays and sands—review. *J. Sediment. Geol.*, **36**, 1085–101.

Overton, H. L. and Timko, D. J. (1969) The salinity principle—a tectonic stress indicator in marine sands, *Log Analyst*, **10**, 34–43.

Parker, C. A. (1974) Geopressures and secondary porosity in the deep Jurassic of Mississippi, *Gulf Coast Assoc. Geol. Socs. Trans.*, **24**, 69–80.

Petroconsultants Ltd (1984) Oil and Gas in Saudi Arabia, **2**, Dublin, Ireland (special consulting report).

Philipp, W., Drong, J. J., Füchtbauer, J., Haddenhorst, J. G. and Jankowsky, W. (1963) The history of migration in the Gifhorn Trough. *Proc. 6th World Pet. Congr., Frankfurt*, Sec. 1, pp. 457–81.

Proshlyakov., B. K. (1960) Reservoir properties of rocks as a function of their depth and lithology. *Geol. Neft. Gaza*, **12**, 24–9.

Powers, M. C. (1967) Fluid-release mechanisms in compacting marine mudrocks and their importance in oil exploration; *AAPG Bull.*, **51**, 1240–54.

Rieke III, H. H. and Chilingarian, G. V. (1974) *Compaction of Argillaceous Sediments*, Elsevier, Amsterdam, 424 p.

Rubey, W. W. and Hubbert, M. K. (1959) Role of fluid pressure in mechanics of overthrust faulting, II. *Geol. Soc. Amer. Bull.*, **70**, 167–206.

Savkevic, S. S. (1971) The possible connection between the leaching-induced porosity of rocks and the primary stage of petroleum generation (migration), *Akad. Nauk SSR Izv. Ser. Geol.*, **6**, 70–7. (In Russian.)

Schmidt, V. and McDonald, D. A. (1979) The role of secondary porosity in the course of sandstone diagenesis, in *Aspects of Diagenesis*, SEPM Spec. Pub. 26, 175–207.

Stephenson, L. P. (1977) Porosity dependence on temperature: limits on maximum possible effect, *A.A.P.G.*, **61**, 407–15.

Tissot, B. (1977) The application of the results of organic geochemical studies in oil and gas exploration, in *Developments in Petroleum Geology—1* (G. D. Hobson (ed.)), Applied Science Publishers Ltd., London, 53–82.

Wedepohl, K. H. (1976) *Handbook of Geochemistry* II-1, New York, Springer-Verlag.

Weller, J. M. (1959) Compaction of sediments, *A.A.P.G.*, **43**, 273–310.

Chapter 15

Size Distribution of Giant Oil and Gas Fields in the World

INTRODUCTION

More than 80% of the world's hydrocarbon reserves are trapped in giant fields (Halbouty, 1970). Production from these giants amounts to 65–70% of the total (Klemme, 1980). In other words, study of the giant fields would be quite valuable in understanding the distribution patterns of petroleum accumulations in the world.

St. John (1980) stated that 'A giant oil field is defined as one containing 500 million barrels of recoverable oil; a giant gas field contains 3 trillion cubic feet of recoverable gas'. He then continued that 'A combination giant oil and gas field with recoverable reserves of 1 billion barrels of oil and 3 Tcf (trillion cubic feet) of gas will have 1·5 billion barrels of oil equivalent (BOE)'.

The size distribution of existing fields in a region may be used in predicting number and sizes of future oil and gas fields in the same region. Refer to White and Gehman (1979) and Magara (1986) for other prediction methods for future fields.

Figure 15.1 shows the oil field distribution of North America (Moody *et al.*, 1970); the recoverable oil volumes are distributed from small to large in this plot. Based on an assumption that the distribution of natural oil deposits is smooth, Moody *et al.*, expected future oil fields where the plot shows significant breaks. Figure 15.2 is a smoothed plot where plus signs infer both sizes and number of future oil fields.

In many sedimentary basins, the field size distributions are known to be log-normal. Haun (1971) showed an example from the Denver Basin as in Fig. 15.3; this frequency plot shows a bell-shaped distribution on semi-log paper (log-normal distribution).

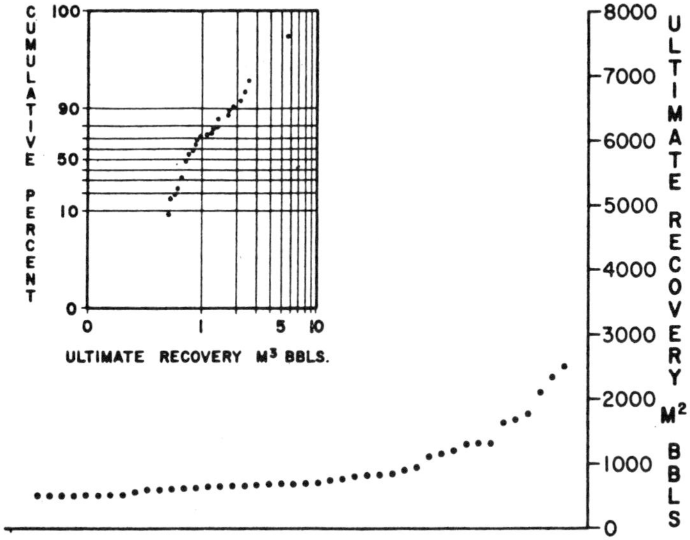

FIG. 15.1. Size distribution of giant fields of North America, in order of magnitude (after Moody *et al.*, 1970). (Reproduced by permission of the American Association of Petroleum Geologists.)

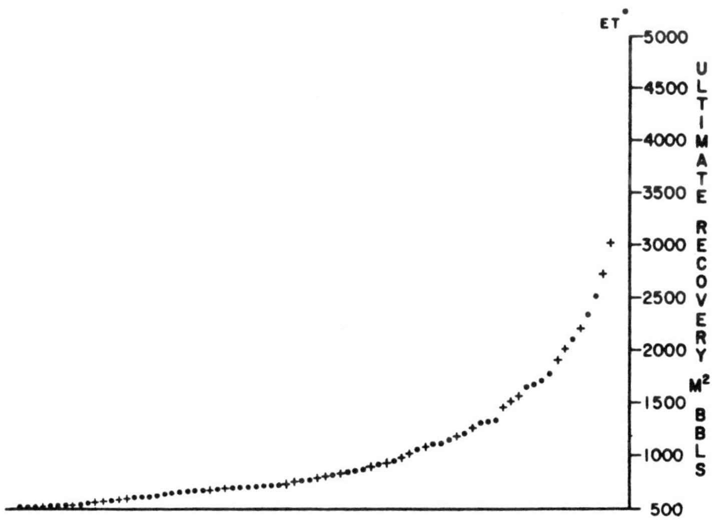

FIG. 15.2. Giant fields of North America in order of magnitude. Circles represent 45 known giants and six fields found but not known to be giants. Pluses represent giants of future (from Moody *et al.*, 1970). (Reproduced by permission of the American Association of Petroleum Geologists.)

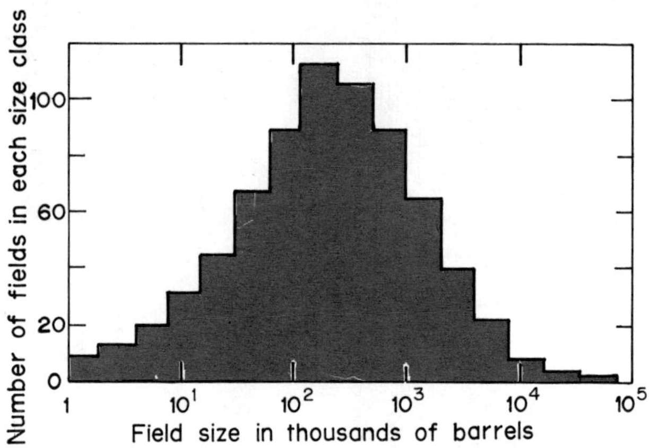

FIG. 15.3. Histogram of oil field sizes in Denver Basin at end of 1969 (from Haun, 1971). (Reproduced by permission of the author.)

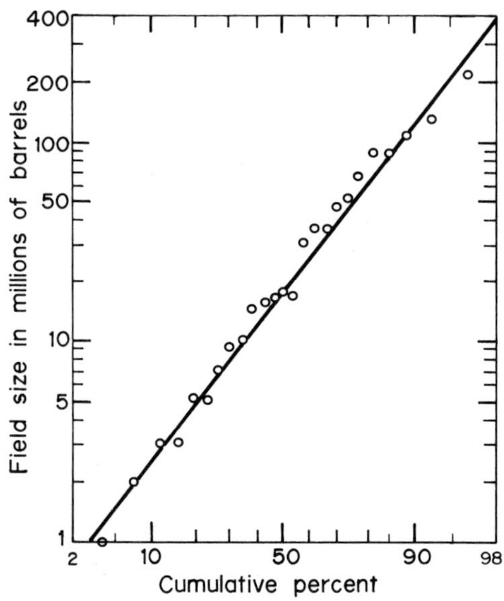

FIG. 15.4. Cumulative frequency distribution of ultimately recoverable oil of north Louisiana fields as of 1946 (from Kaufman, 1963). (Reproduced by permission of Prentice-Hall, Inc.)

Figure 15.4, on the other hand, depicts a cumulative frequency plot of recoverable oil volumes in northern Louisiana on log-probability paper (Kaufman, 1963). A straight-line approximation on this graph suggests the presence of a log-normal distribution of field sizes in this region.

FIELD SIZE DISTRIBUTION OF THE WORLD

Based on the data of the giant oil and gas fields compiled by St. John (1980), eight cumulative frequency plots were made on log-probability paper as follows:

World total	Fig. 15.5
North America	Fig. 15.6
Central and South America	Fig. 15.7
Africa	Fig. 15.8
Europe	Fig. 15.9
Middle East	Fig. 15.10
Asia and Oceania	Fig. 15.11
USSR	Fig. 15.12

The above figures are based on the estimated BOE (barrels of oil equivalent), as explained previously. Each of these figures shows two curved lines; one is based on the recoverable volumes, and another on the number.

The trends in the world total, North America, Europe, Middle East, and USSR plots are relatively straight, suggesting that the fields are mostly distributed log-normally in these regions. The total number and volumes of oil discovered in these regions are relatively large. There are sufficient data to prove the log-normal distributions in these continents. The statistical minimum, most-likely (median), and maximum field sizes may be estimated graphically, by reading the values corresponding to 16, 50 and 84 cumulative %, respectively (see Chapter 12).

The plots of the other regions (Central and South America, Africa, and Asia and Oceania) do not show near-straight lines. This may be due to the fact that the discoveries so far have been relatively small in these regions. More smooth distributions may be established in the future after extensive petroleum exploration activities in these regions.

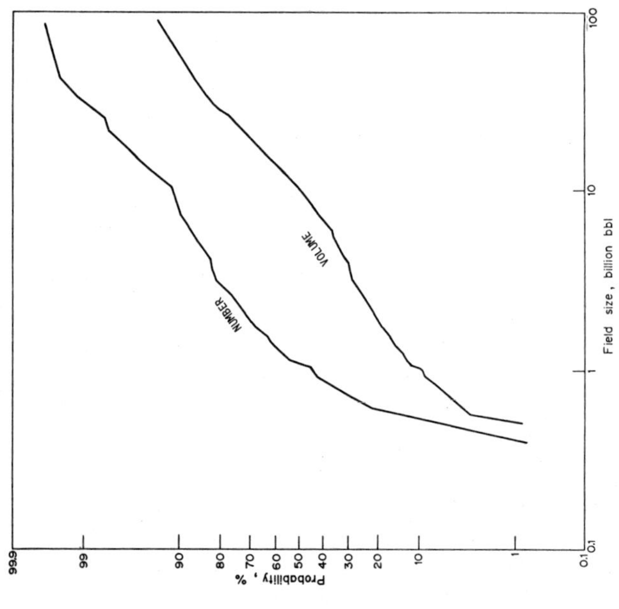

FIG. 15.5. Cumulative frequency percentage plot of giant oil fields of the world (from Huwaidi, 1984).

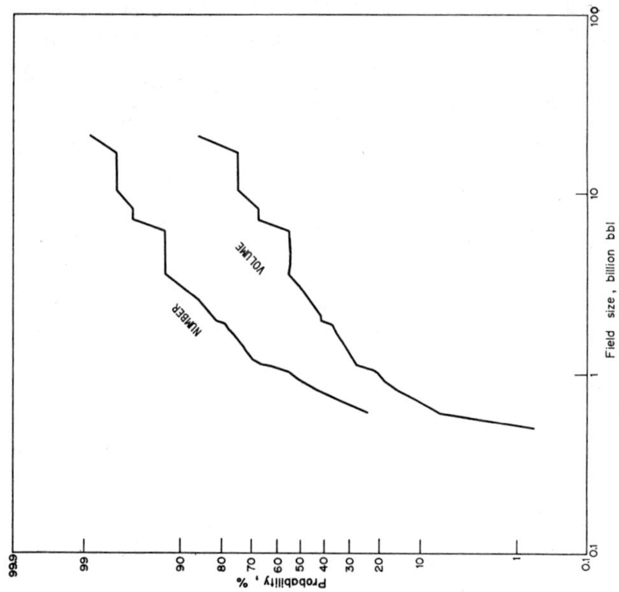

FIG. 15.6. Cumulative frequency percentage plot of giant oil fields of North America (from Huwaidi, 1984).

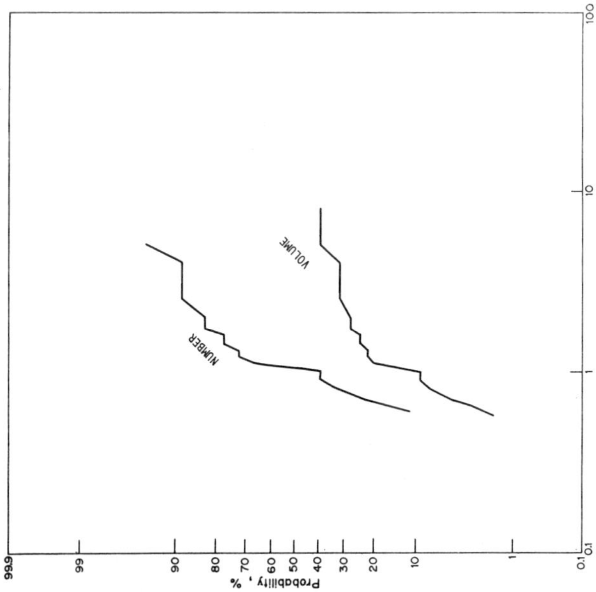

FIG. 15.7. Cumulative frequency percentage plot of giant oil fields of Central and South America (from Huwaidi, 1984).

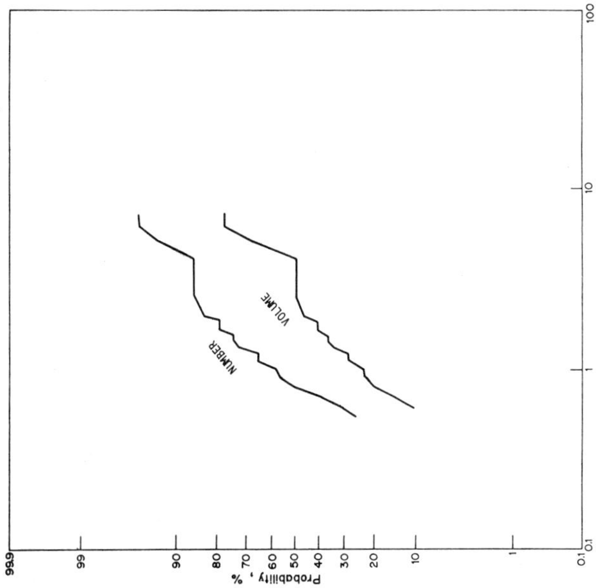

FIG. 15.8. Cumulative frequency percentage plot of giant oil fields of Africa (from Huwaidi, 1984).

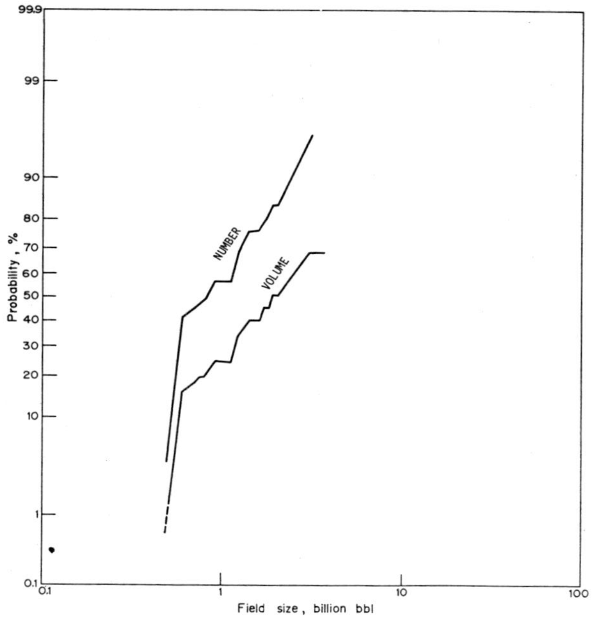

FIG. 15.9. Cumulative frequency percentage plot of giant oil fields of Europe (from Huwaidi, 1984).

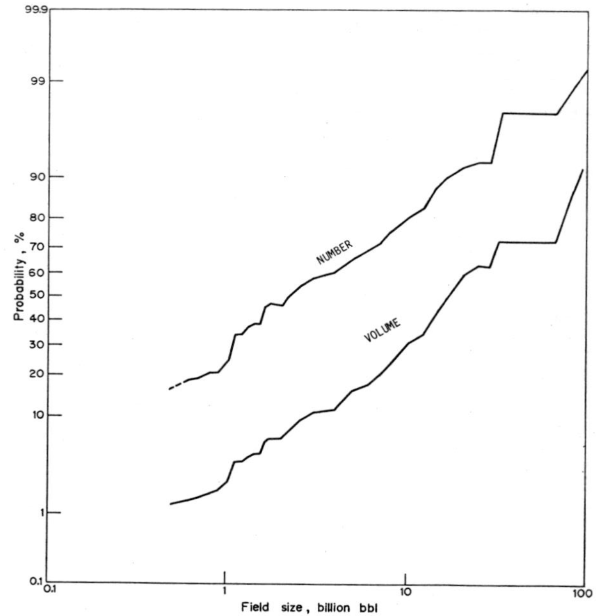

FIG. 15.10. Cumulative frequency percentage plot of giant oil fields of Middle East (from Huwaidi, 1984).

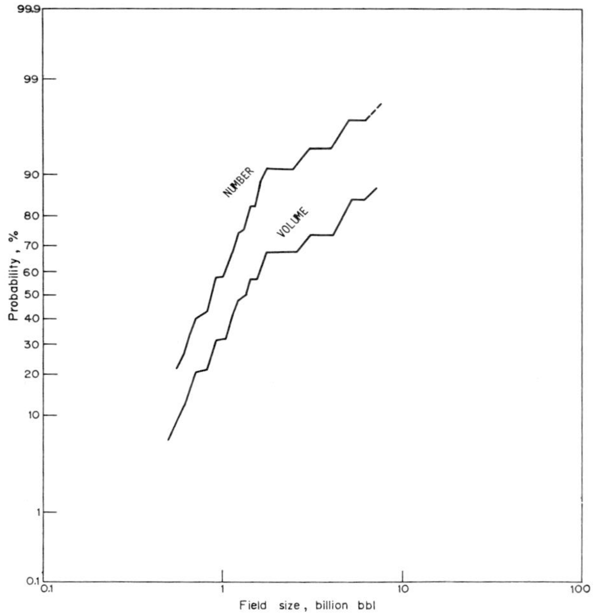

FIG. 15.11. Cumulative frequency percentage plot of giant oil fields of Asia and Oceania (from Huwaidi, 1984).

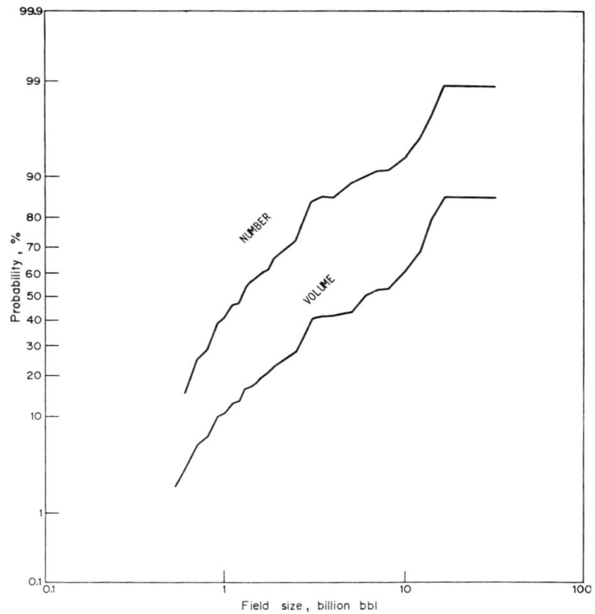

FIG. 15.12. Cumulative frequency percentage plot of giant oil fields of USSR (from Huwaidi, 1984).

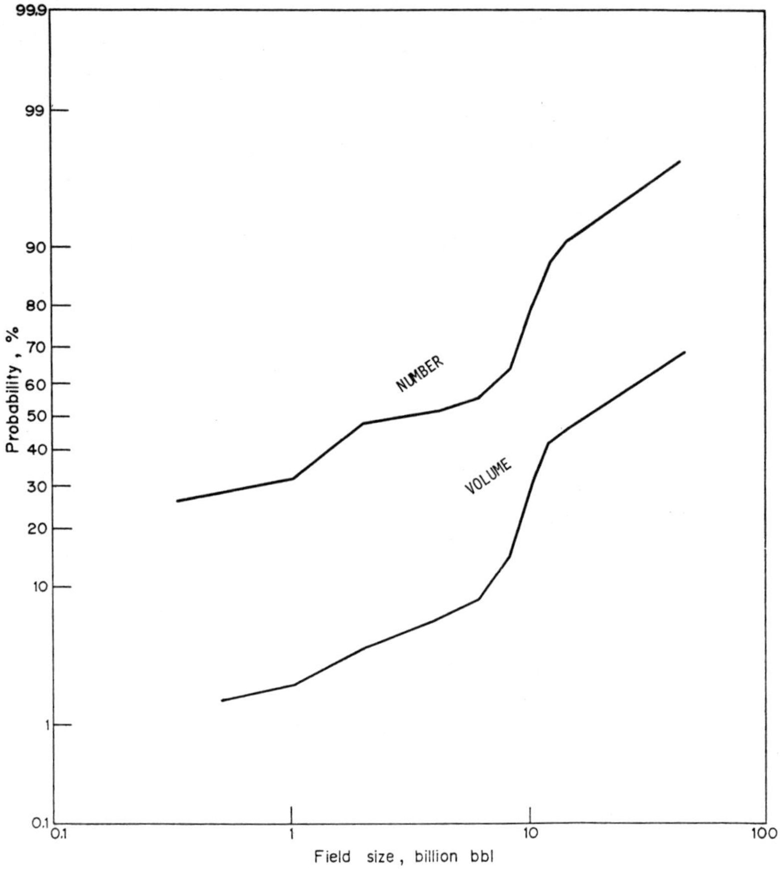

FIG. 15.13. Cumulative frequency percentage plot of giant oil fields of Saudi Arabia (from Huwaidi, 1984).

Figure 15.13 shows a cumulative frequency plot of Saudi Arabia, one of the richest oil nations. This plot shows a significant drop in the trend in the 2–8 billion barrel range. This may suggest that there are potentials for future oil discoveries in this size-range. Using these data, Huwaidi (1984) suggested following future discoveries in Saudi Arabia:

Size range (billion bbl)	*Average size (billion bbl)*	*Number*	*Total oil reserves (billion bbl)*
2–4	3	3	9
4–6	5	4	20
6–8	7	3	21
Total		10	50

The validity of such predictions based on the size distribution must, of course, be checked by other prediction methods using geological, geochemical, and geophysical data and computer statistics.

OTHER STATISTICS OF GIANT OIL FIELDS IN THE WORLD

Other types of the giant oil field data, such as trap type, reservoir lithology, geologic age, and producing depth, were compiled, using the oil field (not gas field) data in Halbouty's book (1970). Percentages based on the number of the giant fields were calculated for each continent (USSR and the Arabian Gulf region are included in Asia).

The results are shown in Tables 15.1, 15.2, 15.3, and 15.4, respectively.

1. Trap Type (Table 15.1)

Anticline and faulted anticline are the major types of traps in all the continents.

2. Reservoir Lithology (Table 15.2)

About $\frac{2}{3}$ of the giant oil field reservoirs are made of sandstone, leaving about $\frac{1}{3}$ for carbonate.

3. Geologic Age (Table 15.3)

According to this summary, the Tertiary is most important in all the continents. In Asia which includes the Arabian Gulf region, the percentages of both the Cretaceous and Jurassic reservoirs are higher than in any other continent.

TABLE 15.1

Percentages of giant oil fields by trap type (data derived from Halbouty, 1970)

Trap type	*Asia*	*Europe*	*Africa*	*North America*	*South America*	*Australia*
Anticline	80	33	37	37	25	50
Faulted anticline	9	33	22	20	10	25
Reef	2	0	12	9	0	0
Salt dome	4	0	0	5	0	0
Other dome	2	0	0	0	0	0
Stratigraphic	0	0	0	4	10	0
Fault	0	0	0	0	15	0
Trancation	0	0	0	7	10	0
Unconformity	0	0	4	0	0	25
Rollover anticline	0	0	19	2	0	0
Other	3	34	6	16	30	0
Total	100	100	100	100	100	100

TABLE 15.2

Percentages of giant oil field reservoirs by lithology (data derived from Halbouty, 1970)

Lithology	*Asia*	*Europe*	*Africa*	*North America*	*South America*	*Australia*
Sandstone	51	50	63	56	64	100
Carbonate	49	50	33	32	30	0
Other	0	0	4	12	6	0
Total	100	100	100	100	100	100

TABLE 15.3

Percentages of giant oil field reservoirs by geologic age (data derived from Halbouty, 1970)

Geologic age	*Asia*	*Europe*	*Africa*	*North America*	*South America*	*Australia*
Tertiary	30	100	57	37	63	100
Cretaceous	40	0	22	9	30	0
Jurassic	18	0	0	17	3	0
Triassic	2	0	0	2	0	0
Paleozoic	10	0	21	35	4	0
Total	100	100	100	100	100	100

TABLE 15.4
Percentages of giant oil field reservoirs by depth range (data derived from Halbouty, 1970)

Depth range (ft)	*Asia*	*Europe*	*Africa*	*North America*	*South America*	*Australia*
–5 000	29	50	19	74	54	32
5 000–7 500	38	50	42	15	17	67
7 500–10 000	23	0	27	5	11	0
10 000–	10	0	12	6	18	1
Total	100	100	100	100	100	100

4. Depth Range (Table 15.4)

Most giant oil field reservoirs have been discovered in the sections above 7500 ft (relatively shallow). This is probably related to two reasons; firstly, petroleum maturation which tends to convert oil into gas at deeper depths, and secondly, depths of petroleum wells which will gradually increase in the future with advanced exploration. In other words, the depth of oil reservoir will increase gradually with advanced exploration and drilling in the future, but will also be limited by the effect of petroleum maturation.

REFERENCES

Halbouty, M. T. (1970) *Geology of Giant Petroleum Fields*, A.A.P.G. Memoir 14, Tulsa, Ok., 555 p.

Haun, J. D. (1971) Potential Oil and Gas Resources: Colorado School of Mines, unpublished progress report, 18 p.

Huwaidi, I. H. (1984) Statistical analysis of oil discovery history in Saudi Arabia—a comparative study of the Kingdom with world-wide data, M.Sc. thesis, Faculty of Earth Sciences, King Abdulaziz University, Jeddah, Saudi Arabia, 247 p.

Kaufman, G. M. (1963) *Statistical Decision and Related Techniques in Oil and Gas Exploration*, Prentice-Hall, Englewood Cliffs, N.J., 307 p.

Klemme, H. D. (1980) Petroleum basins—classifications and characteristics, *Petroleum Geology J.*, **3**, 187–207.

Magara, K. (1986) *Principles of Petroleum Prospecting*, King Abdulaziz University, Jeddah, Saudi Arabia (in press).

Moody, J. D., Mooney, J. W. and Spivak, J. (1970) Giant oil fields of North America. In Halbouty, M. T. (ed.), *Geology of Giant Petroleum Fields*, A.A.P.G. Memoir 14, pp. 8–18.

St. John, B. (1980) *Sedimentary Basins of the World and Giant Hydrocarbon Accumulations*, A.A.P.G., Tulsa, Ok., 23 p.

White, D. A. and Gehman, H. M. (1979) Methods of estimating oil and gas resources. *A.A.P.G.*, **63**, 2183–97.

INDEX